Switched-mode Power Supplies
in Practice

Switched-mode Power Supplies in Practice

Otmar Kilgenstein

Translated by Kenneth G. King

JOHN WILEY & SONS

Chichester · New York · Brisbane · Toronto · Singapore

Originally published under the title 'Schaltnetzteile in der
Praxis' by Vogel Verlag KG, Wurzburg (Bundesrepublik
Deutschland). Copyright by Vogel-Verlag, Wurzburg.

British Library Cataloguing in Publication Data:

Kilgenstein, Otmar
  Switched-mode power supplies in practice.
  1. Electric equipment: Switching power
  convertors
  I. Title
  621.31'37
  ISBN 0 471 92004 5

Phototypesetting by Thomson Press (India) Limited, New Delhi
Printed and bound in Great Britain by Courier International Ltd, Tiptree, Essex

Foreword

All electronic semiconductor circuits need d.c. supplies of relatively low voltage, which in most cases has to be obtained from the 220–240 V mains. Sometimes there is also available a relatively high-voltage d.c. supply, for example a 60 V telephone power supply, which has to be converted to the required low voltage with as little loss as possible. The simplest solution is to use a linear regulator as shown in Figure 1.

Such linear regulators are employed in large numbers, and meet all possible demands as to the constancy and purity of the output: very small variation of the output voltage with variations in mains voltage or load, a wide range of adjustment, simple short-circuit protection and very simple construction with a small number of components. The disadvantages lie in the necessity for an expensive and massive mains transformer, and hence a large equipment volume and weight, and in the relatively low efficiency. The latter factor is particularly important in the case of high d.c. powers, and led to the development of the switching regulator, also known as the 'switched-mode power supply' (SMPS).

In switching regulators, a d.c. supply, obtained from the mains either directly or through a mains transformer, is chopped at a repetition frequency above the audible range, transformed and rectified again. This means a more complicated circuit, however, and the problems of the inherent ripple content of the output, albeit small, and radiated interference must be overcome. The efficiency, on the other hand, reaches values in the region of 70 to 80 per cent.

This book is entirely devoted to switching regulators of all kinds, covering the accurate determination of the characteristics and component requirements to the last detail and comparison with developed and dimensioned examples. Component manufacturers' actual data sheets are used to show the sources of the required data. The reader will learn how to acquire the necessary basic principles and to ensure that a suitably designed circuit

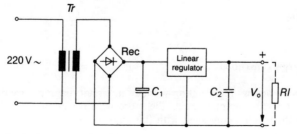

Fig. 1 Power supply with mains isolating transformer, rectifier and linear series regulator providing constant dc output voltage

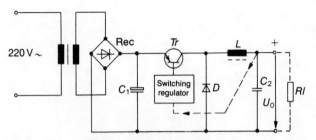

Fig. 2 Power supply with mains isolating transformer, rectifier
and switching series regulator providing constant dc output
voltage (Secondary switching regulator)

will function as desired. Since there is in principle no one optimal circuit—the optimum may be completely different according to the aspect that is optimized—as many variants as possible of the individual sub-circuits are presented. It can also come about that a solution which is in itself optimal in a particular case is either too expensive or not sufficiently reliable.

The book is so structured that from the first circuit considered all the required quantities are evaluated. This means that the reader is always presented with a circuit complete in its essentials, and does not have to gather particular details from different chapters. In later chapters, previously adduced formulae and relationships are in all cases referred to.

A special chapter is devoted to a comprehensive examination of the characteristics, circuits and peculiarities of most of the integrated drive circuits available in the market. Switching frequencies from 20 kHz to about 100 kHz are employed. Higher switching frequencies lead to a reduction in the sizes of transformers and inductors, but result in increased switching losses in core materials, transistors and rectifier diodes. In addition a higher switching frequency makes greater demands on the drive system.

The close association of theory and practice achieved in this book has been made possible only by the availability of the excellent results of a large number of diploma projects carried out at the Georg Simon Ohm Technical College in Nuremberg. The author would therefore like to express his especial thanks to Mr Werner Schaub, Mr Frank Böhner, Mr Dietmar Glang, Mr Anton Fuchs, Mr Rupert Grünzinger and Mr Lutz Metzer. Thanks are also due to Messrs Siemens, SEL Nuremberg, TRW Munich, Hanau Vacuum Smelters and Valvo for their effective support. Without their generous assistance with the supply of components and ancillary materials this undertaking would have been considerably more difficult.

Wei eszett Benohe   Otmar Kilgenstein

NOTE: Voltage arrow convention

Throughout this book arrows representing voltages in circuit diagrams are drawn according to the convention that arrows point towards the more negative point. Equations and algebraic expressions within the text are consistent with this convention.

Contents

vii

1 Introduction and selection of different types of converter

Before a switched-mode power supply is designed the various power converter configurations should be considered to enable selection of the circuit most suitable for the given application. Figure 1.1 provides assistance with initial selection.

In region (a)—up to about 50 V and 50 W—i.e. for relatively low output powers, it is appropriate to use *inductor-coupled* converters (step-down, step-up and flyback converters), so long as the absence of electrical isolation between input and output is acceptable (Figures 1.2 to 1.4). These converters involve a minimum of circuit complexity, but have

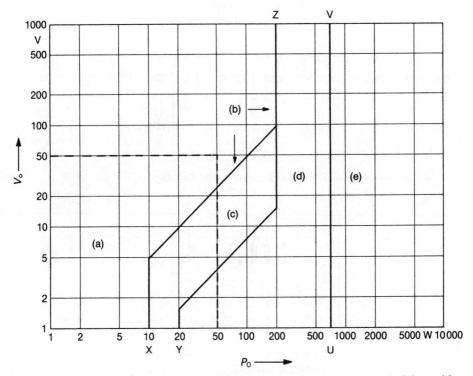

Fig. 1.1 Diagrammatic comparison of different switching regulator principles, with approximate limits of application

1

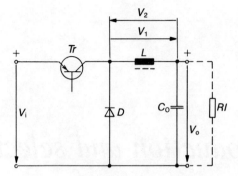

Fig. 1.2 Inductor-coupled step-down
converter

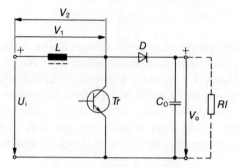

Fig. 1.3 Inductor-coupled step-up
converter

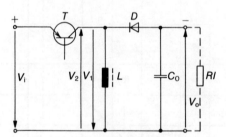

Fig. 1.4 Inductor-coupled flyback
converter

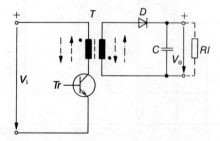

Fig. 1.5 Flyback converter with one
secondary winding; $V_{CEmax} = 2\,V_i$ with
$\delta_T = 0.5$

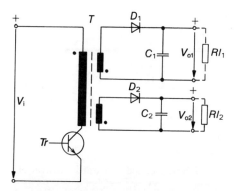

Fig. 1.6 Flyback converter with more than one secondary winding; $V_{\text{CEmax}} = 2V_i$ with $\delta_T = 0.5$

the disadvantage that a relatively large current has to be switched by the transistor. With very high currents there could be difficulties in regard to switching times [9].

In region (b)—to the left of the boundary X–Z—*transformer-coupled flyback converters* are the most frequently used. These converters may provide either a single output or several outputs. In contrast to the operation of the forward converters, in the case of the flyback converter the energy acquired by the transformer during the 'on' time of the transistor is delivered to the output only in the blocking period (Figures 1.5 and 1.6). This can readily be appreciated from the way the dots are marked against the transformer windings. It is assumed, in principle, that all windings are wound in the same direction.

If, for example, during the conducting period of the transistor, current flows from + through the primary winding and the transistor, a voltage in the opposing direction is generated in the primary and secondary windings. Since the secondary winding is connected with reversed polarity, with the diode polarity shown there can be no current flow. When the transistor ceases to conduct, the induced voltage is reversed by the collapse of the magnetic field, and the output capacitor is charged through the diode.

Because flyback converters require only one magnetic component (transformer), they are simpler and cheaper than forward converters. They are therefore used for relatively low powers, especially at low output voltages. Problems arise here in the output transformer, the more so as the output voltage required decreases. The reason for this is the difficulty of controlling the flux leakage. The requirement for very low leakage combined with a high degree of insulation between the primary and secondary windings is not easily satisfied. A further disadvantage of the flyback converter is that the residual ripple in the output voltage is dependent upon the inductance and the equivalent series resistance (ESR) of the output capacitor. For similar reasons, high-frequency noise (spikes) can more easily reach the output.

Region (c) is a transitional region embracing both flyback and single-ended forward converters.

In region (d)—between the boundaries Y–Z and U–V—single-ended forward converters are mostly used (Figures 1.7 and 1.8).

Here again the flux leakage in the transformer presents difficulties with low output voltages, even though such low leakage coefficients as in the flyback converter are not called for. (The reversed polarity relationship between primary and secondary windings in this case should be noted. The intermediate demagnetizing winding need not be

4

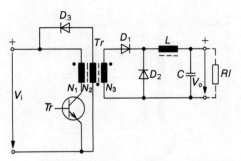

Fig. 1.7 Single-ended forward converter;
$V_{\mathrm{CEmax}} = 2V_i$

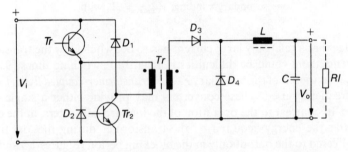

Fig. 1.8 Single-ended forward converter with two transistors;
$V_{\mathrm{CEmax}} = V_i$

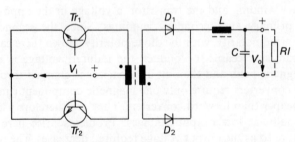

Fig. 1.9 Push–pull forward converter; $V_{\mathrm{CEmax}} = 2V_i$

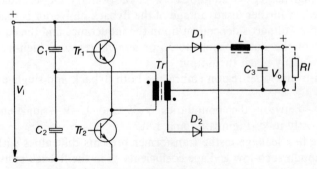

Fig. 1.10 Half-bridge push–pull forward converter;
$V_{\mathrm{CEmax}} = V_i$

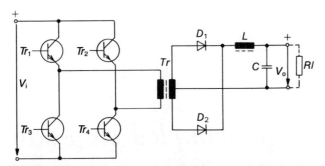

Fig. 1.11 Full-bridge push–pull forward converter;
$$V_{CEmax} = V_i$$

considered at this point.) For very high output powers—in the region to the right of the boundary U–V—it is advantageous to divide the power to be switched on the primary side between a number of transistors in a push–pull arrangement.

In the simple push–pull circuit (Figure 1.9), a high degree of symmetry is necessary in the primary circuit, in order to prevent d.c. magnetization in the transformer. The circuit of Figure 1.10 avoids this undesirable condition by means of the two capacitors. Finally, the very expensive push–pull bridge circuit (Figure 1.11), with four transistors, permits conversion at particularly high power levels. In all push–pull circuits, problems arise in the drive system, because of the requirement that at no time must all the transistors conduct simultaneously.

Since the frequency of the ripple components superimposed on the d.c. output is doubled in the push–pull converter, the residual ripple at the output is correspondingly reduced. It is also an advantage that the output transformer is magnetized in both directions, not in one direction only as in the case of the single-ended converter.

The power switches are shown as bipolar transistors in all cases in Figures 1.2 to 1.11 for the sake of simplicity. All the circuit variants have, however, been constructed both with bipolar power transistors and with power MOSFETs, including the associated drive circuits. The measured quantities of interest are indicated in [ ] alongside the calculated values.

2 Inductor-coupled step-down converters (forward converters)

2.1 BASIC PRINCIPLES

Figure 2.1 shows the circuit (without the protective elements or the integrated drive circuit) of an inductor-coupled step-down converter [1], [7].

The transistor used here is of the *p-n-p* type, since this can be turned on by a supply of negative potential with respect to the emitter. If an *n-p-n* transistor is used, the power supply to the drive IC must be more positive than V_i, which is not possible with only a *single* input voltage. The arrangement shown in Figure 2.2 offers a possible way out of this difficulty, but it can give rise to problems owing to the interrupted negative line (earth).

If the transistor is turned on (to the limit of the low collector–emitter voltage V_{CEsat}), the cathode of the diode D assumes practically the potential V_i. Since V_0 is held constant during the cycle period T by the output capacitor C_0, a voltage $V_i - V_0$ appears across the inductor. The current in the inductor increases approximately linearly. When the transistor is turned off, the voltage induced in the inductance L tends to maintain the current. D now conducts, and a voltage almost equal to the output voltage V_0 appears across the inductor. During the blocking period the inductor current falls linearly back to its original value.

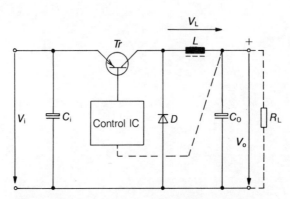

Fig. 2.1 Basic circuit of an inductor-coupled step-down converter with *p-n-p* transistor (control IC connections not shown)

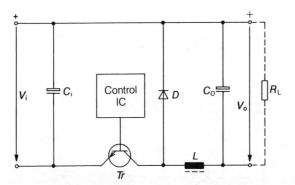

Fig. 2.2 Basic circuit of an inductor-coupled step-
down converter with an *n-p-n* transistor

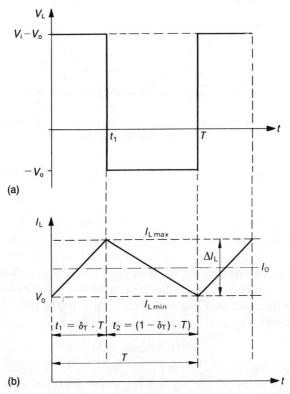

Fig. 2.3 Variation of (a) inductor voltage V_L and (b)
inductor current I_L with time

The duty cycle δ_T denotes the conducting time ('on' time) t_1 of the transistor as a proportion of the cycle period $T = 1/f$.

$$\delta_T = \frac{t_1}{T} \tag{2.1}$$

where t is the 'on' time of the transistor (s) and T is the cycle period (s).

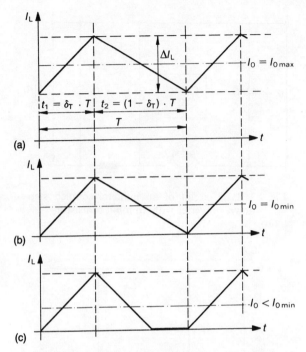

Fig. 2.4 Time–variation of inductor current I_L with load
currents I_0 of different magnitudes

From the relationship $t_1 + t_2 = T$, as shown in Figure 2.3, and Equation (2.1),

$$1 - \delta_T = \frac{t_2}{T} \tag{2.2}$$

Figure 2.3 shows the voltage across and the current in the inductor, while Figure 2.4 shows the waveform of the inductor current with various levels of load current.

The load current may lie between the maximum value I_{0max} and the minimum value I_{0min}. The case illustrated by Figure 2.4(c) (discontinuous-current operation) should not arise, since the output voltage V_0 cannot then be held to the same degree of accuracy.

From Figure 2.4(b),

$$\Delta I_L = 2I_{0min} \quad \text{(A)} \tag{2.3}$$

where I_{0min} is the minimum load current (A) (about 10 to 20 percent of I_{0max}).

The voltage–time areas $(V_i - V_0)t_1$ and $V_0 t_2$ must always be equal (Figure 2.3), and from this follows immediately:

$$V_0 = V_i \delta_T \quad \text{(V)} \tag{2.4}$$

or, with given tolerances applied to the input voltage:

$$V_0 = V_{imin} \delta_{Tmax} = V_{imax} \delta_{Tmin} \quad \text{(V)} \tag{2.5}$$

Equation (2.5) is only theoretically accurate, however, in that it neglects the voltage drops across the diode D and the d.c. resistance R_L of the inductor. Since the output voltage of a forward converter is usually relatively low, these voltage drops are not in fact negligible.

V_0 should therefore always be replaced by the quantity V_0^*.

$$V_0^* = V_0 + V_F + V_L \quad (V) \tag{2.6}$$

where V_0 is the output voltage (V), V_F is the diode voltage drop at I_{Lmax} (V) (approximately 0.5 V in Schottky diodes and about 0.7 to 1 V in epitaxial diodes) and V_L is the voltage drop in the resistance of the inductor (V) (about 0.1 to 0.2 V).

Equation (2.5) then becomes

$$V_0^* = V_{imin}\delta_{Tmax} = V_{imax}\delta_{Tmin} \quad (V) \tag{2.7}$$

(Strictly speaking, Equation (2.6) applies with the transistor blocking: with the transistor conducting the relationship $V_0^* = V_0 + V_L$ must be used, but this corresponds to $V_i - V_{CEsat}$. In fact, the duty cycle remains practically the same.)

The output voltage V_0 is held constant by the control system; if required, it may be adjustable within certain limits. The limits of the duty cycle δ_{Tmax} and δ_{Tmin} are thus determined by the extent of the input voltage variation for a given output voltage.

$$\delta_{Tmax} = \frac{V_0^*}{V_{imin}} \tag{2.8}$$

and

$$\delta_{Tmin} = \frac{V_0^*}{V_{imax}} \tag{2.9}$$

In practice the duty cycles thus calculated will be slightly increased, because the relatively small voltage drop across the transistor has not been taken into account. The ratios ultimately established are determined by the control system to give constant output voltage V_0. These small corrections are allowed for in the provision which is always made for adjustment of the output voltage (in comparison with an internal reference voltage in the controller).

In accordance with Equation (2.8), V_{imin} must in all cases be greater than V_0^*, since δ_{Tmax} has a maximum (theoretical) value of 1; a desirable objective is $V_{imin} \geqslant 2V_0^*$. The minimum possible duty cycle δ_{Tmin} is determined by the switching times of the transistor and the freewheel diode. The limit is given by

$$\delta_{Tmin} T > t_{sw} \tag{2.10}$$

where t_{sw} is the total switching time of the transistor or diode and is given by

$$t_{sw} = t_r + t_s + t_f \quad (s) \tag{2.11}$$

where t_r is the rise time (s), t_s is the storage time (s) and t_f is the fall time (s).

2.2 DIMENSIONING AND DESIGN OF THE INDUCTOR L

From Figures 2.1 and 2.3, and from Equations (2.1) and (2.4) together with Equation (2.7),

$$\Delta I_L L = V_i t_1 = \frac{V_1 \delta_T}{f} = \frac{(V_i - V_0^*)V_0^*}{V_i f}$$

or, from this,

$$L \geqslant \frac{(V_i - V_0^*)V_0^*}{\Delta I_L f V_i} \quad \text{wherein} \quad \frac{V_{imax} - V_{imax}\delta_{Tmin}}{V_{imax}} = 1 - \delta_{Tmin}$$

The worst-case value for L is obtained with V_{imax}.

$$L \geqslant \frac{V_0^*(1 - \delta_{Tmin})}{\Delta I_L f} \quad \text{(H)} \tag{2.12}$$

To obtain the maximum current in the inductor, it is necessary to add $\Delta I_L/2$ to the mean inductor current I_{0max}.

$$I_{Lmax} = I_{0max} + \Delta I_L/2$$

$$= \frac{P_{0max}^*}{V_{imax}\delta_{Tmin}} + \frac{V^{imax}\delta_{Tmin}(1 - \delta_{Tmin})}{2fL} \quad \text{(A)} \tag{2.13}$$

The inductance L should not be too high, however, so that the transient recovery time t_{tr} following a step change in load current is not excessive. According to [3], the transient recovery time may be taken approximately as

$$t_{tr} \approx (5 \text{ to } 20)T \quad \text{(s)} \tag{2.14}$$

This leads to an upper limit to the inductance:

$$L \leqslant \frac{V_0^*}{I_{0max}} t_{tr} \left(\frac{\delta_{Tmax}}{\delta_{T1}} - 1 \right) \quad \text{(H)} \tag{2.15}$$

where I_{0max} is the maximum possible step change in load current (A) $\leqslant 0.9I$ and δ_{T1} is the duty cycle at instant of load change; $\delta_{T1} > \delta_{Tmin}$ (not exactly ascertainable).

Example

Given: $V_0 = 5\,\text{V}$; $I_{0max} = 10\,\text{A}$; $I_{0min} = 1.1\,\text{A}$; $\Delta I_L = 2.2\,\text{A}$

$V_i = 24\,\text{V} \pm 15\% = 20.4\,\text{V}$ to $27.6\,\text{V}$

$f = 20\,\text{kHz}$; $I_{0max} = 5\,\text{A}$; $\theta_{amax} = 40°$

With this low output voltage and the relatively low input voltage, a Schottky diode is the only choice for the free-wheel diode. Estimated forward voltage drop: $V_F = 0.5\,\text{V}$ at I_{0max}; for V_L 0.2 V is assumed, as a value that is certainly attainable.

Thus $V_0^* = 5\,\text{V} + 0.5\,\text{V} + 0.2\,\text{V} = 5.7\,\text{V}$.

From Equations (2.8) and (2.9),

$$\delta_{Tmax} = \frac{5.7\,\text{V}}{20.4\,\text{V}} = 0.28; \quad \delta_{Tmin} = \frac{5.7\,\text{V}}{27.6\,\text{V}} = 0.207;$$

$$\delta_{Tnom} = \frac{5.7\,\text{V}}{24\,\text{V}} = 0.24$$

$$T = \frac{1}{f} = \frac{1}{20 \times 10^3\,\text{s}^{-1}} = 50\,\mu\text{s}$$

$$t_{1min} = \delta_{Tmin}\,T = 0.207 \times 50\,\mu\text{s} = 10.35\,\mu\text{s}$$

$$t_{1max} = \delta_{Tmax}\,T = 0.28 \times 50\,\mu\text{s} = 14\,\mu\text{s}$$

From Equation (2.12),

$$L \geqslant \frac{5.7\,\text{V} \times (1 - 0.207)}{2.2\,\text{A} \times 20 \times 10^3\,\text{s}^{-1}} \quad 102.73\,\mu\text{H};$$

selected: $L = 110\,\mu H$ [$118\,\mu H$]

$$t_{tr} = (5\ldots20) \times T = 250\,\mu s \ldots 1000\,\mu s;\ \text{selected:}\ t = 500\,\mu s$$

δ_{T1} estimated as 0.23.

From Equation (2.15),

$$L \leqslant \frac{5.7\,V}{5\,A} \times 500\,\mu s \times \left(\frac{0.28}{0.23} - 1\right) \leqslant 124\,\mu H,$$

so $L = 110\,\mu H$ is satisfactory.

Inductor design:
The first step is to determine the minimum necessary core size. This can be done in a simple manner with the aid of d.c. magnetization curves for various core patterns (Figure 2.5).
The starting point is the product $(I_{Lmax})^2 L$ in $A^2\,mH$.

$$I_{Lmax} = 10\,A + 0.5 \times 2.2\,A = 11.1\,A,\ \text{or, from Equation (2.13)},$$

$$I_{Lmax} = \frac{5.7\,V \times 10\,A}{27.6\,V \times 0.207} + \frac{27.6\,V \times 0.207 \times (1 - 0.207)}{2 \times 20 \times 10^3\,s^{-1} \times 0.11 \times 10^{-3}\,H}$$

$$= 11\,A$$

$$I_{Lmax}^2 \times L = 123.2\,A^2 \times 0.11\,mH = 13.55\,A^2\,mH$$

From Figure 2.5(b) the most suitable core is found to be the Type E42/20. The copper loss as read off the curve amounts to about 2.5 W; $\mu_e \approx 45$. With a temperature rise of around 40 °C (in the middle of the range indicated by the thick line), the temperature of the inductor will the be

$$\theta_L = 25\,°C + 40\,°C = 65\,°C\ \text{(maximum 80\,°C with}\ \theta_{amax} = 40\,°C).$$

To illustrate the way in which the necessary data for a ferrite-cored coil is obtained from the data manual, Figure 2.6 shows an extract from the manual relating to the selected E42/20 core. The inductance factor A_L, which is required in order to calculate the number of turns in the winding, can be determined immediately from the effective permeability μ_e.

$$A_L = \frac{\mu_e \mu_0}{\Sigma l/A}\ (H) \tag{2.16}$$

where μ_0 is the magnetic constant $= 1.257 \times 10^{-6}\,H/m = 1.257 \times 10^{-9}\,H/mm$ and $\Sigma l/A$ is the magnetic form factor (mm^{-1}) (obtained directly from the data sheet, Figure 2.6, if given: otherwise calculated from l_e and A_e).
For the example,

$$A_L = \frac{45 \times 1.257 \times 10^{-9}\,H\,mm}{0.405\,mm} = 140 \times 10^{-9}\,H\ \text{or}\ 140\,nH$$

From Figure 2.6, $A_L = 140\,nH$ corresponds to a total air gap s of a little over 3 mm. This can be obtained by using two half-cores each with a 1.5 mm air gap in the centre limb. The inductance factors given in the table relate to one half-core with the stated gap in combination with one half-core with no gap. In the present example, two similar half-cores are used, each with a 1.5 mm gap.

12

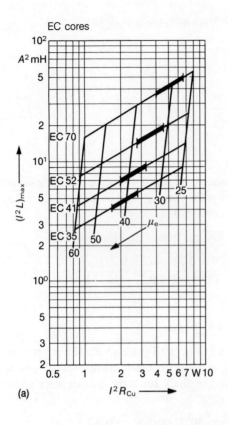

(a)

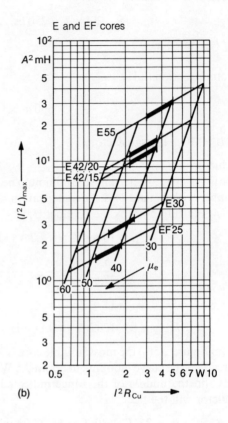

(b)

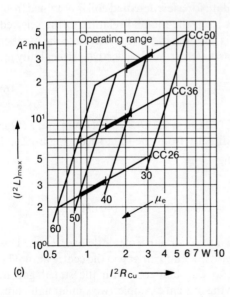

(c)

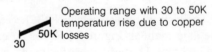

Operating range with 30 to 50K
temperature rise due to copper
losses

Fig. 2.5 D.C. magnetization capability $(I^2 L)_{max}$,
copper loss $I^2 R_{Cu}$, effective permeability μ_e and
temperature rise of EC cores, E and EF cores
and CC cores of Siferrit material (Siemens)

E-K-core   E 42/20

from DIN 41 295

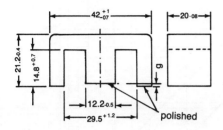

Variation in A_L values with total air gap for one set

1 core $(g≈0)$ and

1 core $(g_X 0)$

or

2 cores $(g > 0)$

Material N 27

Magnetic form characteristic dimensions (per set)

Mag. form factor $\Sigma l/A$	=	0.405	mm^{-1}
Eff. magnetic path length l_e	=	97	mm
Eff. magnetic cross-section A_e	=	240	mm^2
Eff. magnetic volume V_e	=	23300	mm^3

Unit weight ≈ 58g

Accessories: E-cores are individually available by dimension 'S' (foreshortened middle flange). The A_L values in the table are for a core set consisting of the given core and without a fore-shortened middle flange.

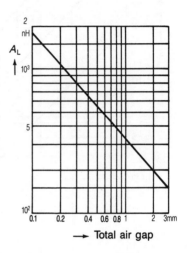

Total air gap

Body of coil

Material	Dimension s (mm)	Tolerance (mm)	A_L value (nH)	Effective permeability (μ_e)
N 27	≈ 0	–	4750$^{+30}_{-20}$%	–
N 27	0.25	± 0.03	≈ 925	≈ 298
	0.50	± 0.05	≈ 560	≈ 180
	1.00	± 0.1	≈ 340	≈ 110
	1.50	± 0.1	≈ 250	≈ 81

Number of chambers	Usable winding area A_N (mm^2)	Mean winding length l_N (mm)	A_R-value ($\mu\Omega$)
1	170	105	21

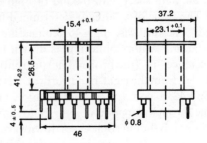

Fig. 2.6 Extract from the data sheet for the E core Type E42/20 (Siemens)

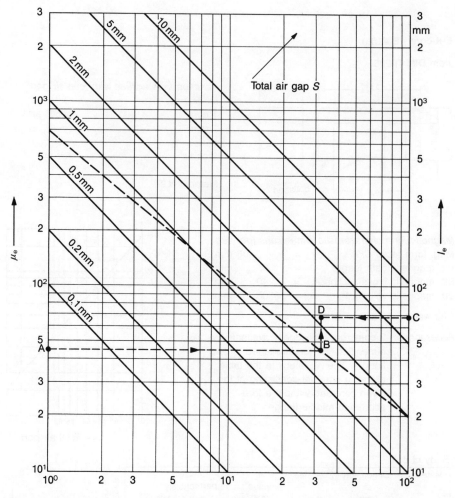

Fig. 2.7 Relationship between effective permeability μ_e, effective path length l_e and total air gap s for ferrite cores with an initial permeability $\mu_i \approx 2000$ (e.g. N27) (Siemens) ------ air-gap reference line

The required air gap s can alternatively be arrived at independently of the values given in the data sheet with the aid of Figure 2.7. The procedure is as follows.

The value of μ_e already determined from Figure 2.5 is found on the left-hand vertical axis—in this case $\mu_e = 45$ (coinciding fortuitously with the point A marked on the diagram). A horizontal line is drawn to the right from this point to meet the 'air-gap reference line' (the broken line) (fortuitously, point B). Now a horizontal line is drawn to the left from the right-hand vertical axis from the value l_e obtained from the data sheet (for the E42/20 core $l_e = 97$ mm), i.e. at 97 mm (not at point C). The intersection of a vertical line through the point of intersection on the air-gap reference line—in this case upwards through B (but in some cases downwards)—with the horizontal line through l_e then gives the air gap s. In the present example this lies between 2 mm and 5 mm: 3 mm is a reasonable choice.

The number of turns required is obtained from

$$N = \sqrt{\frac{L}{A_L}} \tag{2.17}$$

where A_L is the inductance factor (H or nH) and L is the inductance (H or nH respectively).
In the example,

$$N = \sqrt{\frac{110 \times 10^{-6}\,\text{H}}{140 \times 10^{-9}\,\text{H}}} = 28$$

If curves such as those shown in Figure 2.6 are not available, the inductance can be determined by measurement with a test winding (e.g. 20 turns of wire of any convenient cross-section—e.g. 0.2 mm). The value of A_L is determined from Equation (2.17), and μ_e can then be calculated from Equation (2.16). The magnetic form factor must of course be ascertainable from a data sheet. If none of these quantities is given, the mean magnetic path length can be measured on the available core (see Figure 2.6) and the factor A_e obtained from Figure 2.41. The quotient l_e/A_e agrees quite well with the magnetic form factor given in the data sheet. If necessary A_e can also be obtained from the cross-section of the centre limb.

If a core for which details are not given in Figure 2.6 is to be used, the initial choice should be directed towards a magnetic cross-sectional area A_e similar to that of type indicated as suitable in Figure 2.6. In determining the required air gap, an effective permeability μ_e of about 40 to 50 can be assumed (the values according to Figure 2.5 lie in this range) and the gap determined from Figure 2.7. The value of A_L is then calculated from Equation (2.16), based on the next larger *available* air-gap length. It is essential to check, however, that the permissible direct current is greater than the current that actually flows in the worst case. Alternatively a possible air gap may be assumed initially, based on the data sheet for the comparable core, and the d.c. ampere-turns checked by reference to Equation (2.18).

By way of comparison, the latter approach may be tried with the E42/20 core. Assumed: $s = 2$ mm; from Figure 2.6, $A_L = 200$ nH.

$$\mu_e = \frac{200 \times 10^{-9}\,\text{H} \times 0.405\,\text{mm}^{-1}}{1.257 \times 10^{-9}\,\text{H mm}^{-1}} = 64,$$

a somewhat higher value than that obtained from Figure 2.5.
From Equation (2.17),

$$N = \sqrt{\frac{110 \times 10^{-6}\,\text{H}}{200 \times 10^{-9}\,\text{H}}} = 23.5 \approx 24$$

Using Equation (2.18) and Figure 2.9, the figure gives, for $\mu_e = 65$, a value for H of 2500 to 3000 and substituting this in the equation gives

$$I_{\text{dmax}} = \frac{2500\,\text{A/m} \times 97 \times 10^{-3}\,\text{m}}{24} = 10\,\text{A}$$

This is too low, since I_{Lmax} was calculated as 11 A.

The calculation could now be repeated with the next larger air gap of 2.5 mm (1 mm + 1.5 mm); here it will be done straightaway with $s = 3$ mm. $A_L = 150 \times 10^{-9}\,\text{H}$; $\mu_e = 48$

16

(normal value) and $N = 27$. H now becomes approximately 4500 A/m and $I_{\mathrm{dmax}} = 16.2$ A. This represents adequate dimensioning, and results in almost the same values as those found previously.

Some rules follow to assist the accurate reading-off of a logarithmic division—e.g. in Figure 2.7.

- The distance x between two decade points (powers of ten) to a single scale is measured accurately. It should be observed in this connection that occasionally different scales are used in one diagram.
- The distance y between the point to be determined and the first decade point is measured.
- The quotient of the two numbers, y/x, is the required logarithm;

thus $$\mathrm{Log} = \frac{y}{x} < 1.$$

- The number corresponding to the calculated logarithm can be obtained from Table 2.1 or from Figure 2.8. In this way the value is determined accurately.

These rules may now be used to check the previously estimated value for the air gap of 'a little over 3 mm' with the aid of Figure 2.7. (Air-gap lines: $x \rightarrow$ distance from '1 mm' to '10 mm'; $y \rightarrow$ distance from '1 mm' to the ascertained point of intersection.) Distance $x = 36$ mm; distance y (vertical line upwards from B intersecting the line $l_e = 97$ mm) $= 19$ mm.

$$\mathrm{Logarithm} = \frac{y}{x} = \frac{19\,\mathrm{mm}}{36\,\mathrm{mm}} = 0.528.$$

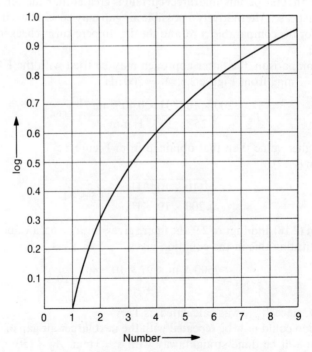

Fig. 2.8 Logarithms of numbers from 1 to 9

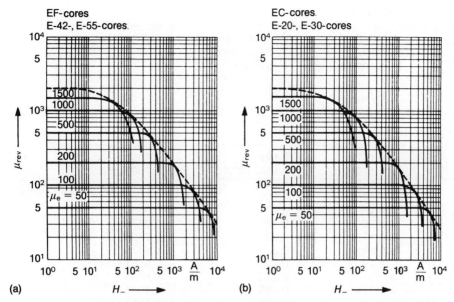

Fig. 2.9 D.C. pre-magnetization in N27 material for EF and EC cores at a temperature of 20 °C with $\hat{B} < 1\,\text{mT}$ (Siemens)

From the table, air gap s (number) = 3.25 to 3.5 mm; from Figure 2.8, more accurately, air gap s (number) = 3.4 mm.

Check: $\log 3.4 = 0.531$.

It remains to be checked that d.c. ampere-turns are not excessive (an excessive d.c. magnetization results in a reduction in the effective permeability).

To this end the maximum value of H must be determined, from Figure 2.9, up to which the curve remains level. With $\mu_e = 45$ the result is approximately $H_{-\text{max}} = 6 \times 10^3\,\text{A/m}$.

The maximum possible direct current is calculated as

$$I_{\text{dmax}} = \frac{H_- l_e}{N}\,(\text{A}) \tag{2.18}$$

where H_- is the magnetic field strength (A/m), I_e is the effective magnetic path length (m) (from data sheet) and N is the number of turns.

In the example, this gives

$$I_{\text{dmax}} = \frac{6 \times 10^3\,\text{A} \times 97 \times 10^{-3}\,\text{m}}{28\,\text{m}} = 20.8\,\text{A} > 11\,\text{A}\,(I_{\text{Lmax}})$$

Next the maximum possible wire thickness is determined. The number of turns that can be accommodated can be calculated from the maximum usable winding area A_N (170 mm^2 from Figure 2.6) only for very fine wires. The procedure here is different. From the wire table (Figure 2.10) and the curves in Figure 2.11 (increase in resistance at high frequencies—skin effect) a wire is selected which exhibits practically no skin effect. The acceptable limit is in the region of $\Delta R/R_d \leqslant 5$ per cent. For $f = 20\,\text{kHz}$ a wire with a nominal diameter of 1 mm $\equiv 1.068$ mm maximum outside diameter is chosen. This results

Wire table

Nominal diameter (= Conductor-⌀) (mm)	Outer diameter of insulated wires (Maximum limits) Easily insulated by lacquering (L) (mm)	D.C. resistance (nominal value at 20°C) (Ω/m)
● 0.1	0.121	2.195
● 0.112	0.134	1.750
● 0.125	0.149	1.405
● 0.14	0.166	1.120
0.15	0.177	0.9756
● 0.16	0.187	0.8575
0.17	0.198	0.7596
● 0.18	0.209	0.6775
0.19	0.220	0.6081
● 0.2	0.230	0.5488
● 0.224	0.256	0.4375
● 0.25	0.284	0.3512
● 0.28	0.315	0.2800
0.3	0.336	0.2439
● 0.315	0.352	0.2212
● 0.355	0.395	0.1742
● 0.4	0.442	0.1372
● 0.45	0.495	0.1084
● 0.5	0.548	0.08781
● 0.56	0.611	0.07000
0.6	0.654	0.06098
● 0.63	0.684	0.05531
● 0.71	0.767	0.04355
● 0.75	0.809	0.03903
● 0.8	0.861	0.03430
● 0.85	0.913	0.03038
● 0.9	0.965	0.02710
● 0.95	1.017	0.02432
● 1	1.068	0.02195

● Denotes preferred types.

Fig. 2.10 Wire table for insulated copper wires of up to 1 mm diameter (extract) (Siemens)

in an increase in resistance of about 4 per cent—well under the prescribed limit. As many wires of 1 mm thickness as possible are wound in parallel. Based on the bobbin dimensions in Figure 2.6, 26.5 mm/1.068 mm = 24.8—rounded down to 24—wires can be laid side-by-side; in terms of depth, 7 mm/1.068 mm = 6.55—i.e. 6 equal layers can be accommodated. (The number must, of course, always be rounded down.) There can thus be 24 × 6 = 144 turns. The largest possible divisor is 144/28 = 5.14, rounded down to 5. It is thus possible to wind five wires in parallel—i.e. all five wires together from the beginning. If sufficient winding accuracy is not attainable, four wires should be used. The 25 per cent higher copper loss is still acceptable.

If high-frequency litz wire is used for the winding, the possible number of layers must be checked.

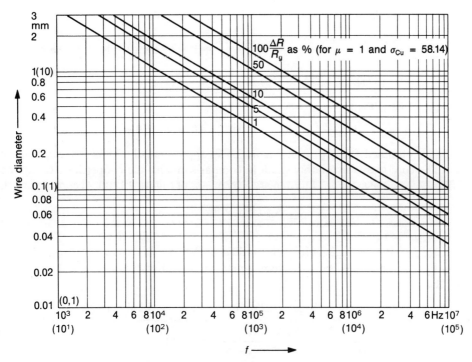

Fig. 2.11 Variation with frequency and wire diameter of resistance increase $\Delta R/R_d$ due to skin effect

Assumed: a winding of three layers is required. The litz wire may then have an outside diameter of 7 mm/3 = 2.33 mm. 26.5 mm/2.33 mm = 11.37, rounded down to 11 turns side-by-side. A total of 11 × 3 = 33 turns is possible. Since only 28 turns are required, it is possible to use this or any other litz wire with a maximum outside diameter of 2.33 mm. If two layers had been assumed, it would have been possible to accommodate only 14 turns (2 × 7) with an outside diameter of 7/2 = 3.5 mm, which would have been much too few.

Calculation of copper loss:

The loss can be calculated approximately from the A_R value given in the data sheet (applicable with $\theta = 70\,°C$).

$$R_{Cu} = A_R N^2 \quad (\Omega) \qquad (2.19)$$

Another possibility commonly used is to calculate the resistance from the mean turn length l_N (data sheet), the number of turns N and the resistance per unit length (Ω/m) (wire table, Figure 2.10). In this case, however, the increase in resistance of the copper wire based on $\alpha_{Cu} = 3.92 \times 10^{-3}/K$ must be allowed for. In accordance with Figure 2.5, a temperature rise of approximately 40 °C will be assumed.

From Equation (2.19), putting $A_R = 21 \times 10\,\Omega^{-6}$,

$$R_{Cu} = 21 \times 10^{-6}\,\Omega \times 28^2 = 0.0165\,\Omega;$$
$$R'_{Cu} = 0.0165\,\Omega\,(1 + 3.92 \times 10^{-3}\,K^{-1} \times 40\,K) = 0.019\,\Omega$$

In general terms, $R_{Cu} = l_N N R'$.

Calculated for the total winding (four windings in parallel):

$$R_{Cu} = \frac{105 \times 10^{-3}\,m \times 28 \times 0.02195\,\Omega/m}{4} \times (1 + 3.92 \times 10^{-3}\,K^{-1} \times 40\,K)$$

$$= 0.0187\,\Omega \; [17\,m\Omega]$$

The copper loss is then

$$W_{Cu} = I_{0max}^2 R_{Cu} = 100\,A^2 \times 0.019\,\Omega = 1.9\,W$$

(The figure obtained for P_{Cu} from Figure 2.5 is approximately 2.5 W—a comparable value.) The core loss is invariably small in comparison with the copper loss. It may be determined from the magnetic flux density B by means of Figure 2.12(a). According to [7],

$$\hat{B} = \frac{\delta_{Tmax} V_{min} \times 10^4}{f N A_e} \quad (T) \tag{2.20}$$

where V_{min} is the voltage across the inductor (V): in this case, for δ_{Tmax}, $V_{min} = V_{imin} - V_0 - V_L$ and A_e is the effective magnetic cross-sectional area (cm^2).

In the example,

$$A_e = 2.4\,cm^2 \quad \delta_{Tmax} = 0.28 \quad V_{min} = 20.4\,V - 5.2\,V = 15.2\,V$$

$$\hat{B} = \frac{0.28 \times 15.2\,V \times 10^4}{20 \times 10^{-3}\,s^{-1} \times 28 \times 2.4\,cm^2} = 32\,mT$$

From Figure 2.12(a), against $f = 20\,kHz$ and $\hat{B} = 32\,mT$, the value of W can be read off as 0.8 mW/g. (Figure 2.12 applied strictly only for a sinusoidal current: in this case it is approximately correct for the triangular current waveform.) With a total mass of 2.58 g, the total core loss is found to be $W_c = 0.8\,mW/g \times 2.58\,g = 0.1\,W$. This is about 1/20 of the copper loss, and thus negligible. (This applies generally to magnetic components with relatively large air gaps.)

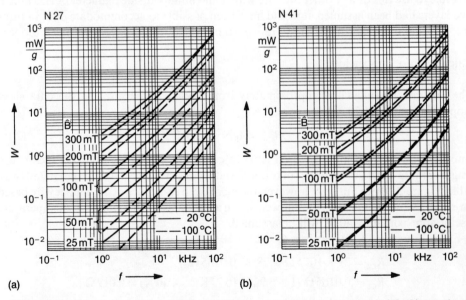

Fig. 2.12 Variation of relative core loss with frequency for N27 and N41 materials (Siemens)

2.3 CALCULATION OF OUTPUT CAPACITANCE C_0

The discontinuous-mode alternating current $\Delta I_L/2$ produces in the (ideal—i.e. loss-free) output capacitor C_0 an alternating charge ΔQ, such that the charges flowing into and out of the capacitor (measured over one cycle) are equal (Figure 2.13).

The discontinuous area is thus given by

$$\Delta Q = \frac{1}{2}\Delta I_L \frac{1}{2}\frac{T}{2} = \Delta I_L \frac{T}{8}$$

Applying the general relationship $Q = CV$,

$$C_0 \geqslant \frac{\Delta Q}{\Delta V_0} = \frac{\Delta I_L}{\Delta V_0}\frac{1}{8f} \quad \text{(F)} \tag{2.21}$$

where ΔI_L is the alternating inductor current (A_{pp}) (peak/peak), ΔV_0 is the maximum permissible output ripple voltage (V_{pp}): normally $\Delta V_0 \approx 0.01\,V_0$ and f is the frequency (Hz) (in most cases the switching frequency: in push–pull converters twice the switching frequency).

The value of capacitance given by Equation (2.21) is accurate only if the capacitor is ideal and has little effective series resistance and inductance. At high frequencies these conditions are met by, for example, polypropylene dielectric capacitors. If electrolytic capacitors are used correction must be made to allow for the parasitic series resistance and inductance. On the assumption that the voltages developed across the ESR and L_C are roughly equal,

$$\text{ESR} \leqslant \frac{\Delta V_0}{2\Delta I_L} \quad (\Omega) \tag{2.22}$$

and

$$L_C \leqslant \frac{\Delta V_0}{2\Delta I_L} \times \frac{\delta_{Tmin}(1-\delta_{Tmin})}{f} \quad \text{(H)} \tag{2.23}$$

If $\omega L_C < \text{ESR}$, this can be simplified to

$$\text{ESR}' \lesssim \frac{\Delta V_0}{\Delta I_L} \quad (\Omega) \tag{2.24}$$

If the ESR (at the relevant operating frequency) is not known, it can be calculated from the loss angle:

$$\text{ESR} = \frac{\tan\delta}{\omega C} \quad (\Omega) \tag{2.25}$$

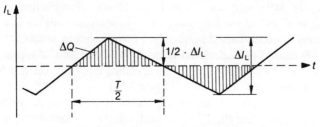

Fig. 2.13 Alternating current in capacitor C_0

Fig. 2.14 Equivalent circuit of an electrolytic capacitor

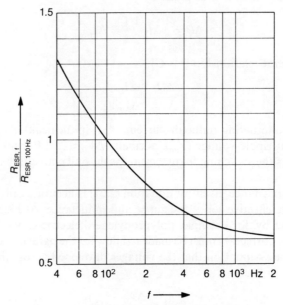

Fig. 2.15 Variation of equivalent series resistance
(ESR) with frequency for a low-voltage electrolytic
capacitor (typical characteristic, relative to f
$= 100\,\text{Hz}$) (Siemens)

Often the total impedance is quoted, in ohms:

$$Z = \sqrt{\text{ESR}^2 + (1/\omega C - \omega L_\text{c})^2} \quad (\omega) \tag{2.26}$$

If the frequency is not too high, the inductive component may if necessary be neglected
(Figure 2.14).

Since than δ increases less than in proportion to frequency (particularly at medium-to-
high ambient temperatures), the series loss resistance ESR (also denoted by R_ESR) must
decrease at higher frequencies in accordance with Equation (2.25) (Figure 2.15).

An extract from the data sheets for electrolytic capacitors suitable for switching duty is
shown in Figure 2.16 (typical values for medium-priced electrolytic capacitors).

The larger the capacitance C_0, the lower becomes the ESR. The capacitance itself then
assumes relatively little significance, and the choice of capacitor is determined by
Equation (2.22) or (2.24). In the frequency range of interest, roughly from 10 to 100 kHz,
ESR $\approx Z$ is almost independent of frequency. In some circumstances the increase in Z at
low temperatures cannot be neglected (Figure 2.17); above 20 °C Z is practically constant.
If a given impedance value is too large for a specified duty, it is often helpful to choose a
capacitor with a higher voltage rating (Figure 2.18).

The simplest, and usually the cheapest, practice is, however, to connect a number of
electrolytic capacitors in parallel. This is particularly satisfactory also in that a single

C_N	V_N	$\tan\delta_{max}$ 100 Hz 20°C	$R_{ESR\,max}$ 100 Hz 20°C	Z_{max} 10 kHz 20°C	$I_{R\,max}$ 5 min 20°C	$I_{C\sim max}$ 100 Hz 85°C	L_{ESL} ca.
(μF)	(V$-$)		(Ω)	(Ω)	(μA)	(mA)	(nH)
47		0.20	6.8	3.3	5	54	20
100		0.20	3.2	1.5	6	92	20
220		0.20	1.5	0.68	8	150	30
470	10	0.20	0.68	0.33	13	300	40
1000		0.20	0.32	0.15	24	460	40
2200		0.24	0.16	0.07	48	840	60
4700		0.28	0.09	0.05	98	1400	60
47		0.16	5.5	2.6	5	70	20
100		0.16	2.6	1.2	7	100	20
220		0.16	1.2	0.55	11	180	40
470	16	0.16	0.55	0.26	19	330	40
1000		0.16	0.26	0.12	36	560	40
2200		0.20	0.13	0.06	74	940	60
4700		0.24	0.08	0.05	150	1500	60
47		0.14	4.7	1.6	6	76	20
100		0.14	2.2	0.75	9	130	30
220	25	0.14	1.0	0.34	15	240	40
470		0.14	0.47	0.16	27	380	40
1000		0.14	0.22	0.08	54	720	60
2200		0.18	0.13	0.05	110	1000	60

Fig. 2.16 Extract from the data sheet for the low-voltage electrolytic capacitor Type B41588 (Siemens)

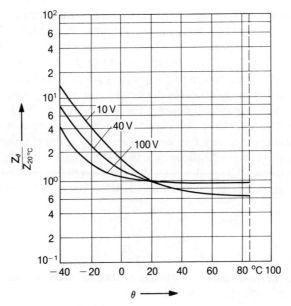

Fig. 2.17 Variation of impedance Z at $f = 10$ kHz with temperature, with operating voltage as parameter (typical characteristics, relative to $\theta = 20\,°$C) (Siemens)

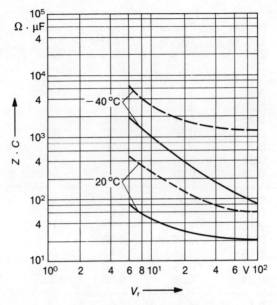

--- Maximum values from DIN 41257
——— Typical values

Fig. 2.18 Variation with rated voltage of time constant $Z \,°\mathrm{C}$ at $f = 10\,\mathrm{kHz}$ (relative to $1\,\mu\mathrm{F}$), with temperature as parameter (Siemens)

capacitor of equivalent capacitance often has a considerable overall height, which may lead to difficulties from the point of view of construction. If the same total capacitance is provided by several capacitors of lower capacitance, the overall height can be kept significantly lower. A larger mounting area is, of course, required, but the use of a number of electrolytic capacitors in parallel, for a given total capacitance, results in a lower overall impedance Z and a higher total alternating current rating. The ultimate choice is naturally influenced also by the cost, and in this regard no general rules can be given.

A further criterion in the selection of the output capacitor C_0 is its alternating current rating. The data sheet (Figure 2.16) quotes the r.m.s. value at $100\,\mathrm{Hz}$ and the maximum permissible ambient temperature (usually $85\,°\mathrm{C}$). What is required, however, is the r.m.s. value at the switching frequency f, in many cases at an ambient temperature other than $85\,°\mathrm{C}$. The alternating current ΔI_L flowing in the output capacitor is, again, given not as an r.m.s. value but in terms of the peak/peak value. This has therefore to be converted into the r.m.s. value. Since ΔI_L is of triangular waveform, the conversion factor to be applied is not $\sqrt{2}$ but $\sqrt{3}$:

$$I_{\mathrm{Crms}} = \frac{\Delta I_L}{2\sqrt{3}} \quad (\mathrm{A}) \tag{2.27}$$

Conversion from the value given for $100\,\mathrm{Hz}$ to that for the switching frequency f—e.g. $20\,\mathrm{kHz}$—is effected with the aid of Figure 2.19.

The conversion from the maximum specified ambient temperature to that given in the data manual (usually $85\,°\mathrm{C}$) can be carried out using the curve of Figure 2.20. The maximum operating life (statistical service life) of an electrolytic capacitor depends

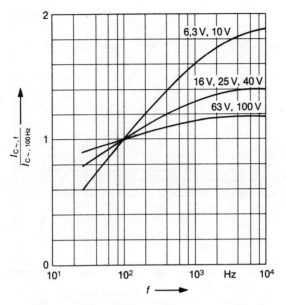

Fig. 2.19 Variation of alternating current rating $I_{C\sim}$ with frequency (relative to $f = 100\,\text{Hz}$) (Siemens)

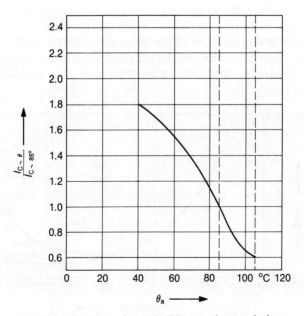

Fig. 2.20 Variation of permissible superimposed alternating current $I_{C\sim}$ with ambient temperature (relative values) (Siemens)

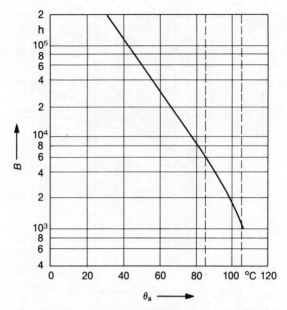

Fig. 2.21 Variation of operating life B with ambient
temperature (Siemens)

strongly upon its operating temperature—i.e. the maximum prevailing ambient
temperature—and the internal heating due to the alternating-current loading
(Figure 2.21).

For high switching frequencies ($> 100\,kHz$) it is advantageous to use the so-called four-
terminal types, in which the connecting leads are taken through the capacitor. The
capacitor foils are welded over the whole length of the wires, with the result that the
impedance remains low even at very high frequencies (Figure 2.22).

In this case the capacitor tables give the permissible r.m.s. alternating current over the
frequency range $10\,kHz$ to $1\,MHz$, for a temperature rise of $10\,°C$. The correction factors

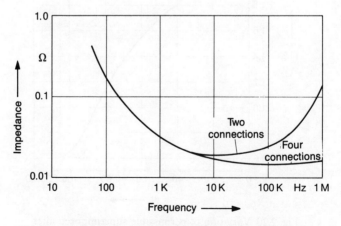

Fig. 2.22 Variation of impedance Z with frequency for two-
terminal and four-terminal electrolytic capacitors (CDE)

Frequency (Hz)	Temperature rise					
	5°C	10°C	15°C	20°C	25°C	30°C
10 K–1 M	0.65	1.00	1.25	1.45	1.70	2.00

Fig. 2.23 Variation with frequency of the temperature rise of a four-terminal electrolytic capacitor (CDE)

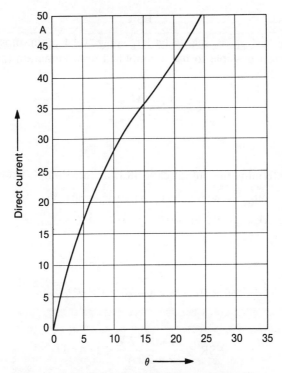

Fig. 2.24 Variation of the temperature rise of a four-terminal electrolytic capacitor with direct current flowing in the connections (CDE)

for different temperature rises can be obtained from Figure 2.23. In these capacitors the direct current flowing in the leads also produces a rise in temperature, which must be taken into account in accordance with Figure 2.24.

With a variable load, and a possible step change in load of $\Delta I_{0\max}$, the capacitor C_0 must be made sufficiently large that the permissible deviation $\Delta V_{0\max}$ in the output voltage up to the end of the transient recovery period t_{tr} (when the inductor current has adjusted to the new load) is not exceeded. This voltage deviation $\Delta V_{0\max}$ (normally up to 0.1 V_0) must not be confused with the much smaller voltage fluctuation ΔV_0 produced by the alternating component of current in the inductor.

$$C_0 \geqslant t_{tr} \frac{\Delta V_{0\max} - \sqrt{\Delta V_{0\max}^2 - \text{ESR}^2 \Delta I_{0\max}^2}}{\text{ESR}^2 \Delta I_{0\max}} \quad \text{(F)} \qquad (2.28)$$

where t_{tr} is the transient recovery time of inductor current (known in connection with the calculation of the inductance) (s), ΔV_{Omax} is the deviation of output voltage V_0 due to a load change (V), ΔI_{Omax} is the maximum load change (A) and ESR is the equivalent series resistance of electrolytic capacitor

$$C_0 \ (\Omega).$$

The largest reasonable value for the capacitance C_0 is that which gives a time constant $\text{ESR}\,C_0 = t_{tr}$. A further increase in C_0 gives no further reduction in ΔV_{Omax}.

$$C_0 \leqslant \frac{t_{tr}}{\text{ESR}} \quad \text{(F)} \tag{2.29}$$

If the permissible ΔV_{Omax} is relatively large (e.g. $0.1\,V_0$), in most cases $\Delta V_{Omax} > \text{ESR}\,\Delta I_{Omax}$. It is then possible to use a simplified approximation (not deducible from Equation (2.28)).

$$C_0 \approx \frac{t_{tr}}{2} \times \frac{\Delta I_{Omax}}{\Delta V_{Omax}} \quad \text{(F)} \tag{2.30}$$

Example

It is required to determine the output capacitor for the example previously considered.

$$\theta_{amax} = 40\,°\text{C}; \quad \Delta V_0 = 0.01\,V_0 = 50\,\text{mV};$$
$$\Delta V_{Omax} = 0.1\,V_0 = 0.5\,\text{V}; \quad \Delta I_0 = 5\,\text{A}.$$

Equation (2.21):

$$C_0 \geqslant \frac{2.2\,\text{A}}{50 \times 10^{-3}\,\text{V} \times 8 \times 20 \times 10^3\,\text{s}^{-1}} = 275\,\mu\text{F}$$

Equation (2.22):

$$\text{ESR} \leqslant \frac{50 \times 10^{-3}\,\text{V}}{2 \times 2.2\,\text{A}} \leqslant 11.4\,\text{m}\Omega$$

Equation (2.32):

$$L_C \leqslant \frac{50 \times 10^{-3}\,\text{V}}{2 \times 2.2\,\text{A}} \times \frac{0.207\,(1 - 0.207)}{20 \times 10^3\,\text{s}^{-1}} \leqslant 93\,\text{nH}$$

From the table of Figure 2.16, two capacitors of $4700\,\mu\text{F}$ each give an ESR value of $50\,\text{m}\Omega/2 = 25\,\text{m}\Omega$ and an L_C of $60\,\text{nH}/2 = 30\,\text{nH}$. This inductance is, however, very much less than the acceptable value, so that Equation (2.24) can be applied. Thus the ESR value calculated above can be doubled; i.e. $\text{ESR} = 22.8\,\text{m}\Omega$. The value obtained from the table is not significantly higher than the permissible value, so that the choice of two capacitors, each of $4700\,\mu\text{F}$, can stand. The required storage capability according to Equation (2.28) is now to be checked.

$$C_0 \geqslant 500 \times 10^{-6}\,\text{s} \times \frac{0.5\,\text{V} - \sqrt{(0.5\,\text{V})^2 - (25 \times 10^{-3}\,\Omega)^2 \times (5\,\text{A})^2}}{(25 \times 10^{-3}\,\Omega)^2 \times 5\,\text{A}}$$

$$\geqslant 2540\,\mu\text{F}$$

In fact the simpler Equation (2.30) could have been used, for $\Delta V_{0max} = 0.5$ V is much larger than $25\,\text{m}\Omega \times 5\,\text{A} = 0.13$ V. Equation (2.30):

$$C_0 \geqslant \frac{500 \times 10^{-6}\,\text{s}}{2} \times \frac{5\,\text{A}}{0.5\,\text{V}} \geqslant 2500\,\mu\text{F, almost the same.}$$

Equation (2.29):

$$C_0 \leqslant \frac{500 \times 10^{-6}\,\text{s}}{25 \times 10^{-3}\,\Omega} \leqslant 20\,000\,\mu\text{F}$$

According to this criterion four capacitors of $4700\,\mu\text{F}$ each might be called for, but this would be excessively expensive and is not in fact necessary.

The alternating current loading is

$$I_{Crms} = \frac{2.2\,\text{A}}{2 \times 1.73} = 0.64\,\text{A}$$

Permissible alternating current loading:

From Figure 2.16 (with $f = 100$ Hz and $\theta_a = 85\,°\text{C}$) $I_\sim = 2.14\,\text{A} = 2.8\,\text{A}$. The correction factor for the much higher switching frequency can be read off from Figure 2.19 as 1.9. The correction factor for the lower temperature of $40\,°\text{C}$ (Figure 2.20) is 1.8.

Total correction factor: $k = 1.9 \times 1.8 = 3.42$.

Total permissible alternating current loading:

$I_{\sim per} = 2.8\,\text{A} \times 3.42 = 9.6\,\text{A}$. The permissible current loading (calculated for a temperature rise of $10\,°\text{C}$) is thus more than ten times the actual loading, and the capacitor C_0 is hardly affected by the alternating current that flows in it.

2.4 DIMENSIONING OF THE POWER-SWITCHING CIRCUIT

To assist in calculating the losses in the transistor and the free-wheel diode, Figure 2.25 shows a complete circuit of the step-down inductor-type converter, including a possible control IC.

Transistor requirements:

$$V_{CER} \approx V_{CE0} \geqslant V_{imax} \quad \text{(V)} \tag{2.31}$$

$$I_{Cmax} \geqslant I_{Lmax} \quad \text{(A)} \tag{2.32}$$

Free-wheel diode requirements:

$$V_R \geqslant V_{imax} \quad \text{(V)} \tag{2.33}$$

$$I_{FAV} \geqslant I_{0max}(1 - \delta_{Tmin}) \quad \text{(A)} \tag{2.34}$$

$$I_F \geqslant I_{Lmax} \quad \text{(A)} \tag{2.35}$$

The selected power transistor (in this case a Darlington transistor with a free-wheel diode in a common housing) has the following data: $V_{CEmax} = 60$ V; $I_{Cmax} = 15$ A. Electrical stress equal for both: $V_{imax} = 27.6$ V < 60 V; $I_{Lmax} = 11$ A < 15 A. The power stage is thus adequately rated.

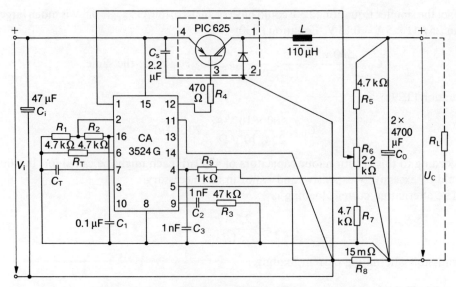

Fig. 2.25 Complete circuit of an inductor-coupled step-down converter including the control IC CA3524G, with possible dimensioning (Unitrode)

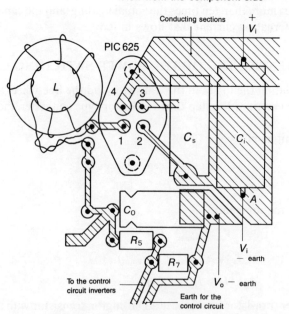

Fig. 2.26 Suggestion of the manufacturers of the PIC625 power integrated circuit for a circuit-board layout to obtain low stray emissions (Unitrode)

A glance at the forward characteristic of the free-wheel diode shows that the power integrated circuit PIC625 (Figure 2.26) contains not a Schottky diode (which would be appropriate for the low output voltage of 5 V) but a simple silicon diode.

Since this makes V_0^* significantly higher, it is necessary to re-calculate the duty cycles.[†] It would, of course, be possible to use another fast Darlington transistor with a separate Schottky free-wheel diode, but that would mean using two components in place of the single component in this instance.

$$V_0^* = 5\,\text{V} + 1.45\,\text{V} + 0.15\,\text{V} = 6.6\,\text{V}$$

$$\delta_{\text{Tmin}} = \frac{6.6\,\text{V}}{27.6\,\text{V}} = 0.24; \quad \delta_{\text{Tmax}} = \frac{6.6\,\text{V}}{20.4\,\text{V}} = 0.32;$$

$$\delta_{\text{Tnom}} = \frac{6.6\,\text{V}}{24\,\text{V}} = 0.28$$

The minimum 'on' time is then

$$t_{1\min} = \delta_{\text{Tmin}} T = 0.24 \times 50\,\mu\text{s} = 12\,\mu\text{s}$$

The switching times, according to the data sheet, are:

$$\text{rise time: } t_r = 0.24\,\mu\text{s; fall time } t_f = 0.6\,\mu\text{s.}$$

The switching times are thus short in comparison with the minimum 'on' time, and the PIC625 is a suitable component.

The losses in the power IC may be calculated for a mean (nominal) duty cycle $\delta_T = 0.28$. This is appropriate in the present case, because the maximum losses occur with δ_{Tmax} in the

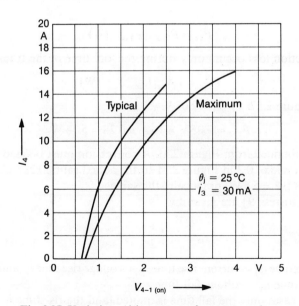

Fig. 2.27 Variation with current of on-state voltage drop in the transistor (Unitrode)

[†] Since the deviation is not too great, it is not necessary to correct the previous calculations.

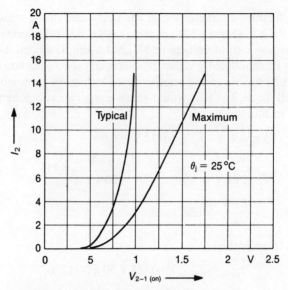

Fig. 2.28 Variation with current of voltage drop
across the free-wheel diode (Unitrode)

transistor and with δ_{Tmin} in the free-wheel diode in the same housing. If the components were separate, the calculations would have to be based on δ_{Tmax} for the transistor and δ_{Tmin} for the diode. No component must be overloaded in the worst case.

The power loss in the transistor consists of the conduction loss P_{Trc} and the switching loss P_{Trs}.

$$P_{\text{Tr}} = P_{\text{Trc}} + P_{\text{Trs}} \quad (\text{W}) \tag{2.36}$$

Since the conduction loss occurs only during the 'on' time of the transistor, in general

$$P_{\text{Trc}} = V_{\text{CEsat}} I_{\text{0max}} \delta_{\text{T}} \quad (\text{W}) \tag{2.37}$$

read off from Figure 2.27: $V_{\text{CEsat}} = 2.05\,\text{V}$.

$$P_{\text{Trc}} = 2.05\,\text{V} \times 10\,\text{A} \times 0.28 = 5.74\,\text{W}$$

A check can be obtained from Figure 2.31. A conversion must be made, however, from the quoted duty cycle of 0.2 in Figure 2.31 to the actual ratio (0.28). Using Figure 2.31, $P_{\text{Trc}} = 4\,\text{W} \times 0.28/0.2 = 5.6\,\text{W}$, practically the same.

The switching losses [9] are given by

$$P_{\text{Trs}} = V_{\text{i}} \frac{t_r + t_f}{2} I_0 f \quad (\text{W}) \tag{2.38}$$

where t_r is the rise time (s) = current rise time t_{ri} + voltage rise time t_{rv} and t_f is the fall time (s) = current fall time t_{fi} + voltage fall time t_{fv}.

Since in many cases only the fall time is quoted (and the rise time is usually not very different from the fall time), the switching loss can also be calculated from

$$P_{\text{Trs}} = V_{\text{i}} t_f I_0 f \quad (\text{W}) \tag{2.39}$$

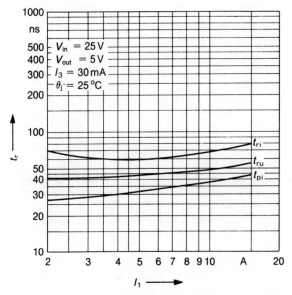

Fig. 2.29 Variation of rise time t_r of the transistor with
current (Unitrode)

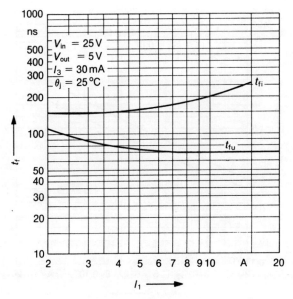

Fig. 2.30 Variation of fall time t_f of the transistor
with current (Unitrode)

34

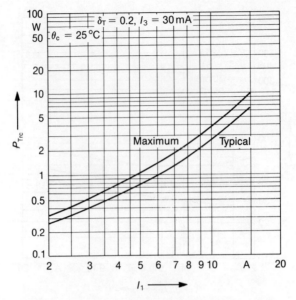

Fig. 2.31 Variation of conduction loss P_{Trc} of the
transistor with current for $\delta_T = 0.2$ (Unitrode)

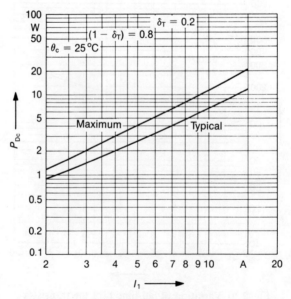

Fig. 2.32 Variation of conduction loss P_{Dc} of the free-
wheel diode with current for $\delta_T = 0.2$ (Unitrode)

The rise time (at 10 A) can be obtained from Figure 2.29:

$$t_r = 70 \text{ ns} + 50 \text{ ns} = 120 \text{ ns}$$

Figure 2.30 gives the fall time:

$$t_f = 200 \text{ ns} + 70 \text{ ns} = 270 \text{ ns}$$

The switching loss is thus

$$P_{Trs} = 24 \text{ V} \times \frac{120 \times 10^{-9} \text{ s} + 270 \times 10^{-9} \text{ s}}{2} \times 20 \times 10^3 \text{ s}^{-1} \times 10 \text{ A}$$

$$= 0.94 \text{ W}$$

Check from Figure 2.33: $P_{Trs} = 0.95$ W (a typical value, because the switching times are quoted as typical values; the maximum is about 1.5 W). As can be seen from Equation (2.39), the switching loss is proportional to, among other factors, the input or collector–emitter voltage.

In inductor-type converters, with their relatively low input voltage, it is thus important to keep the *conduction loss* as low as possible. This implies, however, that the switching transistor should be driven as far into saturation as possible, in order to achieve the lowest possible V_{CEsat}. This has an adverse effect on the switching times, but these are not highly significant, as the figures show. In converters supplied from the mains—e.g. in the forward converter with a transformer—the situation is reversed: with the much higher voltage V_i the switching losses are correspondingly large. In that case it is better to operate the transistor not in the saturated mode, but in the so-called 'quasi-saturated' mode, by means of special drive circuits. The higher conduction loss is then accepted in order to keep the switching losses to a minimum. This increase in conduction loss due to a higher V_{CE} is at

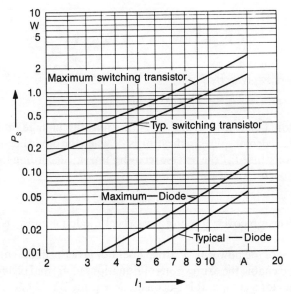

Fig. 2.33 Variation with current of switching losses P_{Trs} in the transistor and P_{Ds} in the free-wheel diode at $f = 20$ kHz (Unitrode)

least partly compensated by the lower primary current. This will be considered in more detail later.

The off-state loss, which should theoretically be taken into account, can in practice be completely neglected.

$$P_{TrR} = V_i I_R (1 - \delta_{Tmin}) \quad (W) \tag{2.40}$$

where I_R is the leakage current (A) at θ_{amax}.

In the example, with $I_R = 10\,\mu A$ at $\theta_c = 100\,°C$:

$$P_{TrR} = 24\,V \times 10 \times 10^{-6}\,A \times 0.76 = 0.18\,mW$$

This is so small in relation to the other losses that in principle the off-state loss can be neglected, even in mains-supplied converters with a voltage ten times higher. The maximum total loss in the transistor is accordingly

$$P_{Tr} = 5.74\,W + 1.5\,W = 7.24\,W$$

The power loss in the free-wheel diode similarly consists of a conduction loss P_{Dc} and a switching loss P_{Ds}.

$$P_D = P_{Dc} + P_{Ds} \quad (W) \tag{2.41}$$

Since the conduction loss in the free-wheel diode occurs only during the 'off' time of the transistor, in general

$$P_{Dc} = V_F I_0 (1 - \delta_{Tmin}) \quad (W) \tag{2.42}$$

V_F forward voltage drop of diode (V) at I_{0max}.

In the example (with $\delta_{Tnom} = 0.28$), Figure 2.28 gives, at $I_0 = 10\,A$, $V_{Fmax} = 1.45\,V$

$$P_{Dc} = 1.45\,V \times 10\,A \times 0.72 = 10.44\,W$$

Check with Figure 2.32:

$$P_{Dc} = 11\,W \times 0.72/0.8 = 9.9\,W$$

which is practically the same.

The switching loss is given by [17]

$$P_{Ds} = Q_{rr} V_R f \quad (W) \tag{2.43}$$

where Q_{rr} is the stored charge (A s) and V_R is the blocking voltage (in this case V_i) (V).

To avoid excessive current spikes in the transistor when the diode begins to block with a delay (reverse recovery time t_{rr}), the reverse recovery time t_{rr}, according to [9], must be less than the rise time of the transistor.

$$t_{rr} \leqslant 0.5\,t_r \quad (s) \tag{2.44}$$

where t_{rr} is the reverse recovery time of free-wheel diode (s) and t_r is the rise time of transistor (s).

The manufacturer of the power IC PIC625 does not give the data required here for the free-wheel diode. To enable the expressions of Equations (2.43) and (2.44) to be estimated, the numerical values for a comparable silicon diode, e.g. the BYV29/300, will be used. The data for this diode are: $I_{FAV} = 9\,A$; $V_{RRM} = 300\,V$; V_F at $10\,A = 1.15\,V$; reverse recovery time: $t_{rr} = 45\,ns$; stored charge: $Q_{rr} = 50\,nA\,s$. Equation (2.44) demands that

$t_{rr} \leqslant 0.5 t_r \leqslant 0.5 \times 120 \times 10^{-9}\,\text{s} \leqslant 60\,\text{ns}$. Equation (2.44) is thus satisfied. (It may be observed that still faster diodes are available.)

Since the mean forward current rating I_{FAV} of this diode is given as 9 A, a check is desirable to see how large the mean direct current in the free-wheel diode can be in the worst case. Equation (2.34):

$$I_{FAVmax} = I_{0max}(1 - \delta_{Tmin}) = 10\,\text{A} \times 0.76 = 7.6\,\text{A} < 9\,\text{A}$$

Switching loss from Equation (2.43):

$$P_{Ds} = 50 \times 10^{-9}\,\text{A} \times 27.6\,\text{V} \times 20 \times 10^3\,\text{s}^{-1} = 27.6\,\text{mW}$$

This figure is so small that it can be entirely left out of account, even if the inbuilt diode has somewhat different characteristics. The maximum diode switching losses at $I_0 = 10\,\text{A}$ can be read off from Figure 2.33 as 55 mW.

From this it may be concluded as a general rule that in inductor-type converters, again because of the relatively low input voltage, the switching losses in the free-wheel diode can be neglected. This is emphasized by the fact that, in the case of the Schottky diodes which are in fact applicable here, a stored charge is hardly ever quoted, because it is so small. These observations apply also to the free-wheel diodes in mains-supplied converters, because the secondary voltages are always relatively low. They do not apply, however, to the possible use of fast diodes on the mains (high-voltage) side, or, necessarily, at much higher switching frequency. The total power loss in the PIC625 power IC is thus

$$P = 5.74\,\text{W} + 1.5\,\text{W} + 10.44\,\text{W} = 17.68\,\text{W}, \text{ or in round figures, } 18\,\text{W}$$

(If a Schottky free-wheel diode were used it would be around 12 to 13 W.)

For the heatsink required to dissipate the heat loss,

$$R_{thca} \leqslant \frac{\theta_j - \theta_a}{W} = R_{thjc} \quad (\text{K/W}) \tag{2.45}$$

R_{thca} is the thermal resistance of heatsink (kW), R_{thjc} is the thermal resistance from semiconductor junction to housing (kW) (data sheet), θ_j is the maximum permissible junction temperature (°C) (data sheet), θ_a is the maximum ambient temperature (°C) and W is the total loss power to be dissipated (W).

For the example calculated above with the PIC625, based on $\theta_{jmax} = 150\,°\text{C}$ and, from the data sheet, $R_{thjc} = 4\,\text{kW}$, the heatsink requirement is given by

$$R_{thca} \leqslant \frac{150\,°\text{C} - 40\,°\text{C}}{18\,\text{W}} - 4\,\text{kW} \leqslant 2.1\,\text{kW}$$

A suitable type with a thermal resistance not exceeding this value must be selected from the heatsink catalogues.

If in place of the power IC Type PIC625 (isolated housing) the Type PIC645, electrically similar but with a non-isolated housing, is used, the heatsink can be made smaller and more conveniently dimensioned, on account of the lower value of R_{thjc} (2 kW). The printed-circuit layout has to be modified accordingly, because Terminal 2 (earth) is in this case connected to the housing, and not to Pin 4 as with the PIC625. (For further details concerning Figure 2.25, see also Chapter 8.)

In the drive IC Type CA 3524 G used here, Terminal 1 is connected to the inverting input of the error amplifier; Terminal 2 is the non-inverting input, and is connected to the voltage reference. Since $V_{ref} = 5$ V (4.6 to 5.4 V) and the error amplifier has a common-mode control voltage range up to only 3.4 V, the reference voltage has to be divided down to less than this value—e.g. halved, to 2.5 V ($R_1 + R_2$, e.g. 4.7 kΩ each). The output voltage is required to be 5 V exactly, so this voltage also must be divided down by R_5 and R_7, similarly of 4.7 kΩ each. The 2.2 kΩ potentiometer R_6 enables the tolerances of the voltage-divider resistances and the reference voltage to be compensated for, and permits precise adjustment of the output voltage.

The reference voltage should be decoupled by means of C_1 (about 0.1 μF); C_2 (1 nF) and R_3 (47 kΩ) are provided for frequency compensation. Resistor R_T (e.g. 3 kΩ) and capacitor C_T (e.g. 20 nF—see data sheet, Chapter 8) determine the frequency $f = 20$ kHz.

Short-circuit current limiting is effected by means of R_8. With a fault voltage of 0.2 V (positive on Pin 4 and negative on Pin 5), and a short-circuit current of e.g. $I_{sc} = 13$ A, R_8 becomes 0.2 V/13 A = 15.4 mΩ (obtained using a link of resistance wire). R_4 carries the control current of 30 mA (according to the data sheet). The value is given by

$$R_4 \leqslant \frac{V_{imin} - V_{BE} - V_{CEsat}}{30 \text{ mA}} \quad (\Omega) \tag{2.46}$$

where V_{BE} is the base-emitter voltage of driver transistor in the PIC625 (V) (about 0.7 V) and V_{CEsat} is the saturation voltage of output transistor in the drive IC (max. 2 V at $I_C = 50$ mA; Terminals 12/11).

R_4 thu becomes

$$R_4 \leqslant \frac{20.4 \text{ V} - 0.7 \text{ V} - 2 \text{ V}}{30 \times 10 \text{ A}} \leqslant 0.59 \text{ k}\Omega$$

The selected standard value is $R_4 = 0.47$ kΩ/1 W.

The input capacitor C_i (e.g. 47 μF/40 V) improves the stability of the circuit and should be connected close to the power IC (see Figure 2.26). Capacitor C_s (e.g. 0.22 to 2.2 μF) reduces r.f. radiated interference originating from the rapid blocking of the free-wheel diode, and should be arranged directly between Terminals 2 and 4 of the PIC625. The most suitable type for this purpose is the multi-layer ceramic capacitor.

A comparison may now be made between the calculated results in the example according to Figure 2.25 and the measured values with nominal input voltage ($V_i = 24$ V), $V_0 = 5$ V and maximum load current $I_{Lmax} = 10$ A. Typical values are assumed; the inductor winding is of four 1 mm-diameter wires in parallel. The measured values are indicated in [ ].

$d_{Tmax} = 0.32$ [0.34]; $\quad \delta_{Tmin} = 0.24$ [0.24]; $\quad \delta_{Tnom} = 0.28$ [0.30];

$V_L = 0.2$ V estimated [0.17 V]; $\quad W_L = I^2 R_{Cu} = 10^2 \text{ A}^2 \times 17 \text{ m}\Omega = 1.7$ W

Total losses:

$P_{tot} = P_{Trc} + P_{Trs} + P_{Dc} + P_{Ds} + P_{R4} + P_L + P_{R8}$

$\quad\quad = 5.74 \text{ W} + 0.95 \text{ W} + 10.44 \text{ W} + 0.055 \text{ W} + 0.42 \text{ W} + 1.7 \text{ W} + 1.54 \text{ W} = 20.85$ W

$P_i = P_0 + W_{tot} = 50 \text{ W} + 20.85 \text{ W} = 70.85 \text{ W}$ [69.8 W]

$$\eta = \frac{P_0}{P_i} = \frac{50 \text{ W}}{70.85 \text{ W}} = 71\% \, [72\%]$$

$L = 110$ H [118 μH]

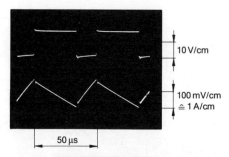

Fig. 2.34 Triangular-waveform current in the inductor (measured by the voltage drop across a $0.09\,\Omega$ resistor in series with the inductor) (lower trace) and the voltage across the inductor (inverted, due to the measuring circuit with its point at the input to the (inductor) (upper trace) in the circuit of Figure 2.25 (oscilloscope d.c.-coupled)

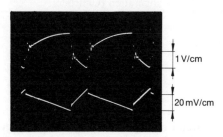

Fig. 2.35 Voltage at the base of the PIC625 (Terminal 3) (upper trace) and output ripple (lower trace) in the circuit of Figure 2.25 (oscilloscope i.c.-coupled)

$\Delta I_L = 2.2\,\text{A}$ [approximately 2 A from Figure 2.34]

$\Delta V_0 \leqslant 50\,\text{mV}_{pp}$ [approximately 28 mV$_{pp}$ from Figure 2.35]

Figures 2.34–2.36 show that the measured values are in very good agreement with the postulated values and calculated results. The deviation in output voltage with input voltage variation ($V_i = 24\,\text{V} \pm 15\%$) at I_{Lmax} and $I_{Lmax}/2$ was a few millivolts; the deviation in output voltage with variation in load current ($I_L = 1$ to 10 A) at $V_i = 24\,\text{V}$ was better than 1 per cent. The inferior result with load variation was presumed to be due to contact resistances. Since, however, the results were in any case quite good, this was not investigated further.

If the converter is to be operated at a significantly higher frequency than 20 kHz (e.g. 100 kHz or higher), the circuit should be modified as shown in Figure 2.36 [9].

By virtue of the catching diode D (1N914), resistors R_5 (56 Ω) and R_{12} (100 Ω), together with the constant-current supply through Tr_2, the total storage time is reduced to around 10 ns. The maximum possible operating frequency, according to [9], can be calculated (for

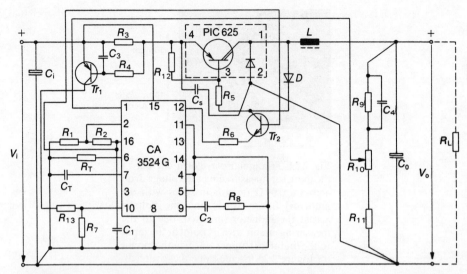

Fig. 2.36 Complete circuit of an inductor-coupled step-down converter, including the CA3524G control circuit with constant-current feed to the power IC, for high-frequency operation (Unitrode)

converters in general) from

$$f_{max} = \frac{0.2\,V_0}{V_{imax}\,t_{smax}} \quad \text{(Hz)} \tag{2.47}$$

where t_{smax} is the maximum storage time (data sheet).

For the PIC625 in the circuit of Figure 2.25, t_{smax} is given in the data sheet as 600 ns:

$$f_{max} = \frac{0.2 \times 5\,\text{V}}{27.6\,\text{V} \times 600 \times 10^{-9}\,\text{s}} = 60\,\text{kHz}$$

With $t_s = 10$ ns, on the other hand, this is increased to 360 kHz.

The resistance R_6 is calculated from

$$R_6 = \frac{V_{ref} - V_{BE/Tr2}}{30\,\text{mA}} = \frac{4.3\,\text{V}}{30\,\text{mA}} = 143\,\Omega; \quad \text{selected: } R_6 = 150\,\Omega$$

Current limiting is here effected in a somewhat different manner from that shown in Figure 2.25. When Tr_1 is turned on (at about $0.7\,\text{V} = V_{BE}$) the shut-down transistor in the drive IC, Terminal 10, is turned on and inhibits the output pulses.

$$R_3 = \frac{0.7\,\text{V}}{I_{sc}} \quad (\Omega) \tag{2.48}$$

where I_{sc} is the short-circuit current (A)—about $1.2 I_{0max}$.

Resistor R_4 (47 Ω), with capacitor C_3 (e.g. 750 pF) prevents an excessively rapid response of the current-limiting circuit. Resistors R_{13} (10 kΩ) and R_7 (2.2 kΩ) limit the current to the shut-down transistor. Resistor R_8 is differently dimensioned here (compared with R_3 in Figure 2.25) at 18 kΩ. Capacitor C_4 (22 nF) improves the control response of the circuit. With $R_T = 2.2$ kΩ and $C_T = 2.2$ nF the circuit operates in this case at $f = 250\,\text{kHz}$. This circuit can be used with a PIC600 up to $I_0 = 2\,\text{A}$ and with a PIC625 up to $I_0 = 5\,\text{A}$; $V_0 = 5\,\text{V}$ in each case. The inductance L and the output capacitance C_0 have naturally to be determined afresh for operation at this high switching frequency. Since these calculations have already been carried out for the converter of Figure 2.25, there is no need to repeat them here.

2.5 CALCULATIONS FOR A DRIVE CIRCUIT FOR AN *n-p-n* POWER TRANSISTOR

The use of *p-n-p* transistors, or the corresponding *p-n-p* power ICs or *p-n-p* Darlington transistors, in accordance with Figures 2.25 and 2.36 makes for simple circuit design. If suitable *p-n-p* transistors are not available, however, the drive to the power stage must be provided by means of an additional driver transistor with a pulse transformer, as shown in Figure 2.37.

In providing the drive for the *n-p-n* power transistor Tr_1 it is necessary to supply at least a certain amount of power, for the achievement of a low saturation voltage V_{CEsat} requires the injection of a sufficient base current. This both eliminates the effect of the spread of current gains B and ensures that Tr_1 does not come out of saturation. This, however, means providing a relatively large series resistance R_B, which drops a voltage several times higher than V_{BEsat}. To facilitate comparison between the various circuits, an output voltage $V_0 = 5\,\text{V}$, $I_{0\text{max}} = 10\,\text{A}$ and $f = 20\,\text{kHz}$ will again be considered. The parameters of this circuit, shown in Figure 2.37, thus correspond to those of Figure 2.25: i.e. the inductor L, the output capacitor C_0 and the connections of the drive IC are the same (not all the circuit elements are shown, in the interests of clarity: the omitted components are as shown in Figure 2.25). The only differences are in the power transistor Tr_1 with its driver stage and the diode D_3. With $I_0 = 10\,\text{A}$, Tr_1 must be a high-current low-voltage transistor: only a few such types are listed in the data manuals. The requirements for Tr_1 are: $V_{\text{CE0}} \geqslant 50\,\text{V}$; $I_{\text{Cmax}} \geqslant 11\,\text{A}$; β_{min} at $I_\text{C} = 10\,\text{A}$: 20 to 30; V_{CEsat} at $I_\text{C} = 10\,\text{A}$ as low as possible; short switching times.

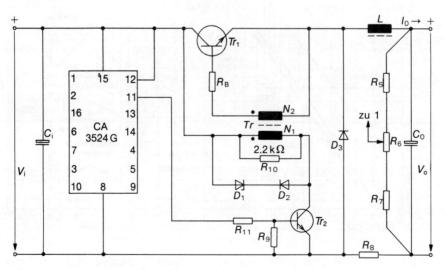

Fig. 2.37 Circuit of an inductor-coupled step-down converter using and *n-p-n* power transistor with drive through a pulse transformer

42

Suitable types are, for example, the Unitrode 2N5658 ($I_{Cmax} = 20$ A; $V_{CE0} = 80$ V; β_{min} at $I_C = 10$ A: 30; V_{CEsat} at $I_C = 10$ A and $I_B = 1$ A: 1 V) and the RCA 2N6686 ($I_{Cmax} = 25$ A; $V_{CE0} = 160$ V; β_{min} at $I_C = 10$ A: 25; V_{CEsat} at $I_C = 20$ A and $I_B = 2$ A: 1.5 V). Since the 2N6686 was readily available, this type was used. Although V_{CEsat} is not quoted at $I_C = 10$ A, it can be assumed that it will not be greater than 1 V, particularly as the typical value is given as only 0.2 V. A base current of 1 A should accordingly be sufficient. t_r and t_f are quoted as $0.15\,\mu s$ and $0.6\,\mu s$ respectively with $I_C = 10$ A and $I_C/I_B = 10$, and so are in accord with the dimensioning arrived at above.

The secondary voltage of the pulse transformer T should on the one hand be high enough to enable the base current to be determined by R_B, but on the other low enough to avoid excessive dissipation in R_B. A secondary voltage of 5 V is chosen as a compromise. This also has the merit that it is within the base-emitter breakdown voltage (about 6 to 7 V). This means that at $I_B = 1$ A the transformer has to transmit a power of about 5 W, which at an estimated efficiency of 80 to 85 per cent corresponds to a primary power of around 6 W.

$$R_B = \frac{5\,V - 1\,V}{1\,A} = 4\,\Omega; \text{ selected standard value: } 3.9\,\Omega/2\,W$$

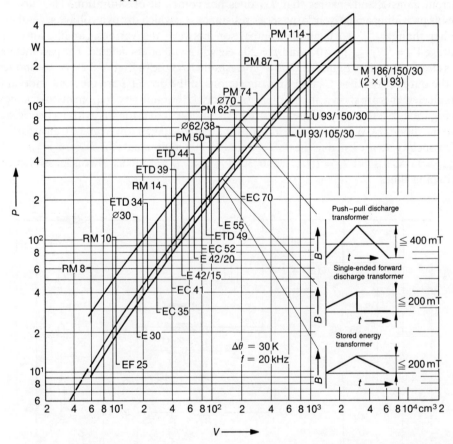

Fig. 2.38 Power-handling capability P and unit volume V of transformers with ferrite cores of N27/N41 material (guidance values) at $f = 20$ kHz for a temperature rise of 30 K in various modes of operation (Siemens)

(The voltage of 1 V subtracted here in the numerator corresponds approximately to V_{BEsat}.) The power loss dissipated in R_B is $W_{\text{RB}} = I_B^2 R_B \delta_{\text{Tmax}} = 1\,\text{A}^2 \times 3.9\,\Omega \times 0.28 = 1.1\,\text{W}$ (using the original value for δ_T). The next step is to choose the core size for T. For this purpose the curves of Figure 2.38, Figure 2.39 or Figure 2.40 may be used.

From Figure 2.38 (middle curve, for the single-ended forward transformer) a component volume of around 4 to 5 cm^3 can be read off by extrapolation for the power of 6 W to be transmitted. The smallest core included in this figure is the Type RM 8, or the EF 25; both are somewhat too large. The component volume can, however, be calculated alternatively from

$$V = \text{core width} \times \text{overall core height} \times \text{largest dimension of bobbin (cm)}^3$$

$$(2.49)$$

Example

For the core Type E42/20, from Figure 2.5,

$$V = 4.22\,\text{cm} \times 4.22\,\text{cm} \times 3.72\,\text{cm} = 65\,\text{cm}^3$$

For this core the value $V = 65\,\text{cm}^3$ can be read off from Figure 2.38.

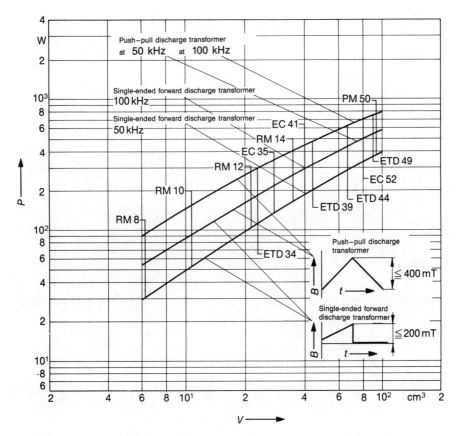

Fig. 2.39 Power-handling capability P and unit volume V of transformers with ferrite cores of N27/N41 material (guidance values) at higher frequencies (50 kHz and 100 kHz) for a temperature rise of 30 K in various modes of operation (Siemens)

For the transformer T under consideration, the EF20 core can now be selected from the data manual, since it can be presumed that it will be of adequate size. Based on the dimensions from the data manual, $V = 2.04 \text{ cm} \times 2.04 \text{ cm} \times 1.4 \text{ cm} = 5.8 \text{ cm}^3$. From Figure 2.38 a power capability of about 10 W can be read off by extrapolation, which is sufficient.

The power-transmitting capability can alternatively be

$$P = Cf\Delta BJf_{\text{Cu}}A_{\text{N}}A_{\min} \times 10^{-4} \quad \text{(W)} \tag{2.50}$$

$C = 1$ for push–pull operation

$= 1/2\sqrt{\delta_{\text{T}}}$ for single-ended forward operation (about 0.71 for $\delta_{\text{T}} = 0.5$ and about 1 for $\delta_{\text{T}} = 0.24$)

$= 0.61$ for single-ended stored-energy operation (flyback converter)

where J is the current density (A/mm^2 or A/cm^2) (see table in Figure 2.41), ΔB is the permissible flux-density excursion (see Figures 2.38 to 2.40) (T), f_{Cu} is the copper space factor (see Figure 2.42); an average value of 0.5 is assumed in Figures 2.38 to 2.40 A_{N} is the usable cross-sectional area of winding (cm^2) and $A_{\min}$ is the minimum effective magnetic

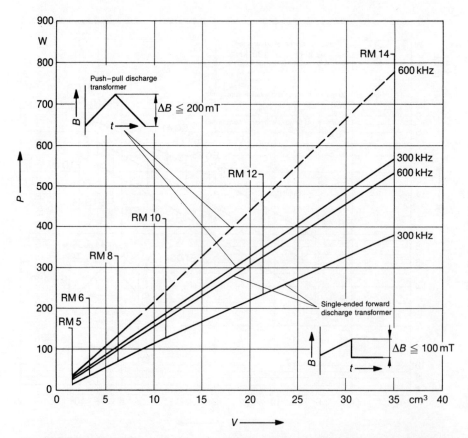

Fig. 2.40 Power-handling capability P and unit volume V of transformers with ferrite cores of N27 material (guidance values) at high frequencies (300 kHz and 600 kHz) for a temperature rise of 30 K in single-ended and push-pull operation (Siemens)

Column 1	2	3	4	5	6	7	8	9	10	11	12	13	14	15	16
Core type	A_e (mm²)	A_{min} (mm²)	V_e (mm³)	A_N (mm²)	V_N (mm³)	R_{th} (K/W)	ΔB (mT)	J (A/mm²)	ΔB_2 (mT)	J_2 (A/mm²)	P (W)	ΔB_2 (mT)	J_2 (A/mm²)	Discharge operation	Energy storage operation
EF12.6	13	12.2	384	11.6	316	77	908	8.6	375	11.8	5.3	188	12.1	2.0	1.7
16	20.1	19.3	754	22.3	758	54	790	6.6	384	8.9	12.3	192	9.3	4.6	3.9
20	33.5	32	1 500	34	1 400	39	670	5.7	382	7.6	27	191	8.0	9.9	8.5
25	52.5	51.5	3 020	56	2 910	28	544	4.7	392	5.8	54	196	6.5	21.3	18.3
E30	60	49	4 000	73	3 360	22	565	4.9	327	6.5	75	163	6.9	28	24
42/15	181	175	17 600	157	13 700	14	361	3.1	361	3.1	254	193	4.1	128	110
42/20	240	229	23 300	170	17 900	14.3	317	2.6	317	2.6	270	191	3.5	153	131
55	354	350	42 500	238	26 900	10	285	2.6	285	2.5	500	198	3.2	307	264
EC35	84.3	71	6 530	97	5 140	18	498	4.4	337	5.6	123	168	6.1	48	41
41	121	106	10 800	134	8 310	15	433	3.8	350	4.5	204	175	5.2	84	72
52	180	141	18 800	212	15 700	11	390	2.2	313	3.8	363	157	4.4	150	129
70	279	211	40 100	469	45 500	7.3	335	2.3	302	2.6	822	151	3.2	359	309
PM50 × 39	340	275	29 600	154	14 900	15	280	2.8	280	2.8	328	162	3.7	173	149
62 × 49	550	460	62 200	270	32 400	11.5	227	2.2	227	2.2	593	167	2.7	380	327
74 × 59	740	615	98 000	423	59 200	9.3	205	1.8	205	1.8	924	166	2.1	620	532
87 × 70	915	700	140 000	657	104 000	7.7	190	1.5	190	1.5	1370	153	1.8	940	808
114 × 93	1730	1340	360 000	1070	225 000	5.7	144	1.2	144	1.2	2560	144	1.2	1817	1561
	Core and winding data					Resistance Temperature Change	$\Delta\theta_{Fe} = 15K$	$\Delta\theta_{Cu} = 15K$	Push—Pull at $B \leqslant 400$ mT			Single-ended forward at $\Delta B \leqslant 200$ mT			

$$\Delta\theta = \Delta\theta_{Fe} + \Delta\theta_{Cu} = 30K$$

Fig. 2.41 Current density J, flux-density excursion B and power-handling capability P of power transformers with ferrite cores at a switching frequency of 20 kHz (Siemens)

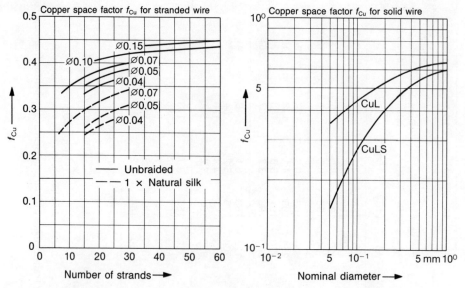

Fig. 2.42 Variation with nominal diameter of copper space factor f_{Cu} for stranded and solid wires (Siemens)

cross-sectional area (if only A_e is given in the case of equal cross-sectional areas, this is the value to be used) (cm^2).

The maximum permissible current density J, the significant parameter in regard to the heating of a transformer, can be obtained from Figure 2.41 in A/mm^2.

Columns 13 to 15 in this figure relate to the single-ended forward converter. They are only applicable, however, to the open construction; with encapsulated windings the loading can be increased, on account of the much better heat dissipation.

Example

The power to be transmitted is to be checked in relation to the EF20 core by means of Equation (2.50).

Core data: $A_N = 0.34\,\text{cm}^2$; $A_e = 0.335\,\text{cm}^2$;

$J = 8\,\text{A/mm}^2 = 800\,\text{A/cm}^2$; $\Delta B = 0.2\,\text{T}$;

$\delta_T = 0.24$: i.e. $C \approx 1$; $f_{Cu} = 0.4$; $f = 20\,\text{kHz}$.

$$P = 1 \times 20 \times 10^3\,\text{s}^{-1} \times 0.2 \times 10^{-4}\,\text{V s/cm}^2 \times 8 \times 10^2\,\text{A/cm}^2$$
$$\times 0.4 \times 0.34\,\text{cm}^2 \times 0.335\,\text{cm}^2 = 14.6\,\text{W}$$

From Figure 2.38 about 9.5 W would be read off for the calculated volume of 5.8 cm^3; Figure 2.41 indicates 9.9 W.

The number of primary turns is calculated on the basis of Faraday's law:

$$N_p \geqslant \frac{V_i \delta_T \times 10^4}{A_{min} f \hat{B}_{max}} \tag{2.51}$$

where A_{min} is the minimum magnetic cross-sectional area (cm^2) (not quoted for cores with equal cross-sections; A_e is then to be substituted) and $\hat{B}_{max}$ is the maximum permissible flux density (T) (for N27, $\hat{B}_{max} = 0.2\,T$).

The factor 10^4 is necessary, since

$$[T] = \frac{V\,s}{m^2} = \frac{V\,s}{10^4\,cm^2}.$$

The product $V_i v_T$ may be represented either by $V_{imax}\delta_{Tmin}$ or $V_{imin}\delta_{Tmax}$. For the transformer in Figure 2.37,

$$N_p \geqslant \frac{27.6\,V \times 0.207 \times 10^4}{0.335\,cm^2 \times 20 \times 10^3\,s^{-1} \times 0.2\,V\,s/cm^2} \geqslant 43\,turns$$

Rounded up:

$$N_p = 45\,turns$$

The maximum possible wire diameter (the maximum value including insulation) is obtained from the winding cross-sectional area:

$$d \leqslant \sqrt{\frac{4A_N f_{Cu}}{N\pi}} \quad (mm) \tag{2.52}$$

A_N is the maximum winding cross-sectional area (mm^2) and f_{Cu} is the copper space factor (about 0.5).

The winding space $A_N = 34\,mm^2$ should be divided equally between the primary and secondary windings:

$$A_N/2 = 17\,mm^2$$

Equation (2.52):

$$d_p \leqslant \sqrt{\frac{4 \times 17\,mm^2 \times 0.5}{50\,\pi}} \leqslant 0.47\,mm \text{ (overall diameter)}$$

From Figure 2.10, the maximum nominal diameter is $d_{pmax} = 0.4\,mm$. Selected: $d_p = 0.355\,mm$; overall diameter: $0.395\,mm \approx 0.4\,mm$.

In the maximum winding width of 12.1 mm the possible number of turns per layer is $12.1\,mm/0.4\,mm = 30$. Two layers are thus required for the 50 turns, giving a winding depth of $2 \times 0.4\,mm = 0.8\,mm \approx 1\,mm$. The transformation ratio is $n = N_s/N_p = V_s/V_p = 5\,V/24\,V = 0.208$. N_s thus becomes $0.208 \times 50 = 10.4$, rounded up to 11 turns. A more accurate value for n is $11/50 = 0.22$.

Secondary wire diameter:

$$d_s = \sqrt{\frac{4 \times 17\,mm^2 \times 0.5}{11\pi}} = 0.99\,mm \text{ overall diameter}$$

The maximum possible nominal diameter is 0.9 mm; selected: $d_s = 0.85\,mm$.

The overall diameter of a wire of 0.85 mm nominal diameter is 0.913 mm. With this relatively thick wire the calculation of the winding space cannot be based on the winding cross-sectional area, but on the winding depth. The bobbin dimensions given in the data manual indicate a winding width of 12.1 mm and a total winding depth of 3.15 mm.

The 11 turns with a maximum overall diameter of 0.913 mm give a winding width of $11 \times 0.913 = 10.04$ mm. Since 12.1 mm is available, the single layer of the secondary winding is easily accommodated. Calculated on an alternative basis: a winding width of 12.1 mm max. and a maximum overall wire diameter of 0.913 mm give a maximum number of turns equal to 13.4, or 13 turns rounded down. Of the total winding depth of 3.15 mm only $1 \text{ mm} + 0.913 \text{ mm} = 1.913 \text{ mm}$ is required. There remains $3.15 \text{ mm} - 1.913 \text{ mm} = 1.237$ mm, which is sufficient for insulation between the primary and secondary windings and a protective covering.

The primary and secondary copper cross-sectional areas are respectively $a_p = 0.0990 \text{ mm}^2$ and $a_s = 0.57 \text{ mm}^2$. The primary current is calculated as

$$I_p = I_s n + I_M \quad \text{(A)} \tag{2.53}$$

where I_s is the secondary current (A) and I_M is the primary magnetizing current (A).

The magnetizing current [7] is given by

$$I_M = \frac{V_i \delta_T}{L_p f} \quad \text{(A)} \tag{2.54}$$

Here again it does not matter which combination of V_i and δ_T is adopted.

From Equation (2.17), $L_p = N_p^2 A_L$ and putting $A_L = 1300$ nH (with no air gap, from the data sheet), the primary inductance is calculated as

$$L_p = 50^2 \times 1300 \times 10^{-9} \text{ H} = 3.25 \text{ mH}$$

$$I_M = \frac{20.4 \text{ V} \times 0.28}{3.25 \times 10^{-3} \text{ H} \times 20 \times 10^3 \text{ s}^{-1}} = 88 \text{ mA}$$

The total primary current thus becomes

$$I_p = 1 \text{ A} \times 0.22 + 0.088 \text{ A} = 0.31 \text{ A}$$

and the current density is then

$$J_p = \frac{0.31 \text{ A}}{0.099 \text{ mm}^2} = 3.13 \text{ A/mm}^2 \quad \text{and} \quad J_s = \frac{1 \text{ A}}{0.57 \text{ mm}^2} = 1.75 \text{ A/mm}^2$$

According to Figure 2.41, the permitted figure is 8 A/mm^2; the transformer will thus become barely warm.

A Schottky diode is chosen for D_3 on account of its low voltage drop: hence the original duty cycle $\delta_T = 5.7 \text{ V}/V_i$. The stressing of the diode amounts to

$$V_{imax} = 27.6 \text{ V, or, rounded, 30 V}; \quad I_{max} = 11 \text{ A}$$

As possible types may be considered the BYS24-25 (Siemens), with the two anodes connected together, or the SD21 (TRW).

The power loss in the free-wheel diode (only the conduction loss is considered, since the switching loss is negligible at this low blocking voltage) is, from Equation (2.42),

$$P_{Dc} = V_F I_0 (1 - \delta_{Tmin}) = 0.6 \text{ V} \times 10 \text{ A} \times 0.793 = 4.76 \text{ W}$$

(At the average duty cycle $\delta_T = 0.24$, $W_{Dc} = 4.56$ W.)

The loss in the transistor is, from Equation (2.37),

$$P_{Trc} = V_{CEsat} I_{Cmax} \delta_{Tmax} = 1 \text{ V} \times 10 \text{ A} \times 0.28 = 2.8 \text{ W}$$

(At the average duty cycle $\delta_T = 0.24$, $W_{Trc} = 2.4$ W.)

Equation (2.39):

$$P_{Trs} = V_i I_0 t_f f = 27.6 \text{ V} \times 10 \text{ A} \times 10^{-6} \text{ s} \times 20 \times 10^3 \text{ s}^{-1}$$
$$= 3.31 \text{ W}$$

(At the nominal voltage of 24 V $W_{Trc} = 2.88$ W.)

The losses at the nominal voltage $V_i = 24$ V thus total: 4.56 W + 2.4 W + 2.88 W + 1.7 W (W_L) + 1.54 W (W_{R8}) = 13 W. To this amount must be added the drive power loss of 24 V × 0.31 A × 0.24 = 1.79 W. The loss in the circuit of Figure 2.37 under average conditions is therefore W = 13 W + 1.79 W = 14.8 W ≈ 15 W.

In the comparable converter with the PIC625 power IC the figure was 17.7 W—about 52 per cent higher. The lower losses in the present case are attributable first to the Schottky diode, with its lower forward voltage drop, and secondly to the lower saturation voltage of the separate transistor in comparison with the Darlington transistor. The lower losses in the version of Figure 2.37 have to be set against the lower circuit complexity in that of Figure 2.25. Whether the optimum should be sought in terms of minimum losses or minimum complexity (and therefore lower cost) is a question of the application requirement, and cannot be decided here.

The connection of the Zener diode D_1 in series with a normal silicon diode D_2 across the primary winding of T shown in Figure 2.37 serves to reset the transformer during the blocking period. The Zener voltage of D_1 is equal to the input voltage V_i. This means that the maximum voltage applied to the transistor T_2 is equal to twice the input voltage; the maximum current has already been calculated. Resistor R_{10} prevents the occurrence of overvoltages when T_2 switches off. Since the stray inductance of the whole circuit has a bearing on this, the required value of R_{10} must be determined experimentally. R_{11} is calculated so that sufficient base current is provided with β_{min} in Tr_2 and V_{imin}, in order to ensure adequate collector current. For the drive transistor, Type BD139-10 is selected ($\beta_{min} = 70$; $I_{Cmax} = 2$ A; $V_{CE0} = 80$ V). The maximum voltage appearing across Tr_2 is $V_{CEmax} = V_{imax} + V_{zmax} + 20$ per cent (for overshoot) = 27.6 V + 25 V + 10.5 V = 63 V, for which the rating is adequate. The maximum loss is $W_{Trc} \approx 0.8$ V × 0.31 A = 70 mW. The transistor therefore needs no heatsink, since its permissible loss without one is specified as about 750 mW.

The drive-circuit loss already calculated as 1.79 W is dissipated mainly in R_B. To ensure that the saturation voltage of Tr_2 remains below 1 V it is overdriven by a factor of three.

$$I_{B/Tr_2} = \frac{I_C}{\beta_{min}} \times 3 = \frac{0.31 \text{ A}}{70} \times 3 = 13.3 \text{ mA}$$

$$R_{11} \geqslant \frac{20.4 \text{ V}}{13.3 \text{ mA}} \leqslant 1.53 \text{ k}\Omega; \quad \text{selected: } R_{11} = 1.3 \text{ k}\Omega, R_9 = 1 \text{ k}\Omega$$

The energy $LI_M^2/2$ stored in the transformer T is converted into heat in the Zener diode during the blocking period. The loss is calculated as

$$P_z = \frac{LI_M^2 f}{2} \quad \text{(W)} \tag{2.55}$$

$$P_{vZ} = 0.5 \times 3.25 \times 10^{-3} \text{ H} \times 0.09^2 \text{ A}^2 \times 20 \times 10^3 \text{ s}^{-1} = 0.26 \text{ W}$$

This is somewhat high for a small Zener diode; the Type ZX24, which can dissipate 1 W without cooling, is therefore chosen. With this the calculations for all the components in

50

the circuit have been performed; all other aspects have been illustrated by means of the examples.

2.6 DRIVE CIRCUIT DESIGN FOR AN *n*-CHANNEL MOSFET POWER TRANSISTOR

If MOSFET power transistors are to be used instead of bipolar transistors, these also must be controlled through a transformer-coupled drive circuit for reasons of polarity (see Figure 2.25 for the drive IC connections not shown).

The transformer T used in this case is most suitably designed with a ratio $n = 1$ and supplied on the primary side at a voltage of about 12 V. This voltage can be obtained from the input voltage in a simple manner by means of a 7812 series regulator. The voltage applied to the gate must be sufficient to ensure that the MOSFET is positively turned on, but at the same time must not exceed the maximum permissible gate voltage (usually ± 20 V). Figure 2.44 shows a typical transfer characteristic, while Figure 2.45 shows a typical family of output characteristics.

To facilitate comparison of the MOSFET circuit with the other versions previously investigated, an input voltage $V_i = 24$ V ± 15 per cent $= 20.4$ V to 27.6 V, a frequency $f = 20$ kHz, an output voltage $V_0 = 5$ V and a maximum load current $I_0 = 10$ A will again be considered. Except for the power switch, the circuit is similar, and need not be discussed further.

A suitable MOSFET transistor is, for example, the Siemens Type BUZ10. For $I_{0max} = 10$ A, corresponding to $I_{Lmax} = 11$ A, according to Figure 2.44 a gate voltage of at least 6 V is required. According to Figure 2.45, however, the gate voltage must be at least 7 V to ensure a low residual voltage. If the drive circuit of Figure 2.43 is supplied at 12 V on the primary side, with $n = 1$ the secondary voltage is similarly 12 V. The two diodes D_1 and D_2

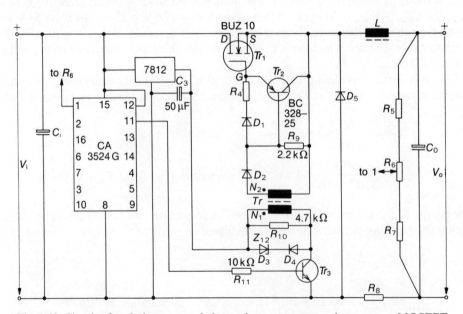

Fig. 2.43 Circuit of an inductor-coupled step-down converter using a power MOSFET with drive through a pulse transformer

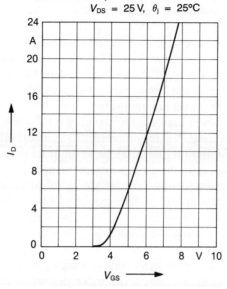

Fig. 2.44 Typical transfer characteristic
$I_D = f(V_{Gs})$ for the BUZ10 MOSFET
power transistor (Siemens)

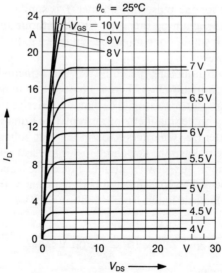

Fig. 2.45 Typical output characteristics
$I_D = f(V_{Gs})$ with V_{Gs} as parameter for the
BUZ10 MOSFET power transistor
(Siemens)

reduce this by about 1.5 V, so that 10.5 V is available. In principle, the pulse transformer could of course be supplied with an input voltage of 24 V, making $n = V_s/V_p = 0.5$. The nominal drive voltage would then again be $12\,V - 1.5\,V = 10.5\,V$, but at the possible minimum input voltage $V_{imin} = 20.4\,V$ the voltage applied to the gate would be only 10.2 V $-1.5\,V = 8.7\,V$ This seems barely sufficient for positive and rapid control. It also has to be borne in mind that Figure 2.44 shows 'typical' characteristics, without maximum limits. The data sheet merely indicates that the threshold voltage—i.e. voltage at which the FET just begins to conduct, shown as 3 V in Figure 2.44—is typically 3 V, but may be as much as 4 V. This means that the maximum applicable characteristics are shifted to the right in Figure 2.44 and downwards in Figure 2.45. In that case 8.7 V is so marginal that positive switching may not be achieved. A gate voltage of 10.5 V, on the other hand, is adequate in all cases, and the transformation ratio may be so adjusted that the range of secondary voltage is shifted upwards. With the given variation of input voltage V_i the ratio could be so chosen that the maximum permissible gate voltage of 20 V was not exceeded, but the higher the gate voltage, the longer is the time required for the relevant capacitances (in the nF region) to discharge, so that the MOSFET switches somewhat less rapidly. From the point of view of optimum drive, the additional complication introduced by the 7812 regulator seems reasonable.

The protective network comprising D_3 and D_4, which comes into effect when the driver transistor switches off, allows the pulse transformer to reset, as in Figure 2.37, and limits the voltage on Tr_3 to 24 V. Resistor R_4 (10 to 100) limits the charging current into the input capacitance of Tr_1 and prevents a tendency to instability. When Tr_3 is turned off, the input capacitance of Tr_1 is discharged rapidly through the conducting transistor Tr_2. Because of the high current gain of Tr_2 the discharge current reaches about 200 mA. D_1, D_2 and D_4 should be fast-switching diodes, e.g. Type BAW76. A Schottky diode is used for D_5 on account of the low output voltage $V_0 = 5\,V$ (see the circuit of Figure 2.37). D_3, as will be seen, can be a small Zener diode with a Zener voltage of 12 V.

The design procedure for the pulse transformer T is similar to that for the circuit of Figure 2.37. The core may again be the Type EF20 (without an air gap, in view of the resetting circuit), or a still smaller one might be suitable, e.g. EF16. The latter possibility will not be checked, however.

Since in the step-down inductor-type converter the principal loss in the MOSFET power transistor is the conduction loss, there is no need to minimize the switching times, which are in any case short, through special design of the derive transformer (with a small air gap, and consequently higher magnetizing current) (see Equation (2.38)). In the case of a mains-supplied converter (see Chapters 5 to 7) the input voltage V_i is an order higher, and it is then worthwhile to keep the switching losses low as well. With $\delta_{Tmax} = 0.28$ and $\hat{B}_{max} = 0.1\,T$ (the maximum flux density should not exceed about 0.1 T if good pulse characteristics are to be achieved), the number of primary turns (= the number of secondary turns if $n = 1$) is calculated from Equation (2.51) as

$$N_p \geqslant \frac{12\,V \times 0.28 \times 10^4}{0.335\,\text{cm}^2 \times 20 \times 10^3\,\text{s}^{-1} \times 0.1\,V\,\text{s/cm}^2} \geqslant 50.2\ \text{turns}$$

Selected: $N_p = N_s = 60$ turns.

The data sheet gives $A_L = 1300 \times 10^{-9}\,H$ with $s = 0$. The primary inductance, from Equation (2.17), is

$$L = N^2 A_L = 60^2 \times 1300 \times 10^{-9}\,H = 4.7\,\text{mH}$$

The magnetizing current, from (2.54), is

$$I_M = \frac{12\,\text{V} \times 0.28}{4.7 \times 10^{-3}\,\text{H} \times 20 \times 10^3\,\text{s}^{-1}} = 36\,\text{mA}$$

Since the drive to a MOSFET power transistor, in contrast to the bipolar transistor, requires only a short pulse of charging current (see Figure 6.21), the average secondary-side power is negligible. The total current in Tr_3, including the small currents in R_{10} and R_9 and the average charging current into the gate of Tr_1, should certainly not exceed something like 50 mA. An exact calculation is not necessary.

In determining the maximum possible wire diameter, the total number of turns $N_p + N_s$ is simply put into Equation (2.52), since in this case the primary and secondary windings are similar. To simplify the calculation of the winding space, it may be based on about threequarters of the available winding cross-sectional area of 34 mm² —i.e. about 25 mm². This leaves sufficient room for the uncritical insulation.

Equation (2.52):

$$d \leqslant \sqrt{\frac{4 \times 25\,\text{mm}^2 \times 0.5}{120\,\pi}} \leqslant 0.36\,\text{mm}$$

According to Figure 2.10 the nearest possible wire diameter (nominal) $d = 0.315$ mm, with a maximum overall diameter of 0.352 mm. The winding length of 12.1 mm will accommodate 12.1 mm/0.352 mm = 34 turns per layer. The 60 turns per winding thus require two layers—four layers in total. The total depth of the two windings is then $h = 4 \times 0.352$ mm = 1.41 mm. There is thus ample room for the insulation between primary and secondary windings and the outer covering.

Since Figure 2.11 shows that with the wire diameter chosen the skin effect is negligibly small, this wire can be adopted. In view of the much lower power compared with the bipolar version, the calculation of winding resistance and the total losses can be dispensed with.

The low primary current, at most about 50 mA, enables a low-power Zener diode to be used—e.g. the ZPY12 among others. Transistor Tr_3 can be a low-power type—e.g. BC238. The use of a ring core of similar size (e.g. R_{16} with an outside diameter of 16 mm and $A_L = 2770$ nH, or R_{25} with an outside diameter of 25 mm and $A_L = 4400$) would have resulted in a smaller number of turns. These cores are, however, more awkward to wind.

It would have been more satisfactory to choose a higher switching frequency (e.g. 50 to 100 kHz), which is possible without any modification because of the very short switching times of the MOSFET (about 60 ns). The 20 kHz switching frequency is retained here, however, to facilitate comparison of the various versions of the circuit.

The significant quantity in the case of the MOSFET power transistor is the maximum on-state resistance $R_{DS(on)}$.

The conduction loss is given by

$$P_{Trc} = R_{DS(on)max} I_D^2 \delta_{Tmax} \quad \text{(W)} \tag{2.56}$$

where $R_{DS(on)}$ is the on-state resistance (Ω) (from Figure 2.46, maximum value at θ_{jmax}, or at least $\theta_j = 100\,^\circ\text{C}$).

For the BUZ10 used in the circuit of Figure 2.43, $R_{DS(on)max} = 0.15\,\Omega$ at $\theta_j = 100\,^\circ\text{C}$. Other types are available with much lower on-state resistance. The conduction loss in the

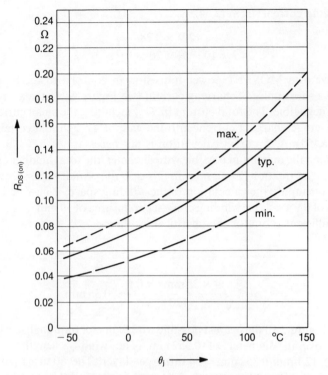

Fig. 2.46 On-state resistance $R_{DS(on)} = f(\theta_j)$ with tolerance range for the BUZ10 MOSFET power transistor (Siemens)

example is then $P_{Trc} = 0.15\,\Omega \times 100^2\,A^2 \times 0.24 = 3.6\,W$ (here again $\delta_T = 0.24$ for comparison). The switching loss is obtained from Equation (2.39) with $t_f = 60\,ns$:

$$P_{Trs} = 24\,V \times 60 \times 10^{-9}\,s \times 10\,A \times 20 \times 10^3\,s^{-1} = 0.29\,W$$

The switching loss is thus very small; even at $f = 100\,kHz$ it would still be less than the conduction loss. Finally, the total transistor loss is

$$P_{Tr} = 3.6\,W + 0.3\,W = 3.9\,W \text{ (for a suitable heatsink see Equation (2.45))}$$
$$P_{tot} = 3.9\,W + 4.56\,W + 1.7\,W + 1.54\,W = 11.7\,W \approx 12\,W$$

Comparing the three variants, the losses, rounded-up, are as follows:

- circuit with power IC Type PIC625: $W = 21\,W$
- circuit with n-p-n power transistor: $W = 15\,W$
- circuit with MOSFET power transistor: $W = 12\,W$

The high losses in the first circuit are due to the fact that the built-in diode is not of the most suitable type. In the second variant the drive power and the switching loss are significant factors. The third variant is clearly the best. To what extent the (ostensibly) higher price stands in the way of a wider use of power MOSFETs is another question.

The transformer-coupled circuits of Figures 2.37 and 2.43 require a transistor in the drive IC in which the collector and emitter connections are brought out externally. This is not the case in all drive ICs. Sometimes only the collector connection is brought out, so that the output transistor is 'active low'. That is, the power transistor is turned on when the

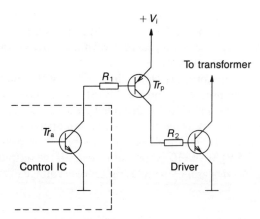

Fig. 2.47 Circuit variant with the output transistor of the control IC switching to earth

output transistor of the drive IC switches in the direction of earth. If a drive IC of this type is to be used, a further small p-n-p transistor Tr_p is necessary, as shown in Figure 2.47, which then drives the driver transistor in Figures 2.37 and 2.43.

While the various kinds of drive IC (controller) have not been discussed so far, since they are considered in detail in Chapter 8, attention must be drawn to possible difficulties that could be encountered with the CA3524G IC used in the present example (and also with other push–pull drive ICs). In the cited example, with $V_i = 24\,\text{V} \pm 15$ per cent and $V_0 = 5\,\text{V}$, the maximum duty cycle has been calculated as 0.28. This presents no difficulty. If the input voltage had been only 10 V, however, corresponding to a maximum duty cycle of just 0.5, then, depending on the relationship of the tolerances, the required output voltage could be obtainable with low load currents but not at high currents. This is easily appreciated from the following consideration.

The CA3524G is conceived as a push–pull controller. According to the data sheet the duty cycle can be varied between 0 and about 90 per cent. A duty cycle greater than 50 per cent can only be obtained by connecting the two outputs in parallel. (The same applies to all push–pull controllers—e.g. the Siemens Types TDA4700, TDA4714, TDA4716 and TDA4718 among others.) Although in theory the duty cycle does not vary with changes in load, in practice it does increase somewhat with increasing load, since the output voltage tends to fall as a result of the finite internal resistance. If δ_T just reaches 0.5, it must become somewhat larger if the load increases, but because of the blanking periods always present between the two outputs in push–pull controllers δ_T cannot increase, and the controller cannot respond. With a materially larger duty cycle this difficulty does not arise. If operation is required with $\delta_T = 0.5$, therefore—e.g. in flyback converters, or when it is necessitated by the given input and output voltages—the choice is restricted to single-ended controllers (e.g. the Valvo TDA 1060).

2.7 CIRCUIT EXAMPLE

Step-down inductor-type converter, Figure 2.48 (brief description): with the given input voltage range of 36 to 62 V, the duty cycle varies between 0.4 and 0.7; using the TDA4716 push–pull drive IC, therefore, since the switching ratio exceeds 0.5, the two outputs Q_1 and Q_2 must be connected in parallel. Difficulties may arise when V_i exceeds about 50 V and

$\delta_T = 0.5$. The power supply to the CMOS inverter is at approximately 8.5 V ($V_z = 10\,\text{V} - V_{BE} - V_D = 10\,\text{V} - 1.5\,\text{V} = 8.5\,\text{V}$), which appears across the 0.33 μF capacitor. The inverter (Figure 2.49) is so arranged that with a positive signal on Terminal I the output Q is at signal 0 and no positive voltage is applied to the gate of the BUZ72A power transistor. When the outputs Q_1/Q_2 in the TDA4716 switch, the input I of the 4049B is driven to a low potential, and the MOSFET is turned on.

The power supply to the TD4716A is provided by a 12 V Zener diode in conjunction with a series resistor R_s. So long as the 'ON' switch is not operated, the drive IC operates at its oscillator frequency of 50 kHz (C_T, R_T) and its input 13 (overvoltage monitoring) is kept at a level below V_{ref} (2.5 V) by the 10 kΩ potentiometer. If the output is to be switched off, this can be done, as in this case, by simulating an overvoltage (Terminal 13 to +12 V).

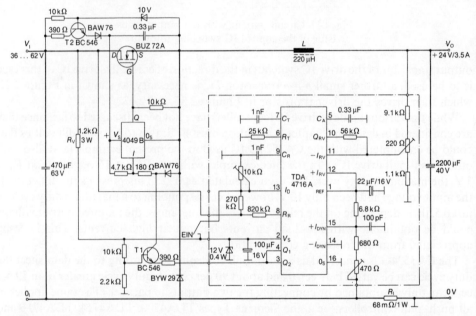

Fig. 2.48 Inductor-coupled step-down converter using the BUZ72A MOSFET power transistor and the TDA4716A control circuit with overcurrent sensing in the neutral line (Siemens)

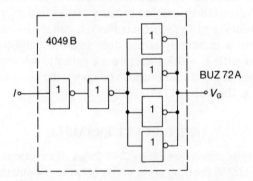

Fig. 2.49 Connections of the 4049B CMOS inverter in the circuit of Figure 2.48 (Siemens)

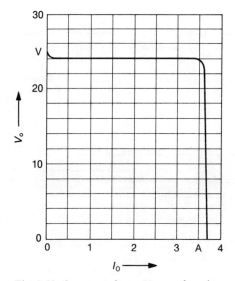

Fig. 2.50 Output voltage V_0 as a function
of load current I_0 (complete characteristic
including current limiting) (Siemens)

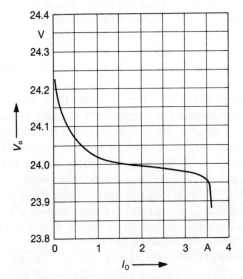

Fig. 2.51 Output voltage V_0 as a function of
load current I_0 (part characteristic from
$I_0 = 0$ to $I_0 = 3.5A$) with expanded voltage
scale (Siemens)

Open-loop control through the $820\,\text{k}\Omega$ resistor to Terminal 8 (R_R) holds the output
voltage V_0 constant to within ± 5 per cent with $V_{\text{inom}} = 48\,\text{V}$ varying down to $V_i = 36\,\text{V}$
and up to $V_i = 62\,\text{V}$. The regulation with load is as shown in Figures 2.50 and 2.51. The
output voltage is divided down at Terminal 11 to approximately $V_0/10 = V_{\text{ref}}$ by means of
a voltage divider. Overcurrent protection is effected by means of R_1, the voltage drop up to

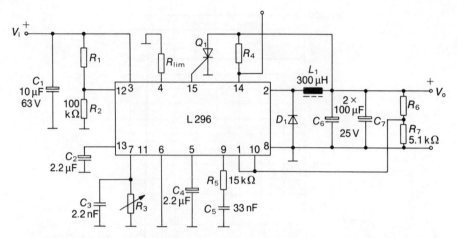

Fig. 2.52 Inductor-coupled step-down converter with the L296 power/control circuit (SGS) for $I_{0max} = 4\,A$ and $V_i = 10$ to $50\,V$

a load current of 3.5 A being compensated by a positive increment to V_{ref}. The inductor L consists of an RM12 core, core material N47, with $A_L = 250$ and $\mu_e = 80$ with an air gap $s = 0.65$ mm. The winding consists of 30 turns of 100×0.1 mm litz wire. The maximum permissible direct current, from Equation (2.18) and Figure 2.9, is approximately 4.8A.

Step-down inductor-type converter, Figure 2.52 (brief description): this circuit includes both the drive stage and the power stage in one integrated circuit, Type L296 (SGS). The maximum load current can be up to 4 A. If Terminal 4 is left unconnected, the maximum short-circuit current is 5 A; connecting a resistor $R_{lim} = 33$ kΩ reduces this to 2.5 A. The switching frequency can be adjusted at Terminals 7 and 11; with $R_3 = 4.7$ kΩ and $C_3 = 2.2$ nF the frequency is exactly 100 kHz. For this frequency the inductance L_1 is calculated to be 300 μH, based on a minimum load current of 0.1 A. Capacitor C_4, with a capacitance of 2.2 μF, provides a starting delay of 20 ms (soft start). At the same time the mean short-circuit current I_{scAV} (not the peak value) is limited by this means to 0.5 A. The connection of R_5 and C_5 to terminal 9 effects frequency compensation for the error amplifier. The internal reference voltage is 5.1 V; since the non-inverting input of the error amplifier is connected directly to V_{ref}, the minimum output voltage V_{0min} is also 5.1 V (R_6 being bypassed). Higher output voltages must be so divided down by R_6 and R_7 that exactly 5.1 V is applied to the inverting input of the error amplifier (Terminal 10). If the overvoltage protection Terminal 1 is connected to Terminal 10, the output 15 delivers a current of 100 mA to trigger a crowbar thyristor as soon as an overvoltage of + 20 per cent is reached at the output. The thyristor then reduces the output voltage V_0, in a maximum of 10 μs, to the on-state voltage of about 2 V, which in turn causes the short-circuit protection system to operate. The input voltage is monitored with respect to V_{imin} through Terminal 12; if the input falls below the response threshold of 5 V, the output 14 (open collector) is activated, and falls to 0.2 V with a loading of 16 mA. This can be used, for example, to turn on an LED as an undervoltage signal. Capacitor C_2 delays this response by 0.1 s. Because of the relatively large inductance in this example, the minimum load current is quite low, and the output capacitors C_6 and C_7 can have proportionally low values, but the transient recovery time in the event of load changes is then fairly long. According to the manufacturer's information the load regulation ΔV_0 with $V_0 = 5.1$ V and

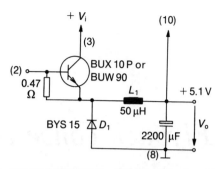

Fig. 2.53 Extension of the circuit of Figure 2.52 for a maximum load current of 15 A

$I_L = 1$ A to 4 A is 10 mV; the output voltage deviation with an input voltage variation $\Delta V_i = \pm 15$ per cent is 15 mV (with $V_0 = 5.1$ V and $I_L = 3$ A). This circuit can be used with input voltages of from 10 to 50 V. The maximum power loss in the L296 regulator is 6 W with $V_i = 35$ V and $I_L = 4$ A; $R_{thjc} = 3$ kW and $\theta_{jmax} = 150\,°C$. This provides the basis for calculating the heatsink requirement. If a Schottky diode is used as the free-wheel diode, the efficiency with $V_0 = 5.1$ V and $I_L = 1$ to 4 A lies between 70 and 80 per cent; with $V_0 \geqslant 20$ V it can be as high as 90 per cent. If the semiconductor element temperature rises above 150 °C, an internal overtemperature protection device switches off the circuit, resetting with a hysteresis of 20 °C. To prevent the appearance of noise voltage at the output, a 2.2 nF ceramic capacitor should be connected between Terminals 3 and 8 by the shortest possible route.

Extension of the circuit of Figure 2.52 for higher load currents: Figure 2.53. With this additional circuit the maximum load current of the circuit of Figure 2.52 can be increased up to 15 A. The L296 functions here as a power driver for the BUX10P OR BUW90 transistor. Extension of the circuit in this way requires new calculations for the diode D_1, the inductor L_1 and the output capacitor.

3   Flyback converters

The flyback converter operates on the blocking-converter principle [7], [1], [25]: that is, energy is transferred to the output during the blocking, or non-conducting, period of the switching transistor. Energy is stored in the inductor when the transistor is turned on, and then delivered to the output when the transistor is again turned off.

3.1 BASIC PRINCIPLES

The relationship between output voltage and duty cycle in the flyback converter is different from that in the step-down inductor-type converter:

$$|V_0| = \frac{\delta_{Tmax}}{1 - \delta_{Tmax}} V_{imin} = \frac{\delta_{Tmin}}{1 - \delta_{Tmin}} V_{imax} \quad (V) \tag{3.1}$$

From this the duty cycles can be calculated as follows:

$$\delta_{Tmin} = \frac{|V_0^*|}{|V_0^*| + V_{imax}^*} \tag{3.2}$$

and

$$\delta_{Tmax} = \frac{|V_0^*|}{|V_0^*| + V_{imin}^*} \tag{3.3}$$

(The use of the absolute values V_0 signifies that only the magnitude, without the (negative) sign, is to be inserted into the equation.)

To obtain realistic results, the input voltage V_i must be considered in conjunction with the voltage drops introduced by the switching transistor (V_{CEsat}), the inductor (V_L) and the current-measuring resistor (V_{R1}). The input voltage V_i^* that remains available is then

$$V_i^* = V_i - V_{CEsat} - V_L - V_{R1} \quad (V) \tag{3.4}$$

In obtaining the output voltage $|V_0|$, the voltage drop across the diode (V_F), and again V_L and V_{R1} must be taken into account: thus

$$|V_0^*| = |V_0| + V_F + V_L + V_{R1} \quad (V) \tag{3.5}$$

where V_F is the forward voltage drop of diode at I_0 (V).

As can be seen from Equations (3.2) and (3.3), the flyback converter can deliver an output voltage which may be either higher or lower than the input voltage: all that is required is an appropriate duty cycle. In practice V_i^* is subject to a lower limit of around $|V_0^*|/4$, since the duty cycle then approaches unity, and no controller IC can attain a ratio greater than about 0.9. For input voltages higher than $|V_0^*|$ the limit is not so critical: it is only necessary that the 'on' time $\delta_{\mathrm{Tmin}} T$ should be long in comparison to the switching time of the transistor. Input voltages of $4|V_0^*|$ or more can certainly be catered for, particularly if MOSFET power transistors are used.

The formula for calculating the inductance can be written in the same form as for the step-down converter: only in this case ΔI_L is not equal to $2I_{0\mathrm{min}}$, but larger, because of the higher inductor current.

$$L \geqslant \frac{|V_0^*|(1-\delta_{\mathrm{Tmin}})}{\Delta I_L f} \quad \text{(H)} \qquad (3.6)$$

(handwritten): $\dfrac{|V_0^*|(1-\delta_{\mathrm{Tmin}})^2}{2 I_{0\mathrm{min}} f}$

$$\Delta I_L = \frac{2I_{0\mathrm{min}}}{1-\delta_{\mathrm{Tmin}}} \quad \text{(A)} \qquad (3.7)$$

The mean current in the inductor is significantly higher than the maximum load current $I_{0\mathrm{max}}$, and is given by [7]

$$I_L = \frac{P_{0\mathrm{max}}^*}{V_{i\mathrm{min}}^* \delta_{\mathrm{Tmax}}} \quad \text{(A)} \qquad (3.8)$$

In the design of the inductor the inductor current given by Equation (3.8) must be augmented by $\Delta I_L/2$:

$$I_{L\mathrm{max}} = \frac{P_{0\mathrm{max}}^*}{V_{i\mathrm{min}}^* \delta_{\mathrm{Tmax}}} + \frac{V_{i\mathrm{min}}^* \delta_{\mathrm{Tmax}}}{2fL} \quad \text{(A)} \qquad (3.9)$$

3.2 FLYBACK CONVERTER WITH BIPOLAR POWER TRANSISTOR

Example

A flyback converter is to be designed to the following requirements:

$V_i = 12$ to $15\,\mathrm{V}$

$V_0 = -15\,\mathrm{V};\quad I_{0\mathrm{max}} = 2\,\mathrm{A};\quad I_{0\mathrm{min}} = 0.17\,\mathrm{A}$

Permissible output voltage ripple: $\Delta V_0 = 0.01|V_0| = 150\,m V_{\mathrm{pp}}$.
Maximum permissible deviation of $|V_0|$ due to a load change:

$$\Delta V_{0\mathrm{max}} = 0.1|V_0| = 1.5\,\mathrm{V}$$

Maximum ambient temperature: $\theta_{a\mathrm{max}} = 40\,°\mathrm{C};\ f = 25\,\mathrm{kHz}$.

The circuit is as shown in Figure 3.1 and the measured values are indicated in [ ].
To enable the maximum current I_L to be estimated, it is necessary first to calculate the theoretical value of δ_{Tmax}; the actual value should not depart too far from this.

$$\delta_{\mathrm{Tmax}}(\text{theoretical}) = \frac{15\,\mathrm{V}}{15\,\mathrm{V} + 12\,\mathrm{V}} = 0.56$$

Then to a first approximation

$$I_L = \frac{15\,\mathrm{V} \times 2\,\mathrm{A}}{12\,\mathrm{V} \times 0.56} = 4.5\,\mathrm{A}$$

Since the actual value may be somewhat higher, the voltage drop V_{CEsat} will be taken from the data sheet for the PIC625 power Darlington transistor for a current $I_L = 5\,\mathrm{A}$.

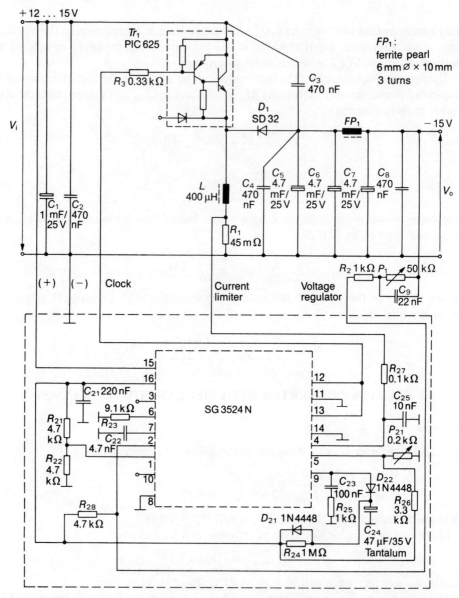

Fig. 3.1 Complete circut of an inductor-coupled flyback converter with the PIC625 power Darlington transistor and the SG3524N control IC

From Figure 2.27, a typical voltage drop at $I_C = 5\,\text{A}$ is 0.8 V, and the maximum value 1.2 V. The calculation may be made with an average value of 1 V. The input voltages V^*_{imin} and V^*_{imax} can now be calculated. Since in this case the diode incorporated in the PIC625 is not used, in the interests of good efficiency, a suitable Schottky diode can be provided, giving a forward voltage drop of about 0.5 V.

$$V^*_{\text{imin}} = 12\,\text{V} - 1\,\text{V} - 0.25\,\text{V} - 0.25\,\text{V} = 10.5\,\text{V}$$
$$V^*_{\text{imax}} = 15\,\text{V} - 1\,\text{V} - 0.25\,\text{V} - 0.25\,\text{V} = 13.5\,\text{V}$$
$$|V^*_0| = 15\,\text{V} + 0.5\,\text{V} + 0.25\,\text{V} + 0.25\,\text{V} = 16\,\text{V}$$

The actual values of the duty cycles and currents can now be calculated.

$$\delta_{Tmin} = \frac{16\,V}{16\,V + 13.5\,V} = 0.54; \quad \delta_{Tmax} = \frac{16\,V}{16\,V + 10.5\,V} = 0.60\,[0.6]$$

Equation (3.8):

$$I_L = \frac{16\,V \times 2\,A}{10.5\,V \times 0.6} = 5.08\,A\,[5.3\,A]$$

(calculated from the voltage drop across R_1 observed on an oscilloscope).

Calculation of the inductor current variation ΔI_L from (3.7):

$$\Delta I_L = \frac{2 \times 0.17\,A}{1 - 0.54} = 0.74\,A\,[0.89\,A]$$

Calculation of inductance from Equation (3.6):

$$L \geqslant \frac{16\,V \times (1 - 0.54)}{0.74\,A \times 25 \times 10^3\,s^{-1}} = 398\,\mu H \approx 400\,\mu H$$

Maximum current from (3.9):

$$I_{Lmax} = \frac{16\,V \times 2\,A}{10.5\,V \times 0.6} + \frac{10.5\,V \times 0.6}{2 \times 25 \times 10^3\,s^{-1} \times 400 \times 10^{-6}\,H}$$

$$= 5.08\,A + 0.315\,A = 5.4\,A\,[5.7\,A]$$

The inductor could now be dimensioned as described in Section 2.2. In this case the Type ZKB455/049-02-H2 from VAC (Vacuum-Schmelze GmbH) will be used, having been kindly made available for this purpose.

The data for this inductor, from the data sheet, are: $L = 400\,\mu H$; maximum permissible direct current: $I_{max} = 6.3\,A$; d.c. resistance $= 50\,m\Omega$.

The inductor thus meets the postulated requirements. The current I_{Lmax} could also have been calculated from the relationship $I_{Lmax} = I_L + \Delta I_L/2$.

$$I_{Lmax} = 5.08\,A + 0.5 \times 0.74\,A = 5.45\,A,$$

practically the same value.

The inductor current I_L has been calculated here for V_{imin} and δ_{Tmax}. If it is calculated for V_{imax} and the associated δ_{Tmin}, the result is a lower value of 4.4 A. The dimensioning of the inductor must of course be based on the higher value, which may also be calculated from Equation (3.8).

A rearrangement of Equation (3.8) gives equally

$$P^*_{0max} = 32\,W = 5.08\,A \times 10.5\,V \times 0.6 = 4.4\,A \times 13.5\,V \times 0.54$$

If now both sides of the equation are divided by the frequency, the energy per cycle is obtained. The energy stored in the inductor is in all cases equal to that delivered to the output.

The output capacitor C_0 (Figure 1.4: in Figure 3.1 the two parallel-connected capacitors C_6 and C_7) must on the one hand be of at least a certain minimum size, and on the other must not have too high an equivalent series resistance (ESR), in view of the prescribed maximum permissible output voltage ripple ΔV_0. It is also necessary to check that the selected type is adequately rated for the alternating current loading. From [4],

$$C_0 \geqslant 50 \times T \times \frac{I_{0max}}{|V_0|} \quad \text{(F)} \tag{3.10}$$

A very short correction time results (load change from no load to I_{0max}):

$$T_0 \approx 1.25\,T \quad \text{(s)} \tag{3.11}$$

Since in the flyback converter energy is supplied to the output only during the non-conducting period of the switching transistor, the diode current jumps from zero to I_{Lmax}. This current I_{Lmax} is divided between the load current I_{0max} and the capacitor current. The difference current $I_{Lmax} - I_{0max}$ flowing into the capacitor produces a voltage drop across the ESR. During the conducting period of the transistor the diode current is zero; the capacitor supplies the load current I_{0max}. This again produces a voltage drop across the ESR, of the opposite polarity. The two voltage drops together constitute the output voltage ripple ΔV_{opp} (peak-to-peak ripple).

$$\Delta V_{opp} = (I_{Lmax} - I_{0max}) \times \text{ESR} + I_{0max} \times \text{ESR} = I_{Lmax} \times \text{ESR}$$

From this the maximum permissible ESR can be determined:

$$\text{ESR} = \frac{\Delta V_0}{I_{Lmax}} \quad (\Omega) \tag{3.12}$$

where ΔV_0 is the permissible output ripple voltage (V_{pp}) and I_{Lmax} is the maximum inductor current (A).

In calculating the r.m.s. alternating current I_{Crms} in the capacitor, it is only necesary to consider the most unfavourable case. According to [21]

$$I_{Crms} = 1.3 I_0 \quad \text{(A)} \tag{3.13}$$

According to [4] the multiplication factor would be 1.1. The output capacitance $C_0(C_6 + C_7)$ is calculated for the example from Equation (3.10):

$$C_0 \geqslant 50 \times 40 \times 10^{-6}\,\text{s} \times \frac{2\,\text{A}}{15\,\text{V}} \geqslant 267\,\mu\text{F}$$

(This relatively low value applies only for an ideal capacitor, with zero ESR. Where polypropylene capacitors are used, at very high switching frequencies, it could be correct, but not for electrolytic capacitors.)

$$\text{ESR} \leqslant \frac{150\,\text{mV}}{5.4\,\text{A}} \leqslant 28\,\text{m}\Omega$$

$$I_{Crms} = 1.3 \times 2\,\text{A} = 2.6\,\text{A}$$

The capacitors may be selected on the basis of the table of Figure 2.16. Possible arrangements would be two electrolytic capacitors each of $4700\,\mu\text{F}/16\,\text{V}$, or $2 \times 2200\,\mu\text{F}/25\,\text{V}$, which latter would give a greater safety margin over the maximum operating voltage. The permissible alternating current loading of these two pairs of capacitors would be respectively

$$I_{C\sim} = 2 \times 1.5\,\text{A} \times 1.4 \times 1.8 = 7.56\,\text{A}$$

and

$$2 \times 1\,\text{A} \times 1.4 \times 1.8 = 5.04\,\text{A}$$

(see Figures 2.16, 2.19 and 2.20).

Since the r.m.s. alternating current that actually flows is only 2.6 A, these rated currents are not a factor in the selection. As in most cases, here it is the value of the ESR or Z that is critical, and the requirement in this respect can be met with only two capacitors, giving $50\,\text{m}\Omega/2 = 25\,\text{m}\Omega$. (In the work recorded in [25], and in Figure 3.1, three capacitors each of 4700 F/25 V were used.) From the point of view of output ripple, C_7 is ample. The ferrite beads affect the steeply rising voltage of the superimposed spikes, but not the 25 kHz alternating voltage. Since, as shown by Figure 2.16, for a given capacitance the impedance Z ($=$ ESR) decreases with increasing operating voltage, it is safe to assume that the total ESR will not be much more than 15 mΩ. Hence the measured value of $\Delta V_0 = 40\,\text{mV}_{\text{pp}}$ was very much less than the prescribed 150 mV. The two multi-layer ceramic capacitors C_4 and C_8 provide better high-frequency characteristics than the electrolytic capacitors alone and attenuate spikes.

Free-wheel diode D_1

The requirements for this diode are as follows:

$$V_\text{R} \geqslant |V_0| + V_{\text{imax}} \quad \text{(V)} \tag{3.14}$$

where V_R is the minimum necessary reverse voltage rating (V).

$$I_{\text{FAV}} \geqslant (0.5 \text{ to } 1)I_\text{L}(1 - \delta_{\text{Tmin}}) \quad \text{(A)} \tag{3.15}$$

$$I_\text{F} \geqslant I_{\text{Lmax}} \quad \text{(A)} \tag{3.16}$$

$$I_{\text{Lmax}} \text{ from Equation (3.9)}$$

The reverse recovery time and the stored charge should be as small as possible. For reverse voltages up to about 45 V (exceptionally up to 90 V) this points to the use of Schottky diodes; for higher voltages, up to about 200 V, fast epitaxial silicon diodes. Since the reverse leakage current I_R is significantly higher in Schottky diodes than in silicon pn diodes, the reverse losses must be taken into account. The very small stored charge in Schottky diodes (often not even quoted) means, however, that the switching losses given by Equation (2.43) can be ignored. The conduction loss is calculated in the same way as for the step-down converter:

$$P_{\text{Dc}} = V_\text{F}I_\text{L}(1 - \delta_{\text{Tmin}}) \quad \text{(W)} \tag{3.17}$$

where I_L is the mean inductor current at I_{0max} (A) (see (3.8)) and V_F is the forward voltage drop at I_L (V).

The reverse loss is given by

$$P_{\text{DR}} = V_\text{R}I_{\text{Rmax}}\delta_{\text{Tmax}} \quad \text{(W)} \tag{3.18}$$

where V_R is the maximum applied voltage (V) and I_{Rmax} is the maximum reverse current (A) at θ_{jmax}.

The conduction loss can alternatively be estimated in another way:

$$P_{\text{Dc}} = V_\text{F}^* I_0 + R_\Delta I_{\text{Fr.m.s.}}^2 \quad \text{(W)} \tag{3.19}$$

where V_F^* is the threshold voltage as shown in Figure 3.5 (the diode characteristic being

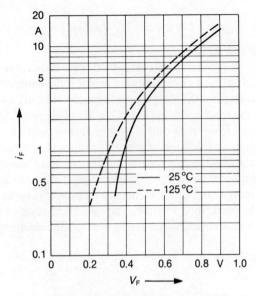

Fig. 3.2 Forward characteristic $I_F = f(V_F)$ of the BYS24-45 Schottky power diode (Siemens)

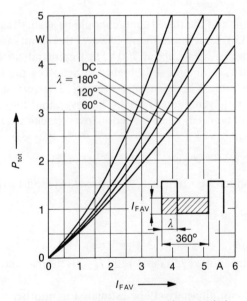

Fig. 3.3 Conduction loss characteristics $P_{Dc} = f(I_{FAV})$ for the BYS24-45 Schottky power diode with conduction period as parameter (Siemens)

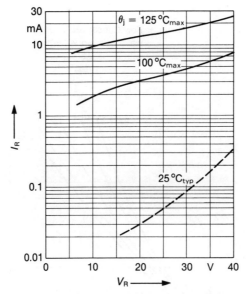

Fig. 3.4 Reverse current I_R as a function of reverse voltage V_R for the BYS24-45 Schottky power diode with junction temperature as parameter (Siemens)

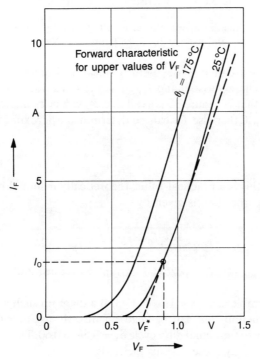

Fig. 3.5 Forward characteristic $I_F = f(V_F)$ of a silicon diode for the determination of incremental resistance R (Valvo)

68

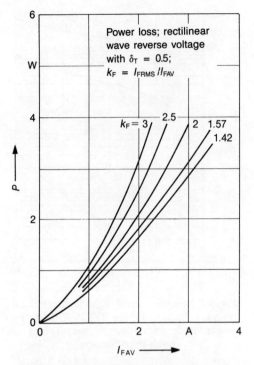

Fig. 3.6 Characteristics for the determination of conduction loss W_{Dc} $(W) = f(I_{FAV})$ for rectilinear-wave reverse voltage with $\delta_T = 0.5$ and form factor k_f as parameter (Valvo)

represented by a back-emf in series with a linear resistance), R_Δ is the incremental resistance (slope of characteristic) and $I_{Fr.m.s.}$ is the r.m.s. value of diode current.

If a graph such as that of Figure 3.5 is available, R_Δ can be estimated from it (in this case, $0.15\,\text{V}/2\,\text{A} = 75\,\text{m}\Omega$). Otherwise R_Δ can be determined approximately from

$$R_\Delta \approx \frac{2V_T}{I_0}$$

where $V_T \approx 50\,\text{mV}$ (this is a practical value; theoretically it should be about 30 mV). The r.m.s. diode current, from [4], is given by

$$I_{Fr.m.s.} = k_f I_0 = I_0 \sqrt{2 + \frac{2}{3p^2}} \quad \text{(A)} \tag{3.20}$$

where $p = P_{0max}/P_{0min}$ for a trapezoidal current waveform and $p = 1$ for a triangular waveform.

In the least favourable case ($p = 1$) $k_f = 1.63$. If a diagram such as that of Figure 3.3 is available, the conduction loss can be read off directly for $\delta_T = 0.5$ ($\lambda = 180°$), which corresponds roughly to the actual duty cycle $\delta_T = 0.54$ to 0.60. The power loss in the diode will now be calculated by these various methods.

Equation (3.14):

$$V_R = 15\,\text{V} + 15\,\text{V} = 30\,\text{V}$$

Since the reverse voltage is relatively low, it is possible to use a Schottky diode. The BYS24-45 (Siemens) or a similar type is suggested. Figures 3.2 to 3.4 relate to this type.

Data for BYS24-45: $I_{FAV} = 2 \times 5\,A$; $V_{Rmax} = 45\,V$.

This diode thus fulfils the given requirements very adequately.

Equation (3.17):
$$P_{Dc} = 0.5\,V \times 5.08\,A \times (1 - 0.54) = 1.2\,W$$
(V_F at 2.5 A from Figure 3.2—the two diodes connected in parallel)

Equation (3.19):

R_Δ is obtained approximately from Figure 3.2 with a voltage variation $\Delta V_F = 0.1\,V$ and a corresponding current variation $\Delta I_0 = 2\,A$ (from 2 A to 4 A), whence $R_\Delta = 0.1\,V/2\,A = 0.05\,\Omega$.
 Or by calculation:
$$R_\Delta = \frac{50\,mV \times 2}{2\,A} = 0.05\,\Omega$$

$I_{Fr.m.s.}$ with $p = 1$, from Equation (3.20),
$$I_{Fr.m.s.} = 1.63 \times 2\,A = 3.26\,A$$

V_F^* cannot be read off directly from the log scale of Figure 3.2, but is less than V_F at I_0, and is accordingly estimated at 0.4 V.
$$P_{Dc} = 2 \times (0.4 \text{ to } 0.7)\,W = 0.8 \text{ to } 1.4\,W$$

Even though all the premises were not quite the same for the different methods, the agreement in the results is quite good.

Equation (3.18):
$$P_{Dr} = 30\,V \times 5 \times 10^{-3}\,A \times 0.6 = 0.09\,W$$

The total diode loss $W_D = 1.2\,W$ (using the average value for W_{Dc}).
 Transistor requirements:
$$V_{CER} \approx V_{CEO} \geqslant |V_0| + V_{imax} \quad (V) \tag{3.21}$$

$$I_{Cmax} \geqslant I_{Lmax} \quad (A) \tag{3.22}$$

Switching times should be as short as possible—i.e. the transition frequency should be as high as possible. In the conducting state the base should be overdriven sufficiently to minimize V_{CEsat}.

Transistor losses

Conduction loss:
$$P_{Trc} = V_{CEsat} I_L \delta_{Tmax} \quad (W) \tag{3.23}$$

Taking $V_{CEsatmax}$ from Figure 2.27,

$$P_{Trc} = 1.2\,\text{V} \times 5.08\,\text{A} \times 0.6 = 3.66\,\text{W}$$

Switching losses:

Since the maximum loss P_{Trc} has been calculated on the basis of δ_{Tmax}, the same operating condition is assumed for the switching losses—i.e. V^*_{imin}.

$$P_{Trs} = (V^*_{imin} + |V^*_0|)\frac{t_r + t_f}{2} I_L f \quad \text{(W)} \tag{3.24}$$

Putting $t_r = 105\,\text{ns}$ at $I_L = 5\,\text{A}$ from Figure 2.29 and $t_f = 230\,\text{ns}$ from Figure 2.30,

$$P_{Trs} = 26.5\,\text{V} \times \frac{105\,\text{ns} + 230\,\text{ns}}{2} \times 5.08 \times 25 \times 10^{-3}\,\text{s}^{-1} = 0.56\,\text{W}$$

(typical value).

The switching losses could alternatively be obtained from Figure 2.33, with the necessary conversion from $20\,\text{kHz}$ to $25\,\text{kHz}$:

$$P_{Trs} = \frac{0.4\,\text{W} \times 25\,\text{kHz}}{20\,\text{kHz}} = 0.5\,\text{W, a similar value.}$$

The total power loss in the transistor is

$$P_{Tr} = 3.66\,\text{W} + 0.56\,\text{W} = 4.22\,\text{W}$$

Since the housing of the PIC power Darlington is electrically isolated, the diode D_1 can be mounted with it on a common heatsink. The cathode of D_1 is connected to the flanged end of its housing. If it is desired to connect the heatsink to earth (to reduce radiated interference), the diode must be insulated by means of a mica washer ($R_{th} \approx 0.1\,\text{kW}$).

Dissipation in transistor and diode:

$$P_{Tr} + P_D = 4.22\,\text{W} + 1.35\,\text{W} = 5.57\,\text{W} \approx 5.6\,\text{W}$$

Specification of heatsink in accordance with Equation (2.45):

$$\text{PIC625:} \quad \theta_{jmax} = 125\,°\text{C}; \quad \theta_{thjc} = 4\,\text{kW}$$
$$\text{BYS24-45:} \quad \theta_{jmax} = 125\,°\text{C}; \quad \theta_{thjc} = 3\,\text{kW}$$

(If the maximum permissible junction temperatures are not the same the calculation must be based on the lower figure; the less favourable (larger) value should be assumed for R_{thjc}.)

$$R_{thca} \leqslant \frac{125\,°\text{C} - 40\,°\text{C}}{5.6\,\text{W}} - 4\,\text{kW} \leqslant 11\,\text{kW};$$

selected, $R_{thca} = 8\,\text{kW}$.

Calculation of resistance R_3, from Equation (2.46) (R_4 in that equation):

$$R_3 \leqslant \frac{12\,\text{V} - 0.7\,\text{V} - 2\,\text{V}}{30 \times 10^{-3}\,\text{A}} = 310\,\Omega; \text{ selected, } R = 330\,\Omega$$

(since $2\,\text{V}$ is a high figure for V_{CEsat} at $I_C = 50\,\text{A}$ in the drive IC).

$$P_{R3} = I^2 R = (30 \times 10^{-3}\,\text{A} \times 0.6)^2 \times 330\,\Omega = 0.1\,\text{W}$$

The calculation of the efficiency must also take account of the losses in the inductor and the current-measuring resistor R_1 and in the drive IC. Inductor resistance $(50\,\text{m}\Omega) + R_1 = 95\,\text{m}\Omega$. The total inductor loss is

$$P_L = (5.08\,\text{A})^2 \times 95 \times 10^{-3}\,\Omega = 2.45\,\text{W}$$

The SG3524N drive circuit used in Figure 3.1 draws an operating current of about 10 mA.

$$P_{IC} = 15\,\text{V} \times 10 \times 10^{-3}\,\text{A} = 0.15\,\text{W}$$

The total power loss in the circuit is then

$$P_{tot} = P_{Tr} + P_D + P_{R3} + P_L + P_{IC}$$
$$= 4.22\,\text{W} + 1.2\,\text{W} + 0.1\,\text{W} + 2.45\,\text{W} + 0.15\,\text{W} = 8.1\,\text{W}$$

In general terms the efficiency is given by

$$\eta = \frac{P_0}{P_0 + W} \tag{3.25}$$

$$\eta = \frac{30\,\text{W}}{30\,\text{W} + 8.1\,\text{W}} = 78.7\%\,[78.9\%]$$

Calculation and measurement are thus in very good agreement.

The efficiency would be improved if a simple transistor with a high current gain were used in place of the Darlington transistor Tr_1. It would be necessary then to replace the SG3524N by the more powerful Type SG3524A (maximum output current 200 mA as against 100 mA for the SG3524N).

Some further observations concerning Figure 3.1 are as follows.

Because the circuit produces a negative output, the output voltage must be applied to the non-inverting input 2 of the error amplifier (see also Figure 2.25). Since the duty cycle is at all times greater than 0.5, the two outputs are connected in parallel.

Capacitor C_3 reduces spikes at the output; the ferrite beads FB_1, with C_7 and C_8, are for the same purpose. In place of the BYS24-45 diode suggested in the calculations, Type SD32 was used in [25]; this type (kindly supplied by TRW) is considerably over-rated.

The work leading to [25] was intended, among other things, to permit a comparison between the various drive ICs, and all the circuits were accordingly built with three different types (SG3524N, MC34060P and TDA4716A).

It is worth noticing that in the version with the Motorola MC34060P, Figure 3.7, in contrast to Figure 3.1, the negative output voltage is applied to the inverting input 2 of the error amplifier. The reason for this is that this controller IC operates in a 'reversed' mode (see also Chapter 8).

As shown by the evaluation of the final measurements according to Figure 3.8 ($|V_0| = f(V_i)$), Figure 3.9 ($|V_0| = f(I_0)$, part characteristic) and Figure 3.10 ($|V_0| = f(I_0)$, complete characteristic), the stability of the output voltage V_0 is very good, and practically independent of the drive IC used.

The minimum load current (limit of continuity) provided for in the example as 0.17 A was measured as 0.13 A. By virtue of the high gain of the control circuits, however, the output voltage was held constant through the region of discontinuous-current operation (easily recognized with the oscilloscope) down to no load.

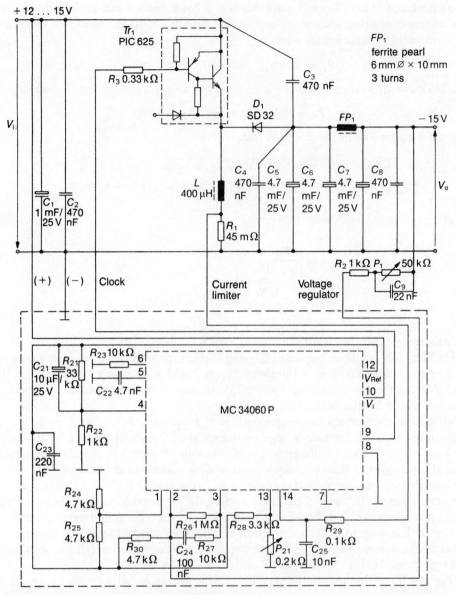

Fig. 3.7 Complete circuit of an inductor-coupled flyback converter with the PIC625 power Darlington transistor and the MC34060P control IC

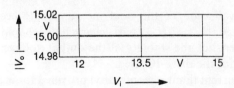

Fig. 3.8 Output voltage $V_0 = f(V_i)$ at $I_0 = 1\,A$ for the inductor-coupled flyback converter of Figures 3.1, 3.7 or 3.11

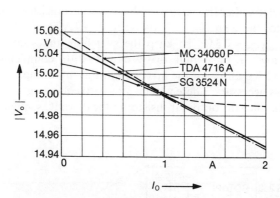

Fig. 3.9 Output voltage $V_0 = f(I_0)$ at $V_i = 13.5\,\text{V}$ (part-characteristic) for the inductor-coupled flyback converter of Figures 3.1, 3.7 or 3.11 with various control ICs

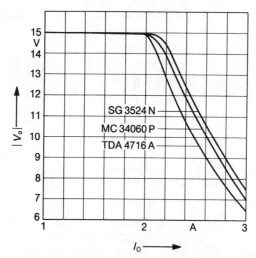

Fig. 3.10 Output voltage $V_0 = f(I_0)$ at $V_i = 13.5\,\text{V}$ (complete characteristic) for the inductor-coupled flyback converter of Figures 3.1, 3.7 or 3.11 with various control ICs

3.3 FLYBACK CONVERTER WITH MOSFET POWER TRANSISTOR

To clarify the question, whether it is more satisfactory to use a bipolar power transistor (e.g. PIC625) or a MOSFET transistor, the converter described above was also built with a MOSFET power stage, with the same input and output conditions. Since the switching times of MOSFET power transistors are generally less than those of bipolar transistors (at least in converters without complicated drive circuits), the switching frequency was increased from 25 kHz to 65 kHz.

In consequence of the higher switching frequency the inductance was considerably lower (160 μH), as were also the dimensions of the inductor. A VAC inductor was again used, namely Type ZKB422/059-02-H2, with a maximum permissible current of 6.3 A— amply dimensioned. Figure 3.11 shows the complete circuit with the SG3524N drive IC.

74

Two International Rectifier IRF9530 MOSFET power transistors were used in parallel. The essential data for these are: $V_{DS} = -100\,V$; $I_D = -7\,A$ at $\theta_c = 100\,°C$; $V_{GS} = \pm 20\,V$; $\theta_{jmax} = 150\,°C$; $R_{DS(on)max} = 0.3\,\Omega$. To keep the switching times, and consequently the switching losses, low, a push–pull drive circuit with complementary low-power MOSFETs was used. To allow for the inversion in this drive stage, a further low-power MOSFET (Tr_1) was necessary. There was no alternative to using a p-channel MOSFET as the power switch, since the output transistor in the drive IC switches towards frame in the 'on' periods.

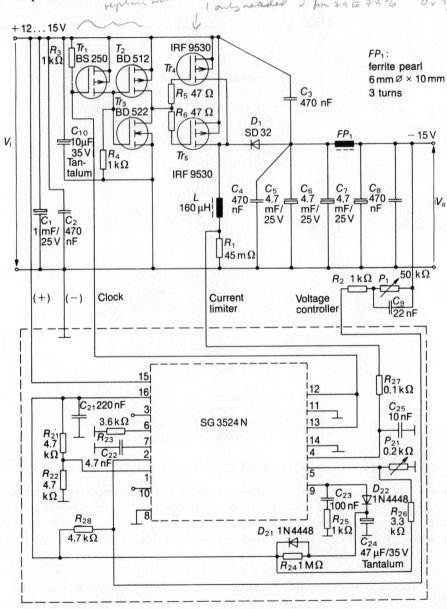

Fig. 3.11 Complete circuit of an inductor-coupled flyback converter with a p-channel power MOSFET and the SG3524N control IC

Although the 160 μH inductor used here has a d.c. resistance of only 25 mΩ, V_i, V_0 and the duty cycles are affected only minimally, so that the previously calculated values can be retained. The conduction loss must now be calculated using Equation (2.56) (substituting I_L for I_D). With the two power transistors connected in parallel, $R_{DS(on)} = 0.3\,\Omega/2 = 0.15\,\Omega$. For the increased temperature, from Figure 2.46, a correction factor of about 1.5 ($\theta_j = 100\,°C$) can be estimated.

$$P_{Trc} = 0.23\,\Omega \times 5^2\,A^2 \times 0.6 = 3.45\ W$$

Equation (3.24): ($t_r = t_f = 140$ ns from the data sheet)

$$P_{Trs} = 26.5\ V \times \frac{140\,ns + 140\,ns}{2} \times 5\,A \times 65 \times 10^3\,s^{-1}$$

$$= 1.2\ W$$

$$P_{Tr} = 3.42\ W + 1.2\ W = 4.65\ W$$

The power loss in the inductor is somewhat reduced by the lower inductor resistance. The loss in $R_1 + R_L$ is

$$P'_L = (45 + 25) \times 10^{-3}\,\Omega \times 5.08^2\,A^2 = 1.8\ W$$

Since the turning-on of the output transistors in the drive IC drives the gate of Tr_1 to ground, the gates of Tr_2 and Tr_3 receive the full positive potential. R_3 and R_4 are thus subjected to the full input voltage for the duration of the 'on' period.

Loss in $R_3 + R_4$: ($T_{R3max} = I_{R4max} = 15\ V/1\ k\Omega = 15\ mA$):

$$P_R = 2 \times 1 \times 10^3\,\Omega \times (15 \times 10^{-3}\,A \times 0.6)^2 = 0.16\ W$$

The loss in the diode D_1 is the same as before (1.2 W), as is that in the IC (0.15 W). The total loss is then

$$P_{tot} = 4.65\ W + 1.2\ W + 0.16\ W + 1.8\ W + 0.15\ W$$

$$= 7.96\ W \approx 8\ W$$

$$\mu = \frac{30\ W}{30\ W + 8\ W} = 79 \text{ per cent } [81.4 \text{ percent}]$$

Comment on the circuit of Figure 3.11:
This circuit is designed for the highest possible efficiency, without regard to complexity. It would have been possible, however, without any other changes, to use a single power transistor, Type IRF9530 or the cheaper IRF9531, with $V_{DS} = -60\ V$. Furthermore, this could have been driven directly from the output transistors of the drive IC, with simply a resistor of about 150 Ω from the gate to $+V_i$ in place of the transistors Tr_1 to Tr_3. The current of 0.1 A which then flows should not increase the switching times significantly; if necessary the current could be increased to 0.2 A by using the 3524A drive IC. The loss in this 150 Ω or 75 Ω resistor would be greater than that in $R_3 + R_4$, and the conduction loss W_{Trc} would be doubled. A rough calculation shows that an efficiency in the region of 75 per cent would be attained, with much reduced complexity.

It can be said by way of summary that there is no significant difference between the versions with bipolar transistor and MOSFET. As was to be expected, there was also no difference in respect of constancy of output voltage.

4 Inductor-coupled step-up converter

The inductor-coupled step-up converter [25], [7], constitutes a blocking converter, in that the energy is conveyed to the output during the non-conducting periods of the transistor.

4.1 BASIC PRINCIPLES

The relationship between the output voltage and the switching ratio in the step-up converter is different again:

$$V_0 = \frac{1}{1 - \delta_{Tmax}} V_{imin} = \frac{1}{1 - \delta_{Tmin}} V_{imax} \quad \text{(V)}$$

(4.1)

By rearranging the equation, the duty cycles can be calculated; as before, if practically useful results are to be obtained, the quantities to be entered are the output voltage V_0^*, allowing for the voltage drop in the diode, and the input voltage V_i^* as reduced by the voltage drops in the transistor, the resistance of the inductor and the current-measuring resistor R_1.

$$\delta_{Tmin} = \frac{V_0^* - V_{imax}^*}{V_0^*} \quad \text{(V)}$$

(4.2)

and

$$\delta_{Tmax} = \frac{V_0^* - V_{imin}^*}{V_0^*} \quad \text{(V)}$$

(4.3)

The total output voltage V_0^* to be applied is thus

$$V_0^* = V_0 + V_F \quad \text{(V)}$$

(4.4)

where V_F is the forward voltage drop of diode at I_L (V).

The effective input voltage V_i^* is calculated as for the flyback converter in accordance with Equation (3.4), and the fluctuation ΔI_L in the inductor current similarly in accordance with Equation (3.7).

A different relationship applies to the inductance L, however:

$$L \geqslant \frac{V_0^* \delta_{Tmin} (1 - \delta_{Tmin})}{\Delta I_L f} \quad \text{(H)}$$

(4.5)

76

The mean direct current in the inductor is again considerably higher than the maximum load current I_{0max}, and reaches its highest value with V^*_{imin} and P^*_{0max}.

$$I_L = \frac{P^*_{0max}}{V^*_{imin}} \quad (A) \tag{4.6}$$

The maximum inductor current—important in the dimensioning of the inductor—is given by

$$I_{Lmax} = \frac{P^*_{0max}}{V^*_{imin}} + \frac{V^*_{imin}\delta_{Tmax}}{2fL} \quad (A) \tag{4.7}$$

$$P^*_{0max} = V^*_0 I_{0max}$$

Transistor requirements:

$$V_{CER} \approx V_{CE0} = V^*_0 \quad (V) \tag{4.8}$$

I_{Cmax} in accordance with Equation (3.22)

Requirements for free-wheel diode D_1:

$$V_R \geqslant V_0 + V_{CEsat} \quad (V) \tag{4.9}$$

I_F in accordance with Equation (3.16)

I_{FAV} in accordance with Equation (3.15)

4.2 STEP-UP CONVERTER WITH BIPOLAR POWER TRANSISTOR

Example

An inductor-coupled step-up converter is to be designed on the basis of the following data:

Output voltage $V_0 = 30\,\text{V}$; maximum load current $I_{0max} = 2.5\,\text{A}$; input voltage $V_i = 12\,\text{V} \pm 15\% = 10.2$ to $13.8\,\text{V}$; switching frequency $f = 25\,\text{kHz}$; minimum load current $I_{0min} = 0.4\,\text{A}$; permissible output voltage ripple:

$$\Delta V_0 = 0.01\,V_0 = 300\,\text{mV}_{p-p}; \quad \theta_{amax} = 40\,°\text{C}.$$

Circuit as shown in Figure 4.1; measured values in [ ].

To enable the voltage drop across the transistor to be estimated approximately, I_L is first determined theoretically:

$$I_L = \frac{30\,\text{V} \times 2.5\,\text{A}}{10.2\,\text{V}} = 7.35\,\text{A}.$$

In view of the losses to be expected this may be rounded up to about 8.5 A.

For the power transistor the Type PIC656 Darlington transistor was chosen. The PIC655 or the PIC635 could equally well have been used (with the same characteristics but different connections), as could a Darlington transistor with short switching times (= higher transition frequency).

From Equations (4.8) and (3.22) the transistor requirements are

$$V_{CE} \geqslant 30\,\text{V}; \quad I_{Cmax} \geqslant 9\,\text{A}$$

PIC656 power IC: $V_{CE} \geqslant 80\,\text{V}$; $I_{Cmax} \geqslant 15\,\text{A}$

The requirements are thus amply met.

In view of the relatively low voltages, the built-in diode in the PIC656 is not used: a Schottky diode is provided instead.

From Equations (4.9) and (3.15) the diode requirements are

$$V_R \geqslant 32 \text{ V}; \quad I_{FAV} \geqslant 1.8 \text{ to } 3.6 \text{ A (see below)}.$$

The BYS24-45 used for the flyback converter, with $V_{Rmax} = 45$ V and $I_{FAV} = 2 \times 5$ A meets these requirements. In the practical investigation the over-rated TRW diode Type SD32 was again used.

Equation (4.4):

$V_F \approx 0.6$ V at $I_F = 4.5$ A, from Figure 3.2 (the two diodes connected in parallel)

$V_0^* = 30 \text{ V} + 0.6 \text{ V} = 30.6 \text{ V}$

Equation (3.4):

V_{CEsat} from Figure 2.27 (corresponds to the selected Type PIC656)

$V_{CEsat} = 1.3$ V at $I_C \approx 9$ A; typical value (1.8 V maximum)

V_L estimated as 0.1 V; $V_{R1} = 0.2$ V

$V_{imin}^* = 10.2 \text{ V} - 1.3 \text{ V} - 0.1 \text{ V} - 0.2 \text{ V} = 8.6 \text{ V}$

$V_{imax}^* = 13.8 \text{ V} - 1.3 \text{ V} - 0.1 \text{ V} - 0.2 \text{ V} = 12.2 \text{ V}$

Equations (4.2) and (4.3):

$$\delta_{Tmin} = \frac{30.6 \text{ V} - 12.2 \text{ V}}{30.6 \text{ V}} = 0.60$$

$$\delta_{Tmax} = \frac{30.6 \text{ V} - 8.6 \text{ V}}{30.6 \text{ V}} = 0.72 \ [0.71]$$

Equation (3.7):

$$\Delta I_L = \frac{2 \times 0.4 \text{ A}}{1 - 0.60} = 2.0 \text{ A}; \quad [I_{0min}, \text{ measured: } 0.34 \text{ A}]$$

Equation (4.5):

$$L \geqslant \frac{30.6 \text{ V} \times 0.6 \times (1 - 0.6)}{2 \text{ A} \times 25 \times 10^3 \text{ s}^{-1}} \geqslant 147 \text{ H} \approx 160 \text{ } \mu\text{H}$$

Equation (4.6):

$$I_L = \frac{30.6 \text{ V} \times 2.5 \text{ A}}{8.6 \text{ V}} = 8.9 \text{ A, rounded to 9 A } [9 \text{ A}]$$

The previous estimate of 8.5 A is thus well substantiated, and the calculations need not be corrected.

Equation (4.7):

$$I_{Lmax} = 8.9 \text{ A} + \frac{8.6 \text{ V} \times 0.72}{2 \times 25 \times 10^3 \text{ s}^{-1} \times 160 \times 10^{-6} \text{ H}}$$

$$= 8.9 \text{ A} + 0.8 \text{ A} = 9.7 \text{ A } [9.67 \text{ A}]$$

It would be possible with these values to draw up a design for an inductor in the manner described in Section 2.2 (core size, air gap, inductance factor, number of turns, wire gauge, winding resistance, flux density). It was decided, however, again to use a commercially available VAC inductor, namely the Type ZKB443/303-02-H2. The particulars for this inductor are: $L = 160 \, \mu H$; $I_{max} = 16 \, A$; $R_{Cu} = 8 \, m\Omega$. These values make the inductor suitable for the given duty.

The *diode losses* are in this case—as with the other inductor-coupled converters with Schottky diodes—for practical purposes only conduction losses. These can be calculated from Equation (3.17).

Equation (3.17):
With $V_F = 0.58 \, V$ at $I_F = 4.5 \, A$, from Figure 3.2 (two diodes in parallel),

$$P_{D1} = 0.58 \, V \times 9 \, A \times 0.4 = 2.1 \, W$$

The *transistor losses* are again a combination of conduction losses and switching losses. The conduction loss is calculated from Equation (3.23).

Switching losses:

$$P_{Trs} = V_0^* \frac{t_r + t_f}{2} I_L f \quad (W) \tag{4.10}$$

Figures 2.29 and 2.30 can be used to determine the switching times, since the PIC656 is similar in this respect to the PIC625.
Putting $t_r = 115 \, ns$ and $t_f = 270 \, ns$ gives

$$P_{Trs} = 30.6 \, V \times \frac{(115 + 270) \times 10^{-9} \, s}{2} \times 9 \, A \times 25 \times 10^3 \, s^{-1}$$

$$= 1.3 \, W$$

$$P_{Tr} = 8.4 \, W + 1.3 \, W = 9.7 \, W, \text{ or approximately } 10 \, W$$
$$\text{(maximum 13 A)}$$

If a common heatsink is to be used for the diode and the transistor, at least one of the two components must be insulated, since the cathode of the diode and the cathode of the diode incorporated in the PIC656 are both connected to the housing flange. The heatsink can be specified in accordance with Equation (2.45), allowing at least 0.1 to 0.2 kW for the insulation. The calculation will not be repeated here.

The losses in the inductor and the current-measuring resistor amount to

$$P_L' = (8 \, m\Omega + 30 \, m\Omega) \times 9^2 \, A^2 = 3.1 \, W$$

The PIC656 transistor requires a base current of about 30 mA at $I_C = 15 \, A$, so that at 9 A about 25 mA is adequate. R_5 is then calculated as

$$R_5 \leqslant \frac{V_{imin} - V_{BETr3} - R_{R1}}{(25 \text{ to } 30) \times 10^{-3} \, A} \quad (\Omega) \tag{4.11}$$

$$V_{BETr3} \approx 1.5 \, V$$

$$R_5 \leqslant \frac{10.2 \, V - 1.5 \, V - 0.25 \, V}{(25 \text{ to } 30) \times 10^{-3} \, A} = 338 \text{ to } 281 \, \Omega;$$

selected: $R_5 = 300\,\Omega$

$$P_{R5} = I^2 R_5 = (30 \times 10^{-3}\,\text{A})^2 \times 300 = 0.3\,\text{W}$$

(The on-state resistance $R_{DS(on)}$ of Tr_1, at $10\,\Omega$ maximum, is negligible in comparison with R_5.)

The value of R_2 is not critical. On the one hand it should be low enough that the input capacitance of Tr_2 is discharged rapidly: on the other hand it should not dissipate excessive power. A value of $1\,\text{k}\Omega$ was chosen.

$$P_{R2} = \frac{V_i^2}{R_2} = \frac{(13.8\,\text{V})^2}{1 \times 10^3\,\Omega} = 0.2\,\text{W}; \quad P_{IC} \text{ is again about } 0.15\,\text{W}$$

Total power loss:

$$P = P_{D1} + P_{Tr} + P_L' + P_{R5} + P_{R2} + P_{IC}$$
$$= 2.1\,\text{W} + (9.7 \text{ to } 13)\,\text{W} + 3.1\,\text{W} + 0.3\,\text{W} + 0.2\,\text{W} + 0.15\,\text{W}$$
$$= 15.6 \text{ to } 19\,\text{W}$$

The efficiency is given by Equation (3.25):

$$\eta = \frac{P_0}{P_0 + P} = \frac{30\,\text{V} \times 2.5\,\text{A}}{30\,\text{V} \times 2.5\,\text{A} + (15.6 \text{ to } 19)\,\text{W}} = 83 \text{ to } 80\% \ [82\%]$$

Since the duty cycles are at all times greater than 0.5, the two outputs of the drive IC must be connected in parallel. For further details see Chapter 8.

Note on Figure 4.1: when the output transistors in the drive IC are turned on, the gate of the p-channel MOSFET Tr_1 is driven to earth potential; it is thus turned on, and base current flows through R_5 to Tr_3, which in turn is also turned on. When the drive IC reverts to the blocking state, Tr_1 is turned off, and at the same time the n-channel MOSFET Tr_2 is turned on by the positive gate potential received through R_2. This connects the base of Tr_3 directly to ground causing rapid turn-off.

The limiting factor in this circuit variant is the maximum permissible gate voltage for Tr_1 and Tr_2 of $\pm 15\,\text{V}$. Since the input voltage has a maximum value of 13.8 V, this presents no difficulty. For higher input voltages the two MOSFETs could be replaced by a bipolar complementary stage, with the value of R_2 chosen appropriately.

Again the choice of output capacitor C_0 is determined entirely by its equivalent series resistance (ESR), to ensure that the prescribed output voltage ripple ΔV_{pp} is not exceeded. Thus only Equation (3.12) is applied.

It has also to be checked that the permissible alternating current loading is not exceeded. For this purpose the most unfavourable value according to Equation (3.13) should be assumed.

For the present example,

$$\text{ESR} \leqslant \frac{300\,\text{mV}}{10\,\text{A}} \leqslant 30\,\text{m}\Omega \quad \text{and} \quad I_{Crms} = 1.3 \times 2.5\,\text{A} = 3.25\,\text{A}$$

Since the output voltage $V_0 = 30\,\text{V}$, electrolytic capacitors with the standard rating of 40 V are appropriate.

Reference should now be made to the table of Figure 2.16. While this table extends only to an operating voltage of 25 V, Figure 2.18 shows that the impedance decreases with

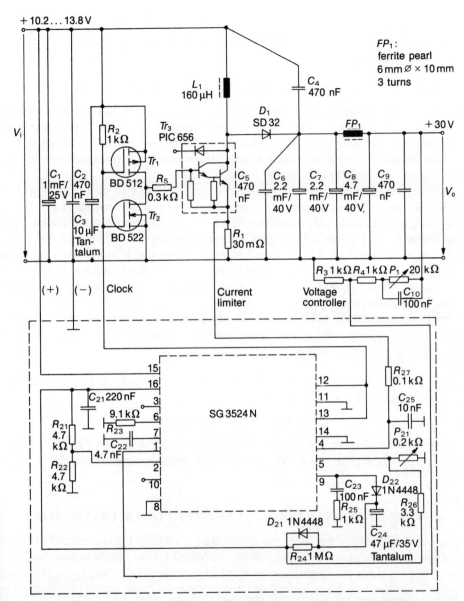

Fig. 4.1 Complete circuit of an inductor-coupled step-up converter with the PIC656 power Darlington transistor and the SG3524N control circuit

increasing operating voltage, and two electrolytic capacitors of 2200 μF each will certainly be adequate; if necessary two 1000 μF capacitors would suffice. Two capacitors of 2200 μF/40 V were decided on. The 4700 μF capacitor C_8 in Figure 4.1 being more than sufficient for the suppression of spikes, it also contributes to the reduction of ripple. The measured value of 75 mV$_{pp}$ is only a quarter of the prescribed magnitude as a result.

Permissible alternating current loading according to the table of Figure 2.16 and the correction factors from Figure 2.19 (1.4) and Figure 2.20 (1.8):

For $C_0 = 2 \times 2200 \,\mu\text{F}/40$ V ($I_{C\sim}$ is higher at $V_{rat} = 40$ V than at $V_{rat} = 25$ V):

$$I_{C\sim} = 2 \times 1 \text{ A} \times 1.4 \times 1.8 = 5.04 \text{ A}$$

For $C_0 = 2 \times 1000 \,\mu\text{F}/40$ V:

$$I_{C\sim} = 2 \times 0.72 \text{ A} \times 1.4 \times 1.8 = 3.63 \text{ A} > 3.25 \text{ A}$$

Thus the permissible alternating current loading of two $1000 \,\mu\text{F}$ units would also be sufficient, since the rating $I_{C\sim}$ is higher than the current I_{Crms} which flows in the least favourable case. Even a slight excess over the rating is not critical, since it does not matter very much whether the capacitors attain a temperature rise of 10 °C (the limit on which the current ratings are based) or somewhat more.

To keep spikes at the output below the prescribed limit of output voltage ripple of 300 mV$_{pp}$, an output filter consisting of a ferrite bead together with capacitors C_8 and C_9 is provided, C_8 being considerably overdimensioned. C_4 also contributes to the reduction of output spikes. To limit spikes in the output voltage to a minimum, the layout of the circuit would have to be optimized, which was not possible in this case. It would also have to be checked, whether an improved performance could be obtained using a higher filter inductance (ferrite beads with more turns or a number of smaller ferrite beads in series) and smaller capacitors.

A better understanding of the circuit may be gained, as in the case of the flyback converter, by comparing the power ratios with the transistor conducting and non-conducting (e.g. with V^*_{imin} and I_{0max}.

Transistor conducting, energy storage in the inductor: voltage across the inductor $V^*_{imin} = 8.6$ V; inductor current $I_L = 9$ A.

$$P_L = 8.6 \text{ V} \times 9 \text{ A} \times 0.72 = 55.7 \text{ W}$$

Transistor non-conducting: applicable voltage difference (30.6 V − 8.6 V) at $I_0 = 2.5$ A.
$$P'_L = (30.6 \text{ V} - 8.6 \text{ V}) \times 2.5 \text{ A} = 55 \text{ W, practically equal.}$$

4.3 STEP-UP CONVERTER WITH MOSFET POWER TRANSISTOR

In a further version the power stage was replaced by a MOSFET power transistor, Type BUZ10. In view of the short switching times of the MOSFET the switching frequency was raised to 65 kHz. Since the operating conditions were unaltered, the duty cycles remained the same; hence the inductance L obtained from Equation (4.5) can simply be converted in the inverse ratio of the switching frequencies. The result is a value of 61.5 μH, rounded up to 63 μH. The VAC Type ZKB455/052-02-H2 was selected.

$$\text{Inductor data: } L = 63 \,\mu\text{H}; \quad I_{max} = 16 \text{ A}; \quad R_{Cu} = 8 \text{ m}\Omega$$

The inductor is thus suitable for the circuit. Figure 4.2 shows the complete circuit in the version with the SG3524N drive IC.

For the calculation of losses, it is only necessary to recalculate the transistor losses W_{Trc} (from Equation (2.56)) and W_{Trs} (from Equation (4.10)). All the other loss components can be taken from the bipolar version.

The chosen transistor Type BUZ10, with $V_{DS} = 50$ V and $I_D = 12$ A, meets the requirements derived from Equations (4.8) (30.6 V max.) and (3.22) (9.7 A max.). The

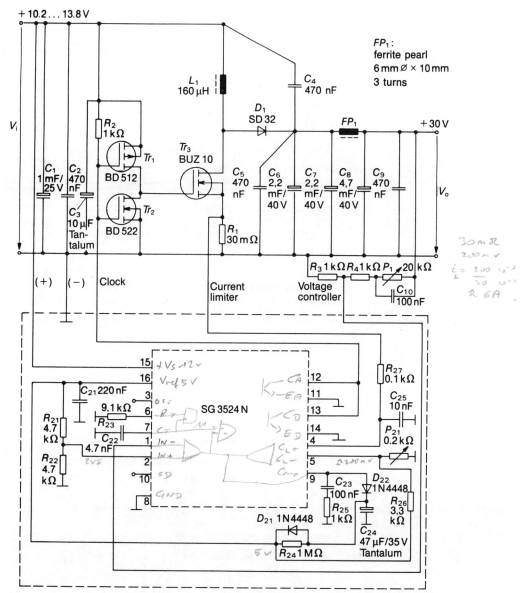

Fig. 4.2 Complete circuit of an inductor-coupled step-up converter with the BUZ10
n-channel power MOSFET and the SG3524N control IC

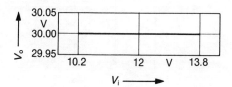

Fig. 4.3 Output voltage $V_0 = f(V_i)$ at $I_0 =$
1 A for the inductor-coupled step-up converter of Figure 4.1 or 4.2

84

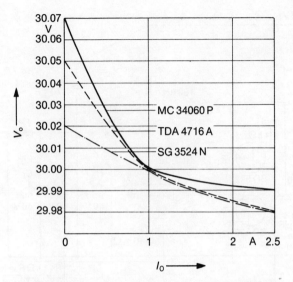

Fig. 4.4 Output voltage $V_0 = f(I_0)$ at $V_i = 12\,\text{V}$ (part-characteristic) for the inductor-coupled step-up converter of Figure 4.1 or 4.2 with various control ICs

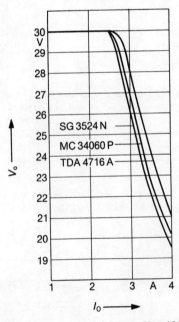

Fig. 4.5 Output voltage $V_0 = f(I_0)$ at $V_i = 12\,\text{V}$ (complete characteristic) for the inductor-coupled step-up converter of Figure 4.1 or 4.2 with various control ICs

permissible gate voltage is ± 20 V, so there are no problems in this regard in the present example. The switching times t_r and t_f are both quoted in the data sheet as 60 ns.

From Equation (2.56) and Figure 2.46 ($\theta_j = 100\,°C$), with $R_{DS(on)} = 0.15\,\Omega$,

$$P_{Trs} = 0.15\,\Omega \times 9^2\,A^2 \times 0.72 = 8.75\,W$$

$$P_{Trs} = 30.6\,V \times \frac{(60 + 60) \times 10^{-9}\,s}{2} \times 9\,A \times 65 \times 10^3\,s^{-1}$$

$$= 1.1\,W$$

$$P_{Tr} = 8.75\,W + 1.1\,W = 9.85\,W$$

Adopting again the previously derived figures for $P_{Dc} = 2.1$ W, $P'_L = 3.1$ W, $P_{R2} = 0.2$ W and $P_{IC} = 0.15$ W, the total power loss becomes

$$P = P_D + P_{Tr} + P'_L + P_{R2} + P_{IC}$$

$$= 2.1\,W + 9.85\,W + 3.1\,W + 0.2\,W + 0.15\,W = 15.4\,W$$

Then from Equation (3.25) the calculated efficiency is

$$\eta = \frac{30\,V \times 2.5\,A}{30\,V \times 2.5\,A + 15.4\,W} = 83\%\ [86.5\%]$$

The measured quantities were $V_i = 10.2$ V and $I_i = 8.5$ A. This gives an input power of 86.7 W, and hence the efficiency of 86.5 per cent and a power loss $P = 86.7\,W - 75\,W = 11.7\,W$. Calculation and measurement are thus in very good agreement.

The constancy of the output voltage with variations in input voltage and load is shown in Figure 4.3, representing $V_0 = f(V_i)$ with $I_0 = 1$ A; Figure 4.4, $V_0 = f(I_0)$ with $V_i = 12$ V (part characteristic); Figure 4.5, $V_0 = f(I_0)$ with $V_i = 12$ V (complete characteristic). In this case again, measurements were carried out with various drive ICs. While the SG3524N, particularly in Figure 4.4, comes out best, the other types can also be said to be very suitable. Figures 4.3 to 4.5 are identical for the two versions (bipolar and MOSFET); this is only to be expected, since the control characteristics are not dependent upon the type of output stage.

Since the calculation of losses is always on a worst-case basis, the measured efficiency is in most cases higher than the calculated figure.

5 Single-ended forward converter for mains input

The single-ended forward converter, the most widely used type for direct mains input (primary-switched—see also Figure 1.7) is shown diagrammatically in Figure 5.1. Protection networks and drive circuitry are not shown.

The a.c. mains supply is connected through the protective resistor R_i and the suppression inductor L_i (usually in the form of a symmetrical inductor divided between the two mains leads, but represented here as a single inductance) and rectified in the bridge rectifier Rec. C_1 is a reservoir capacitor. The resistor R_i must limit the switch-on surge current, due to the presence of the initially discharged capacitor C_1, to such an extent that the rectifier Rec is not overloaded (peak current or I^2t value) and that no (slow) fuse included in the equipment can be blown or (quick-acting) m.c.b. in the mains supply tripped.

The transistor is switched alternately on and off by the drive circuit. The arrows by the transformer T show the direction of the induced e.m.f.s. With the transistor Tr turned on, the transformer draws a current nI_0 ($n = N_3/N_1 < 1$) plus a rising magnetizing current i_m. Because of this rising current a voltage is induced (broken arrow) in opposition to the input voltage V_i. Since the secondary winding N_3 is similarly poled, a positive voltage is applied to the diode, and it conducts.

When the transistor has been turned off, the magnetizing current i_m falls, and a voltage is generated in the reverse direction (chain-dotted). Diode D_2 blocks, and the load current is

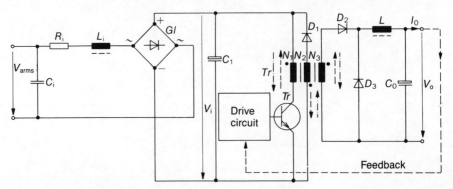

Fig. 5.1 Single-ended forward converter (without protective networks) details of the control circuit or e.m.i. filter

86

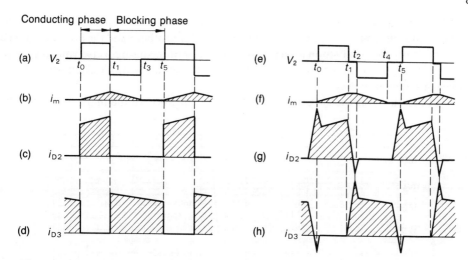

Fig. 5.2 Current and voltage waveforms in the forward converter: (a) to (d) ignoring rectifier and free-wheel diode characteristics; (e) to (h) showing the effects of diode recovery times (Valvo) [18]

transferred to the free-wheel diode D_3, as in the inductor-coupled step-down converter (see Chapter 2). The magnetizing energy stored in T must now be recovered. This is normally done by feeding it back to the input supply V_i by means of the winding N_2 (in most cases with the same number of turns as N_1). At the instant of switching (when the transistor Tr is turned on or off), both diodes D_2 and D_3 conduct simultaneously for a short period (the reverse recovery time t_{rr}), so that an additional current peak flows in the transistor (see also Figure 5.8, where these current peaks are easily seen). Very fast diodes must be used for D_2 and D_3, therefore, to avoid unnecessary loading of the transistor.

Schottky power diodes are frequently used for low output voltages (e.g. 5 to 10 V). These

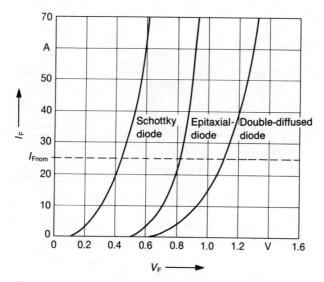

Fig. 5.3 Typical forward characteristics of Schottky, epitaxial and double-diffused diodes (Valvo) [18]

have the lowest possible recovery times, and are available for very high forward currents, but have only relatively low reverse voltage ratings. On the other hand the forward voltage drop (and hence the conduction loss) is relatively low. For medium reverse voltages (up to 200 to 300 V), it is more appropriate to use the so-called epitaxial diodes, and beyond that normal fast silicon diodes.

The calculations for the circuit begin with the rectification of the mains input supply.

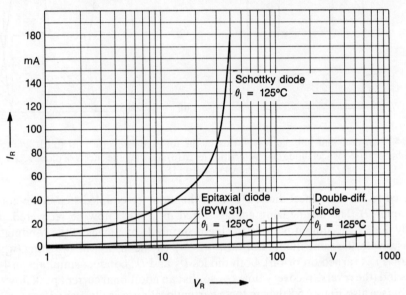

Fig. 5.4 Typical reverse characteristics of Schottky, epitaxial and double-diffused diodes (Valvo) [18]

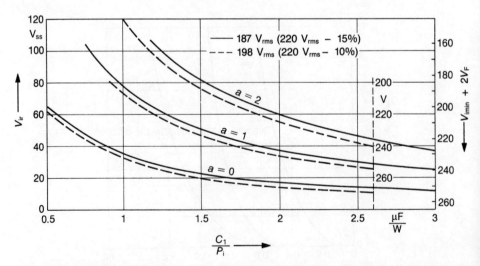

Fig. 5.5 Variation in ripple voltage V_{ir} (V_{pp}) and direct input voltage $V_{imin} + 2V_F$ (V) with reservoir capacitance C_1, relative to input power P_i, with the number of missing mains half-cycles (a) as parameter. Solid lines are for -15% mains voltage (187 V_{rms}) and dashed lines for -10 per cent mains voltage (calculated in accordance with [4])

5.1 MAINS RECTIFICATION

The first step is to determine the capacitor C_1 (high-voltage electrolytic). For this purpose the curves of Figure 5.5 may be employed, converted in accordance with [4].

The soild lines apply to a mains voltage reduction of 15 per cent (220 V r.m.s. $- 15\% =$ 187 V 4.m.s. $= 265$ V peak), the dashed curves to a voltage reduction of 10 per cent (220 V r.m.s. $- 10\% = 198$ V r.m.s. $= 280$ V peak). The left-hand vertical scale shows the peak-to-peak ripple voltage; on the right is shown the minimum converter input voltage $V_{imin} + 2V_F$, at which the converter must continue to operate. The parameter 'a' signifies the extent to which allowance is made for missing half-cycles in the input: $a = 2, 1, 0$ denote respectively two missing half-cycles, one or none.

Example

$a = 0$ and $V_a = 187$ V r.m.s.:

$$\frac{C_1}{P_1} = 1 \ \mu\text{F/W}: \ V_{ir} = 35 \text{ V}, \quad V_{imin} + 2V_F = 230 \text{ V}.$$

Alternatively, $\hat{V}_a = 187$ V $\times \sqrt{2} = 265$ V peak;

$$265 \text{ V} - 35 \text{ V} = 230 \text{ V}.$$

The capacitor should be as small as possible from the point of view of cost and space; on the other hand the ripple voltage should be as low as possible. Depending upon the requirements and cost considerations, a will vary between 0 and 1, and C_1/P_1 between 1 and 2 μF/W. The capacitor must have a rated voltage of at least 220 V $\times$ 1.15 $\times$ 1.41 $=$

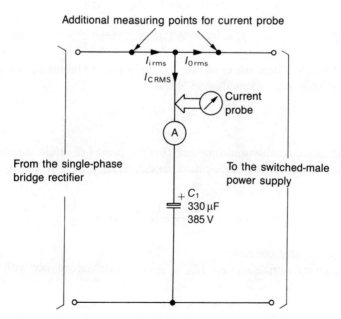

Fig. 5.6 Definitions of alternating capacitor currents with indications of measuring points (Valvo) [4]

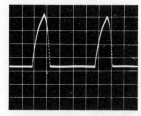

Fig. 5.7 Current wave-form at the input to the capacitor in a 300 W forward-converter power supply unit with a capacitive input filter. Scale: horizontal, 2 ms/division; vertical, 2 A/division (Valvo) [14]

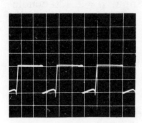

Fig. 5.8 Switching transistor collector current in a 300 W forward-converter power supply unit. Scale: horizontal, 10 μs/division; vertical, 2 A/division (Valvo) [14]

357 V; ratings of 350 V or 385 V are usual. For a given electrolytic capacitor, it remains to check the alternating current loading.

If it is possible to observe the current waveform, e.g. by means of a current probe, the total alternating current in the capacitor (100 Hz input current and switching-frequency output current) can be calculated in accordance with [14]:

$$I_{C1\,rms} = \sqrt{I_i^2 + I_0^2} \quad \text{(A)} \tag{5.1}$$

where I_i is the r.m.s. input alternating current at twice mains frequency (100 Hz)(A) and I_0 is the r.m.s. output alternating current at switching frequency f(A)

$$I_i^2 = \left(\frac{\pi^2}{8\delta_{Ti}} - 1\right) I_{iAV}^2 \quad \text{(A}^2) \tag{5.2}$$

where I_{iAV} is the mean direct current drawn by the converter (input current at minimum mains voltage) according to Equation (5.5) (A)

$$\delta_{Ti} = \frac{\alpha_i}{D} \tag{5.3}$$

where α_i is the diode conduction time and D is the period of diode current cycle: with $f_a = 50$ Hz, $D = 10$ ms for a single-phase bridge rectifier, 3.3 ms for a three-phase bridge rectifier.

$$I_0^2 = \left(\frac{1}{\delta_{Tmin}} - 1\right) I_{iAV}^2 \quad \text{(A}^2) \tag{5.4}$$

(applies only to forward converters).

The mean converter input current I_{iAV} is calculated in accordance with [4] as

$$I_{iAV} \approx \frac{P_i}{\hat{V}_a - 0.5V_r} \quad \text{(A)} \tag{5.5}$$

where $\hat{V}_a$ is the minimum peak mains voltage (V), P_i is the converter input power =

converter output power/η(W), η is the efficiency: about 75 to 80 per cent and V_r is the peak-to-peak ripple voltage (V).

Example

A single-ended forward converter is designed for $P_0 = 300$ W; $\theta_{jmax} = 45\,^\circ$C; minimum mains voltage $V_{arms} = 187$ V; $a = 1$; $C_1 = 1.5\,\mu$F/W (estimated; $V_0 = 5$ V; $I_{0max} = 60$ A; $f = 40$ kHz; $\Delta I_L = 12$ A.

From Figure 5.7, $\alpha_i = 2.8$ ms.

$$\delta_{Ti} = \frac{2.8\,\text{ms}}{10\,\text{ms}} = 0.28$$

$P_i = 300$ W/(0.8 to 0.75) $= 375$ to 400 W; $P_i = 390$ W assumed as a basis for calculation. From Figure 5.5, with $C_1/P_i = 1.5\,\mu$F/W and $a = 1$, $V_{ir} = 50$ V. $V_{imin} + 2V_F = 214$ V, or $V_{imin} = 212.5$ V, say 212 V.

Equation (5.5):

$$I_{iAV} = \frac{390\,\text{W}}{187\,\text{V} \times 1.41 - 0.5 \times 50\,\text{V}} = 1.63\,\text{A} \qquad (1.55\,\text{A with } a = 0)$$

Equation (5.2):

$$I_i^2 = \left(\frac{3.14^2}{8 \times 0.28} - 1\right) \times 1.55^2\,\text{A}^2 = 8.17\,\text{A}^2; \quad I_i = 2.86\,\text{A}$$

Equation (5.4):

$$I_0^2 = \left(\frac{1}{0.28} - 1\right) \times 1.55^2\,\text{A}^2 = 6.18\,A^2; \quad I_0 = 2.49\,\text{A}$$

($\delta_{Tmin} = 0.28$ determined below)
Equation (5.1):

$$I_{C1rms} = \sqrt{8.17\,\text{A}^2 + 6.18\,\text{A}^2} = 3.8\,\text{A}$$

This calculation of the alternating capacitor current is based on observation of the current waveform, e.g. using an oscilloscope with a current probe. If this is not possible, the current can be calculated in accordance with [4].

For this purpose the diode current form factor k_{fD} is required. This factor, with the mean direct current I_{FAV} already calculated, would enable the conduction loss in the rectifier diodes to be calculated, but since there are adequately rated rectifier bridges available, the choice may be restricted to these. The form factor can be used, however, to calculate the r.m.s. current in the reservoir capacitor and the series resistor R_i to a good approximation.

According to [4], the diode current form factor k_{fD} (neglecting R_i, and therefore giving a figure on the high side) is given by

$$k_{fD} = \frac{I_{Frms}}{I_{FAV}} = \frac{1}{X}\sqrt{\pi\left(\frac{\pi}{2} - \sin^{-1}(1 - X) - (1 - X)\sqrt{2X - X^2}\right)}$$

where

$$X = \frac{V_{ir}}{\hat{V}_a} \tag{5.6}$$

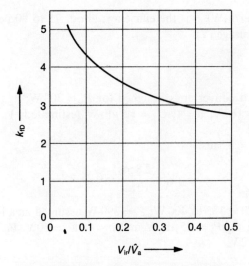

Fig. 5.9 Variation of diode current form factor k_{fD} with ripple voltage (V_{ir}) relative to the peak minimum voltage $\hat{V}_a$ (V) $(\hat{V}_a = 187\,\text{V} \times 2 = 265\,\text{V})$

V_{ir} and $\hat{V}_a$, as peak-to-peak and peak values respectively, are obtained from Figure 5.5 ($\hat{V}_a$ is in most cases $187\,V \times \sqrt{2} = 265\,\text{V}$).

To avoid the complication of Equation (5.6), the form factor k_{fD} has been calculated for a range of conditions, and is represented by the curve of Figure 5.9.

The alternating capacitor current is given by

$$I_{C1rms} = I_{iAV}\sqrt{0.5\,k_{fD}^2 + k_{fi}^2 - 2} \quad \text{(A)} \tag{5.7}$$

In practice the form factor k_{fi} of the converter input current lies between the limits:

$$1.3 < k_{fi} < 1.7 \tag{5.8}$$

The loading of the resistor R_i is given by

$$I_{Rirms} = I_{iAV}\frac{k_{fD}}{\sqrt{2}} \quad \text{(A)} \tag{5.9}$$

For comparison, I_{C1rms} will now be calculated from Equation (5.7) with an average form factor $k_{fi} = 1.5$.

With $\hat{V}_a = 265\,\text{V}$ and $V_{ir} = 50\,\text{V}$ (from Figure 5.5 with $C_1/P_i = 1.5\,\mu\text{F/W}$ and $a = 1$),

$$\frac{V_{ir}}{\hat{V}_a} = \frac{50\,\text{V}}{265\,\text{V}} = 0.19 \ (0.9 \text{ with } a = 0)$$

Reading-off from Figure 5.9, $k_{fD} = 3.6$ (4.4 with $a = 0$).

Equation (5.7):

$$I_{Cirms} = 1.63\,\text{A} \times \sqrt{0.5 \times 3.6^2 + 1.5^2 - 2} = 4.23\,\text{A}$$
$$(5.1\,\text{A with } a = 0)$$

C_a	V_a	V_s	$R_{ESR,max}$ 120 Hz 20·C	Z_{max} 10 kHz 20·C	$I_{R,max}$ 5 min 20·C	$I_{C\sim max}$ 120 Hz 85·C
(μF)	($V-$)	($V-$)	(Ω)	(Ω)	(mA)	(mA)
250			1.100	0.370	0.22	690
430			0.550	0.220	0.36	1100
640	200	250	0.350	0.150	0.53	1500
1400			0.150	0.077	1.1	2600
2500			0.095	0.048	2.0	3500
5500			0.050	0.029	4.4	5500
170			1.200	0.480	0.19	660
350			0.600	0.240	0.37	1000
800	250	300	0.260	0.110	0.82	1800
1200			0.170	0.079	1.2	2400
2000			0.100	0.052	2.0	3400
4500			0.050	0.030	4.5	5500
95			2.300	0.700	0.15	480
190			1.200	0.350	0.29	730
290			0.750	0.240	0.43	1000
420	350	400	0.500	0.170	0.61	1300
650			0.320	0.110	0.93	1800
1100			0.200	0.072	1.6	2400
2500			0.090	0.038	3.5	4100

Fig. 5.10 Extract from a data sheet for mains-input electrolytic capacitors (B43461 series, Siemens)

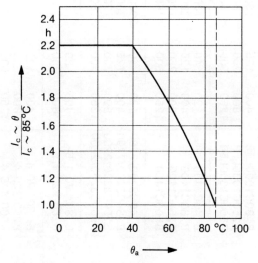

Fig. 5.11 Variation of superimposed alternating current with ambient temperature for electrolytic capacitors listed in Figure 5.10 (Siemens)

Silicon
Single-phase bridge rectifiers

Type	V_{RRM} (V)	V_{VRMS} (V)	C_{max} (μF)	R_{MM} (Ω)	Cooling[1]	I_{FSM} 10 ms, (A)	i^2t T_{vjmax} (A²s)	θ_{vj} max. (°C)	R_{thjc} (°C/W)	R_{thja} (°C/W)
B 40C1000L5B	120	40	5000	0.5	I	50	12.5	150		42
B 80C1000L5B	400	125	1600	1.5	M					
B250C1000L5B	800	250	800	3						
B500C1000L5B	1250	500	400	6						
B 40C1400	120	40	7000	0.4	I	70	24.5	150		30
B 80C1400	400	125	2200	1.1	M					
B250C1400	800	250	1000	2.5						
B500C1400	1250	500	500	5						
B 40C3200/2200	120	40	10000	0.25	I	100	50	150		25
B 80C3200/2200	400	125	3000	0.8	M					
B250C3200/2200	800	250	1700	1.6						
B500C3200/2200	1250	500	800	3						
B 80/ 70-1,5	400	80	—	—	I,M	50	12.5	150		23
B250/220-1,5	800	250	—	—	—					
B 80/ 70-4	400	125		0.5	I,M	150	110	150		13
B250/220-4	800	220		1	M					
B500/445-4	1250	500		2						
15/02A2	200	60		0.15	I	320	500	150	1	12
/04A2	400	125		0.3	M					4.3
/08A2	800	220		0.5						2.7

[1] I Self-supporting or on an insulating plate   M Metal plated with a minimum size of 250 × 250 × 1 mm³, lacquered

Fig. 5.12 Extract from data manual for silicon single-phase bridge rectifiers (Siemens)

The second method gives a result around 20 to 50 per cent higher than the first, based on measurement. The reason for this is the neglect of R_i (the mains impedance, the impedance of the filter and any protective resistance), which leads to higher current amplitudes. Compare the measurement with $a = 0$.

The capacitance required is $C_1 = 1.5 \, \mu F/W \times (375 \text{ to } 400 \text{ W}) = 562$ to $600 \, \mu F/350 \text{ V}$. The factor 2.1 must now be read-off from Figure 5.11 for $\theta_{amax} = 45 \, °C$ and applied to the permissible alternating current loading $I_{C\sim}$ in Figure 5.10. Expressed differently, the given alternating current value of about 4.2 A is divided by 2.1, so that a capacitor must be selected from the table with an alternating current rating $I_{C\sim}$ of 2 A. A single capacitor $C_1 = 1100 \, \mu F/350 \text{ V}$ would be suitable, but is much too large and too expensive; it is therefore necessary to connect several capacitors in parallel. One possibility is to use two capacitors each of $290 \, \mu F = 580 \, \mu F$ giving $I_{C\sim} = 2 \times 1 \text{ A} = 2 \text{ A}$.

Another possibility is a group of three parallel-connected capacitors each of $190 \, \mu F = 570 \, \mu F$ and $I_{C\sim} = 2.2 \text{ A}$. Which solution is adopted is determined by the cost and the dimensions of the capacitors. The permissible alternating current is so defined that it causes a limit temperature rise of $10 \, °C$. A small excess over the rating results in only a slightly higher temperature and hence an insignificant reduction in life. Figure 5.10 shows an extract from the data manual. Reference to other manufacturers and other unit types will reveal further possibilities for capacitors better suited to this application, although it is essential to take the cost into account.

In a commercial 300 W power-supply unit, four $470 \, \mu F$ electrolytic capacitors were connected two-in-series, two-in-parallel, giving a total capacitance of $470 \, \mu F$. Taking the capacitors of Figure 5.10, the alternating current rating would be $2 \times 1.3 \text{ A} = 2.6 \text{ A}$ at $85 \, °C$ or 5.5 A at $45 \, °C$. This combination is thus satisfactory at up to about $\theta_{amax} = 65 \, °C$, from Figure 5.11 (factor 1.6), and embodies a sufficient margin of reserve.

The rating of the bridge rectifier Rec presents no difficulty; bridge rectifier units are available on the market for (nominal) supply voltages of from 40 to 500 V, including three-phase bridges. Figure 5.12 shows an extract from a data sheet covering rectifiers suitable for the present example.

For the 300 W power supply the Type B250C3200/2200 is suitable; it does not even require a heatsink. The permissible mean direct current $I_{FAV} = 2200 \text{ mA} = 2.2 \text{ A}$ is higher than the worst-case calculated figure of 1.63 A. In fact a mean current of 3.2 A is permissible if the unit is mounted on a metal chassis by means of a fixing clip. Since this rectifier is cast into a metal or plastic housing, no problems arise in regard to isolation. From Figure 5.12 the maximum rated reverse voltage is 880 V. If the most unfavourable case is considered ($+ 10$ per cent mains voltage with no load), there results a figure of $220 \text{ V} \times 1.1 \times 1.41 \times 2$ (allowing for the back-voltage of a charged capacitor) $= 683 \text{ V}$, so that there is a safety margin of only about 117 V against transient overvoltages. This is relatively little, even though the absolute no-load case does not actually arise because of the ever-present losses.

One possible way of dealing with such transient voltages is to connect a varistor (rated at 200 V r.m.s.) across the a.c. terminals of the rectifier. Another possibility is to choose a rectifier with a higher reverse voltage rating —e.g. Type B500C3200/2200; this would afford a safety margin of over 560 V, which would be sufficient, since the sharply rising voltage spikes, of relatively low energy, are attenuated by the input filter which is always included.

Figure 5.12 gives a minimum value of 1.6 ohms for resistance R_i with $C_{1max} = 1700 \, \mu F$. Since, however, in the present example the capacitance is only $(570 \text{ to } 580) \, \mu F \times 1.5$

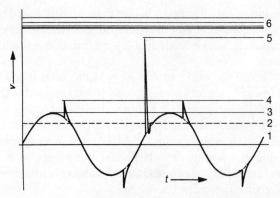

6 Rectifying element breakthrough range

5 Surge overvoltage (surge peak voltage); peak value obtained by, for example, breaking the circuit operation

4 Peak value for periodically occurring overvoltages

3 Peak value for sinusoidal terminal voltage

2 r.m.s. value for sinusoidal terminal voltage

1 Instantaneous value for terminal voltage

Fig. 5.13 Explanation of terms relative to mains supplies

(allowing for the $+50$ per cent tolerance on C_1) $= 855$ to $870\,\mu\text{F}$, i.e. nearly enough half the permissible capacitance, R_i also can be reduced to a half, i.e. to $0.8\,\Omega$, although it remains to be checked that the I^2t rating of the rectifier, i.e. the surge rating, is not exceeded.

The I^2t rating is frequently quoted in the data manual. Alternatively it can be calculated from the permissible 10 ms surge current I_{FSM} in accordance with [11]:

$$I^2t = \int_0^{10\,\text{ms}} i^2\,dt = 0.5 \times 10 \times 10^{-3}\,\text{s} \times I_{\text{FSM}}^2 \quad (\text{A}^2\,\text{s}) \tag{5.10}$$

For the suggested rectifier B250C3200/2200,

$$i^2\,dt = 0.5 \times 10 \times 10^{-3}\,\text{s} \times 100^2\,\text{A}^2 = 50\,\text{A}^2\,\text{s}$$

(see Figure 5.12).

The charging of C_1 draws from the mains supply an amount of energy $C_1\hat{V}_a^2$. On completion of the charge, the energy contained in C_1 is $C_1\hat{V}_a^2/2$. The remaining energy, similarly $C_1\hat{V}_a^2/2$, is converted into heat in R_i. The I^2t integral is then, from [4],

$$\int i^2\,dt = \frac{\Delta E}{R_i} = \frac{1}{2}\frac{C_1\hat{V}_a^2}{R_i} \quad (\text{A}^2\,\text{s}) \tag{5.11}$$

from which

$$R_1 \geqslant \frac{0.5\,C_1\hat{V}_a^2}{i^2\,dt} \quad (\Omega) \tag{5.12}$$

For the B250C3200/2200 bridge rectifier,

$$R_i \geqslant \frac{0.5 \times 1700 \times 10^{-6}\,\text{F} \times (220 \times 1.41\,\text{V})^2}{50\,\text{A}^2\,\text{s}} \geqslant 1.63$$

(see Figure 5.12).

The value obtained here for R_i, identical with that given in the table, is valid for the nominal values of $\hat{V}_a$ and C_1. When unfavourable tolerances are taken into account, R_i has to be increased. In the example,

$$R_i \geqslant \frac{0.5 \times 870 \times 10^{-6}\,\text{F} \times (220 \times 1.1 \times 1.41\,\text{V})^2}{50\,\text{A}^2\,\text{s}} \geqslant 1$$

(based on $C_{1\text{max}} = 870\,\mu\text{F}$).

From Equation (5.9), with the form factor $k_{\mathrm{fD}} = 3.6$ already determined, the current in R_{i} is

$$I_{\mathrm{Rirms}} = 1.63\,\mathrm{A} \times \frac{3.6}{1.41} = 4.16\,\mathrm{A}$$

The loss in R_{i} is then

$$P_{\mathrm{Ri}} = I^2 R = 4.16^2\,\mathrm{A}^2 \times 1.1\,\Omega = 19\,\mathrm{W}$$

Relative to the converter output power of 300 W, around 6.5 per cent is lost in heat in R_{i} alone. The question of the dimensioning of R_{i} is not, however, concerned only with the relatively high power loss. The e.m.i. filter which is always necessary at the input to the converter, with its inductance L_{i} and the capacitor C_1, forms a series-resonant circuit. This must be sufficiently damped by R_{i} that an overswing of the input voltage V_{i} is avoided as far as possible.

The overswing voltage (which can be neglected in the case of the frequently-used current-balance inductors: see Chapter 9) is given according to [4] by

$$\frac{\Delta V_{\mathrm{i}}}{V_{\mathrm{i}}} = \exp\left(-\frac{\pi}{\sqrt{\dfrac{1}{D^2} - 1}} \right) \tag{5.13}$$

where

$$D = \frac{R_{\mathrm{i}}}{2\sqrt{\dfrac{L_{\mathrm{i}}}{C_1}}} = 1\ \text{to}\ 0.5 \tag{5.14}$$

or

$$R_{\mathrm{i}} \geqslant (1\ \text{to}\ 2)\sqrt{\frac{L_{\mathrm{i}}}{C_1}}\quad (\Omega) \tag{5.15}$$

Equation (5.13) is presented graphically in Figure 5.14.

Thus, where possible, R_{i} should be chosen such that $R_{\mathrm{i}} \geqslant \sqrt{L_{\mathrm{i}}/C_1}$ to $2\sqrt{L_{\mathrm{i}}/C_1}$, or

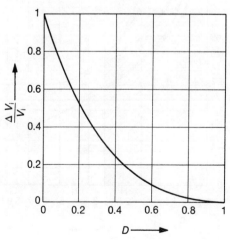

Fig. 5.14 Variation of voltage overshoot $\Delta V_{\mathrm{i}}/V_{\mathrm{i}}$ with damping factor

98

$D \geqslant 0.5$. The overswing is then limited to a maximum of about 16 per cent and can be allowed for in the design of the converter.

In the filter catalogues, L_i is given as about 3 mH for a current of 6 A (a necessary rating, since the current drawn by the 300 W power supply exceeds 4 A).

With $C_1 = 580 \,\mu$F maximum, R_i should therefore be

$$R_i \geqslant \sqrt{\frac{3 \times 10^{-3} \,\text{H}}{580 \times 10^{-6} \,\text{F}}} \geqslant 2.3 \text{ to } 4.6 \,\Omega$$

There is, however, a further aspect to be considered. When the equipment is switched onto the mains supply, the charging current that flows initially is limited only by R_i, the low impedance of the mains (a few tenths of an ohm) and the e.m.i. filter inductance L_i. Even though the charging time-constant amounts to only a few milliseconds, under unfavourable conditions the charging current may cause the operation of a fuse (an equipment fuse, usually with a time-delay characteristic) or the relatively quick-acting miniature circuit breakers in the supply system. It can be seen from the characteristics of various fuses and m.c.b.s that a charging current of 50 to 100 A can be sustained for the very short period in question.

This limit leads to a further possible deciding factor for R_i:

$$R_i \geqslant \frac{\hat{V}_{a\max}}{50 \text{ to } 100 \,\text{A}} \quad (\Omega) \tag{5.16}$$

This value is an absolute quantity, and applies to all converters. Thus with the upper-

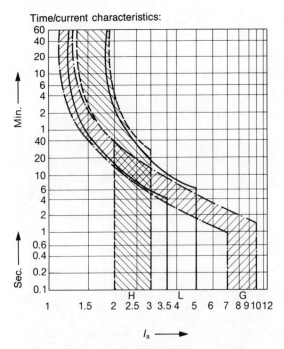

Fig. 5.15 Time/current characteristics of (narrow) panel-mounting miniature circuit breakers

limit mains voltage of $+10$ per cent to $+15$ per cent to be allowed for, R_i becomes

$$R_i \geqslant \frac{220\,\text{V} \times 1.41 \times (1.1 \text{ to } 1.15)}{50 \text{ to } 100\,\text{A}} \geqslant 3.4 \text{ to } 7.1\,\Omega$$

In the 300 W converter, with its calculated input current of 4.16 A r.m.s., such a large protective resistance, even taking the minimum value of $3.4\,\Omega$, would in itself reduce the efficiency by 20 per cent, which would not be acceptable. The determination of R_i may be thus summarized:

● Because of the high inrush current, which can lead to the operation of a fuse or m.c.b., R_i must be larger than about $3.5\,\Omega$.
● R_i must not be below a minimum value in accordance with Equation (5.15).
● R_i must not be below a minimum value given in the table applicable to the rectifier. If necessary this value can be determined from Equation (5.12) with the $I^2 t$ value.
● The maximum permissible surge current of the rectifier must not be exceeded.

As shown by Equation (5.5), the converter input current I_{iAV} is approximately proportional to the input power, and hence to the output power. The same applies to I_{Ri}, since the form factor k_{fD} can vary little with the size of the converter. The power loss dissipated in R_i is, however, proportional to the square of I_{Ri}. This means that the determining equations presented above for R_i lead invariably to acceptable results in small converters, but not in higher-power converters. Considering, for example, a power-supply unit with an output of 60 W—a fifth of the 300 W so far considered—the loss in R_i is only a twenty-fifth. The reduction in efficiency with $R_i = 3.4\,\Omega$ is then not 20 per cent but 20 per cent/5 = 4 per cent, which is quite tolerable. It cannot be stated in advance, therefore, that the determination of R_i is pointless, and that it should be decided straightaway from Equation (5.16). In small converters (which are in any case cheaper than large ones), a value of about $3\,\Omega$ can be used for R_i; in larger converters there is no choice but to make R_i around 5 to $8\,\Omega$ (to ensure absolute security against the operation of fuses or m.c.b.s) and to short-circuit the resistor either by means of a time-delay circuit after a few milliseconds or electronically, by means of a thyristor circuit, when the surge current has decayed.

The relay circuit of Figure 5.16 provides an operating delay of several milliseconds by virtue of the electrolytic capacitor in parallel with the relay winding. The release delay which also occurs is of no interest. The circuit of Figure 5.17 operates in the following manner: capacitor C_1, uncharged, constitutes practically a short circuit, and the whole of the peak mains voltage appears across R_i. This situation, however, lasts less than 1 ms, so that the loss in R_i is not great. Due to the high voltage across R_i the breakover diode Z conducts, transistor Tr is turned on and short-circuits C_2. Thus no triggering pulse is

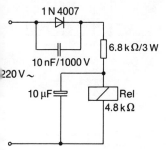

Fig. 5.16 Circuit for a slow-response relay on the 220 V mains for the bridging of R_i

applied to thyristor Th. When the voltage falls to the point where the breakover diode becomes non-conducting, C_2 is charged to about 30 V through R_1 and the thyristor receives a triggering pulse and thereupon conducts. R_i is thereby virtually short-circuited.

The r.m.s. voltage measured across a thyristor of the appropriate power rating amounted to only about 0.25 V, and is thus negligible in any circumstances. The time required for the charging of C_2 with the dimensioning given could only have been about 1 ms or less, and is thus short in relation to the duration of the inrush current pulse. Since when the thyristor is triggered C_2 is not fully discharged, the time period between the partial discharge and the recharging to about 30 V is certainly shorter still. The thyristor turns off after each half cycle of the mains voltage (see Figure 5.7), and has to be triggered afresh. This can be arranged without the capacitor C_2 and the trigger diode D_i, but R_1 has then to be of much lower resistance, since it must supply the minimum triggering current required for the thyristor, and the loss in R_1 is considerably greater. In the circuit of

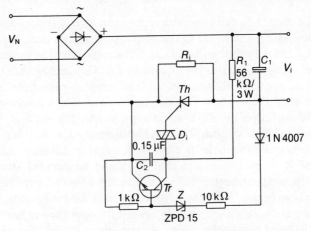

Fig. 5.17 Electronic bridging circuit for R_i depending on the decay of charging current into the initially discharged capacitor C_1

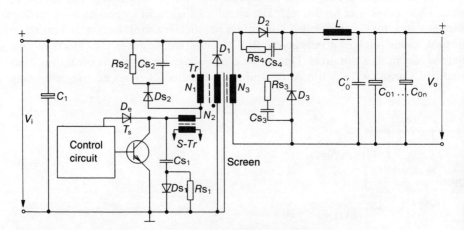

Fig. 5.18 Circuit of a single-ended forward converter, omitting the drive and control circuits and the mains rectification with its e.m.i. filter

Figure 5.17 the resistor R_i must obviously be connected in the d.c. circuit after the rectifier, but from the point of view of the magnitude of the inrush current it is immaterial whether it is placed before or after the rectifier.

The problems of mains rectification having been discussed in some detail, attention may be turned to further aspects of the circuit of a single-ended forward converter. Figure 5.18 shows the circuit again, but without the mains rectification. The control and drive circuits are indicated in block form, since there are a great many drive circuits differing in complexity and power capability, using both bipolar and MOSFET transistors. Drive circuits are discussed more specifically in Section 5.5. The whole of Chapter 8 is devoted to the many available integrated control circuits.

5.2 DESIGN OF THE MAIN TRANSFORMER *T*

As a first step the duty cycles δ_{Tmax} and δ_{Tmin} must be determined. The maximum permissible duty cycle is derived from the following considerations:

- The non-conducting period $(1 - \delta_{Tmax})\,T$ of the transistor must be longer than its conducting period $\delta_{Tmax}\,T$, to ensure that the transformer is reset through its winding N_2.
- The voltage at the input to the inductor must be appreciably higher than V_0, to make control possible.

Both considerations dictate that $\delta_{Tmax} < 0.5$, so, allowing a safety margin, it can be established that

$$\delta_{Tmax} = 0.45 \tag{5.17}$$

As shown previously in connection with the step-down inductor-coupled converter, here again

$$\delta_{Tmax} V_{imin} = \delta_{Tmin} V_{imax}$$

or, rearranged,

$$\delta_{Tmin} = \delta_{Tmax} \frac{V_{imin}}{V_{imax}} \tag{5.18}$$

δ_{Tmin} can in this way be calculated from the given value of $\delta_{Tmax} = 0.45$.

Example

For a forward-converter power supply it is given that $V_s = 220\,V + 10$ per cent, -15 per cent. Hence $V_{imax} + 2V_F$ (voltage drop across the rectifier diodes) is given by

$$V_{imax} + 1.5\,V = 220\,V \times \sqrt{2} \times 1.1 = 341.2\,V \text{ or } V_{imax} \approx 340\,V.$$

$$\delta_{Tmin} = 0.45 \frac{210\,V}{340\,V} = 0.28\,(V_{imin} = 212\,V \text{ for } C_1/P_i = 1.5\,\mu F/W \text{ and } a = 1; \text{ allowing for}$$

$V_{CEsat}, V^*_{imin} \approx 210\,V)$.

The next step is to determine the transformation ratio $n = N_3/N_1$.

$$n = \frac{N_3}{N_1} = \frac{V_{Omax} + V_F + V_{LS}}{\delta_{Tmax} V^*_{imin}} \tag{5.19}$$

where V_{0max} is the maximum output voltage (V), V_F is the forward voltage drop of diodes D_2 and D_3 (V) (approximately 0.5 V for Schottky diodes; up to 1 V for epitaxial diodes) and V_{LS} is the voltage drop across secondary winding and smoothing inductor (V) (about 0.2 V).

For the present example, with $V_0 = 5$ V,

$$n = \frac{5\,V + 0.5\,V + 0.2\,V}{0.45 \times 210\,V} = 0.0603, \quad \text{or} \quad \frac{1}{n} = 16.58$$

Winding N_2 usually has the same number of turns as N_1. The voltage across N_2 is accordingly the same as that across N_1, i.e. up to a maximum of V_{imax}. With the back-e.m.f. V_{imax} induced in N_1 and the input voltage V_{imax}, the collector-emitter voltage immediately after the transistor is turned off is equal to $2V_{imax}$; when the transformer has been reset and the induced voltage has died away, V_{CE} is reduced to V_{imax}.

In principle the number of turns on N_2 can be increased. The voltage V_i would then be reached sooner and the collector-emitter voltage would be less than $2V_{imax}$. This is sometimes recommended in the literature, but rarely put into practice. A doubling or trebling of the number of turns on N_2 certainly results in a reduction of the induced voltage, but introduces considerable problems in the winding of the transformer. Winding N_2 needs to be as closely couple to N_1 as possible, and the higher the ratio, the more difficult this becomes.

The next requirement is to determine the size of the core. A guide to this is given by the curves in Figures 2.38 to 2.40.

For $P_0 = 300$ W, or a transformer rating of about 350 W (in the present case the ratio $V_0/(V_0 + V_F + V_L) = 5\,V/5.7\,V = 0.88$ results straightaway in a loss of 12 per cent; in addition there is a loss of a few percent in the transformer, so that the losses in the output circuit will certainly not be less than 15 per cent) the E55 core, according to Figure 2.38 (single-ended forward converter) is too small; it would be good for about 280 W. The curves of Figure 2.38 are valid, however, for a switching frequency of 20 kHz, and since, from Equation (2.50), the power that can be handled is proportional to frequency, at $f = 40$ kHz the E55 core could handle $2 \times 280\,W = 560\,W$. It is thus very adequate for the power of about 350 W. (Too close a dimensioning of the core is not worthwhile, since the magnetizing losses become greater and problems can arise with the winding space; see also Figure 2.41.) The ultimate decision as to whether the core is suitable or not is only reached when it is established that the necessary insulation can be incorporated (for a test voltage of 2.5 kV). The design may thus be proceeded with on the basis of the E55 ferrite core.

Principal data for the E55 core:

Effective magnetic cross-sectional area: $A_e = 354\,mm^2 = 3.54\,cm^2$ (equal to A_{min}); inductance factor with no air gap: $A_L = 5800$ nH; winding bobbin with one winding section; available winding cross-sectional area: $A_N = 280\,mm^2$; winding depth: 8.5 mm; winding length: 33.4 mm; mean length of turn: $l_N = 113$ mm.

It is best to start by calculating the number of turns for the secondary winding.

$$N_s \geqslant \frac{\delta_{Tmax} V_{imax} n \times 10^4}{f A_{min} \hat{B}_{max}} \tag{5.20}$$

where A_{min} is the minimum cross-sectional area of core material (if not quoted, use A_e) (cm^2) and $\hat{B}_{max}$ is the maximum flux density from Figure 2.38 (T) (in the present case 0.2 T).

Equation (5.20) incorporates not the usual combination of δ_{Tmax} and V_{imin}, but the larger value $\delta_{Tmax} V_{imax}$. This ensures that the maximum flux density is not exceeded in the event of step changes of load.

For the present example,

$$N_s \geqslant \frac{0.45 \times 340\,\text{V} \times 0.0603 \times 10^4}{40 \times 10^3\,\text{s}^{-1} \times 3.54\,\text{cm}^2 \times 0.2\,\text{T}} \geqslant 3.26$$

or about 3 turns. According to Equation (5.20) the next higher number or turns—in this case four—should really be chosen, in view of the $\geqslant$ sign. It was nevertheless rounded down, because step-load changes are relatively infrequent and the normal flux density is much less than 0.2 T.

The number of primary turns then becomes $N_p = N_s/n = 3/0.0603 = 49.75$, or approximately 50 turns. As a check on the maximum working flux density:

$$\hat{B}_{max} = \frac{\delta_{Tmax} V_{imin} \times 10^4}{f A_{min} N_p} = \frac{0.45 \times 210\,\text{V} \times 10^4}{40 \times 10^3\,\text{s}^{-1} \times 3.54\,\text{cm}^2 \times 50}$$

$$= 0.134\,\text{T} < 0.2\,\text{T}$$

Using Equation (2.17), the primary inductance can be calculated from the number of primary turns N_p and the inductance factor A_L:

$$L_p = N_p^2 A_L = 50^2 \times 5800 \times 10^{-9}\,\text{H} = 14.5\,\text{mH}$$

The current flowing in the primary circuit consists of the reflected load current and the magnetizing current I_M:

$$I_{pmax} = (I_{0max} + \Delta I_L/2)n + I_M \quad \text{(A)} \tag{5.21}$$

For a 300 W forward converter with $V_0 = 5\,\text{V}$ and, e.g., $\Delta I_L/2 = 6\,\text{A}$ the primary load current is

$$I_{pmax} - I_M = (60\,\text{A} + 6\,\text{A}) \times 0.0603 = 3.98\,\text{A} \approx 4\,\text{A}$$

The maximum possible magnetizing current with a step change in load is given by [7]

$$I_{Mmax} = \frac{V_{imax} \delta_{Tmax}}{f L_p} \quad \text{(A)} \tag{5.22}$$

In the example,

$$I_{Mmax} = \frac{340\,\text{V} \times 0.45}{40 \times 10^3\,\text{s}^{-1} \times 14.5 \times 10^{-3}\,\text{H}} = 0.264\,\text{A}$$

In normal operation, with δ_{Tmin} at V_{imax}, $I_{Mmax} = 0.16\,\text{A}$.

The magnetizing current is generally between 1 and 10 per cent of $n I_{0max}$. The maximum total collector current on the primary side is thus

$$I_{pmax} = 3.98\,\text{A} + 0.264\,\text{A} = 4.25\,\text{A}$$

Even though in principle no air gap is necessary in this case, because of the resetting circuit, a small air gap of 0.1 to 0.2 mm is nevertheless recommended. Since in the case of the E55 core (as with most small cores) there is no version available with such a small gap, it can be provided by interposing shims of a thickness equal to half the required gap length (since there are gaps on both sides of the core) between the halves of the core. The sprung core clamping device can accommodate the dimensional increase without difficulty.

The introduction of an air gap reduces the inductance factor and hence the primary inductance. This signifies a higher magnetizing current and therefore a somewhat higher primary current. The advantage lies in the quicker magnetization of the core and consequently in a reduction in the transistor switching times. This applies both to the main transformer and also the drive transformer. The reduction in switching times and the consequent reduction in switching losses more than compensates for the slight increase in the conduction loss due to the higher primary current. Since fundamentally the drive power is small in comparison with the total power handled by the circuit, the measure in. that respect hardly affects the total power balance.

As an example, numerical values may be compared for $d = 0$, $d = 0.1$ mm and $d = 0.2$ mm (normal operation):

$$d = 0 \text{ mm: } L_p = 14.5 \text{ mH}; \ I_{Mmax} = 0.16 \text{ A}; \ I_{pmax} = 4.14 \text{ A};$$

$$I_{pr.m.s.} = 2.78 \text{ A}$$

$$d = 0.1 \text{ mm}; \ A_L = 3500 \times 10^{-9} \text{ H}; \ L_p = 8.75 \text{ mH};$$

$$I_{Mmax} = 0.27 \text{ A}; \ I_{pmax} = 4.25 \text{ A}; \ I_{pr.m.s.} = 2.85 \text{ A}$$

$$d = 0.2 \text{ mm}; \ A_L = 2000 \times 10^{-9} \text{ H}; \ L_p = 5 \text{ mH};$$

$$I_{Mmax} = 0.48 \text{ A}; \ I_{pmax} = 4.46 \text{ A}; \ I_{pr.m.s.} = 3 \text{ A}.$$

The increase in I_{pmax} and $I_{pr.m.s.}$ thus amounts to about 8 per cent with $d = 0.2$ mm and only 3 per cent with $d = 0.1$ mm, which is quite acceptable.

The r.m.s. current loading of the primary winding is given, from [4], by

$$I_{pr.m.s.} = I_{pmax} \sqrt{\delta_{Tmax}} \quad \text{(A)} \tag{5.23}$$

In the example,

$$I_{pr.m.s.} = 4.25 \text{ A} \times \sqrt{0.45} = 2.85 \text{ A}$$

The winding length of $33.4 \text{ mm} - 2 \times 1.5 \text{ mm}$ (cheek thickness) $\approx 30 \text{ mm}$, with the assumed two layers, leads to a wire diameter (including insulation) of $30 \text{ mm}/25 = 1.2 \text{ mm}$. This corresponds to a nominal diameter of about 1.1 mm. With $f = 40 \text{ kHz}$, however, the skin effect in a wire of this thickness results in an increase in resistance of about 20 per cent (Figure 2.11). A wire with a nominal diameter of 0.5 mm is therefore chosen (skin effect about 1 per cent and thus negligible; maximum outside diameter 0.548 mm). Two windings are applied together in each winding and connected in parallel. The depth of the primary winding is then

$$h_p = 2 \times 0.548 \text{ mm} = 1.1 \text{ mm}$$

With a mean length of turn of $113 \text{ mm} = 0.113 \text{ mm}$, each of the parallel-connected windings requires a wire length of $0.113 \text{ m} \times 50 = 5.65 \text{ m}$. The copper resistance, based on a resistance per unit length of $0.0878 \ \Omega/\text{m}$ is then

$$R_{Cu} = 0.0878 \ \Omega/\text{m} \times 5.65 \text{ m}/2 = 0.25 \ \Omega$$

The primary loss is therefore

$$P_p = 2.85^2 \text{ A}^2 \times 0.25 \ \Omega = 2.03 \text{ W or approximately 2 W.}$$

The resetting winding N_2 is loaded only to the extent of the small magnetizing current I_M. It can also be wound with a 0.5 mm diameter wire, which means a single layer. The nett winding depth of the two windings N_1 and N_2 is thus 1.1 mm + 0.55 mm = 1.65 mm. Given the available winding depth of 8.5 mm there remains 8.5 mm − 1.65 mm = 6.85 mm.

If a single layer is allocated to the three secondary turns and h.f. litz wire of 5 mm diameter is used, 1.85 mm is left for all the insulation and screens, which is enough for a large number of insulation layers. Concerning the winding construction it should be said that, to preserve close coupling between N_1 and N_2, one layer of N_1 should first be wound, then N_2 (with insulation interposed), and then the second layer of N_1. Next, in view of the necessary isolation of the output from the mains, as much insulation should be applied as will allow the secondary winding N_3 to be accommodated. It is usually necessary to incorporate a screen between the primary and secondary windings (either a complete winding of thin wire, of, say, 0.1 mm diameter, or a thin copper foil, which must not form a short-circuited turn), or even two screens. These screens must, of course, be wound between layers of insulation. Where there are two screens, one is earthed on the primary side (usually to the collector of the switching transistor) and the other is earthed on the secondary side; if there is only one screen it is necessary to determine the most suitable earthing point.

Wire of circular cross-section or strip cannot be used for the high-current secondary winding, since with the large cross-sectional area required the skin effect would be much too great.

HF litz wire of 5 mm diameter, with a copper space factor of about 0.42 (see Figure 2.42), may contain about 0.82 mm^2 of copper. The required length of wire can be estimated here as 3 × 0.15 m = 0.45 m. This leads to a secondary resistance of 1 mΩ. The secondary loss is then

$$P_s = (40.25 \text{ A})^2 \times 1 \times 10^{-3}\,\Omega = 1.62 \text{ W}$$

The core losses are obtained from Figure 2.12(a) (the curves are valid for sinusoidal excitation; with a rectilinear waveform and $\delta_T \approx 0.5$ the estimate should be based on half the value—i.e. about 70 mT) as

$$P_{core} = 8 \text{ mW/g} \times 216 \text{ g} = 1.73 \text{ W}$$

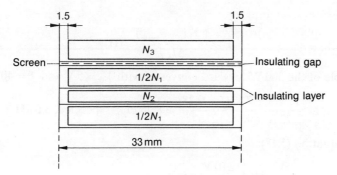

Fig. 5.19 Possible construction of the power transformer T in
Figure 5.18 (E55 core)

The total transformer loss is then

$$P_T = 2.03\,\text{W} + 1.62\,\text{W} + 1.73\,\text{W} = 5.38\,\text{W} \approx 5.4\,\text{W}$$

The current densities are as follows:

Primary:

$$A_p = 2\frac{\pi d^2}{4} = \frac{2 \times 3.14 \times 0.5^2\,\text{mm}^2}{4} = 0.39\,\text{mm}^2$$

$$J_p = \frac{2.85\,\text{A}}{0.39\,\text{mm}^2} = 7.3\,\text{A/mm}^2$$

Secondary:

$$A_s = 8.2\,\text{mm}^2 \quad I_{\text{r.m.s.}} = 60\,\text{A} \times \sqrt{0.45} = 40.25\,\text{A}$$

$$J_s = \frac{40.25\,\text{A}}{8.2\,\text{mm}^2} = 4.9\,\text{A/mm}^2$$

The current densities calculated here are considerably higher than would be permissible for the E55 core according to Figure 2.41 (about $3.2\,\text{A/mm}^2$). In the present instance, however, the transformer was resin-cast, because of the high insulation voltage required between the primary and secondary windings, and a fan was also provided.

The values given in Figure 2.41 are for the most unfavourable case (without resin casting and with natural convection). The higher loading of the transformer in the example cited was compensated by the considerably better heat dissipation.

It is clear from the results evaluated here that with an unconstrained design of the transformer based on Figure 2.41—simply on account of the heat dissipation—a larger core would have to be selected, even though the core characteristics were adequate in all other respects.

5.3 DESIGN OF THE SECONDARY CIRCUIT

Calculations for the inductor L are exactly as for the inductor-coupled step-down converter (see Chapter 2).

The inductance is given by the same equation.

Equation (2.12):

$$L \geqslant \frac{V_0^*(1 - \delta_{\text{Tmin}})}{\Delta I_L f} \quad \text{(H)}$$

In the example of the 300 W forward converter with $V_0 = 5\,\text{V}$ and $f = 40\,\text{kHz}$,

$$L \geqslant \frac{(5\,\text{V} + 0.7\,\text{V})(1 - 0.28)}{12\,\text{A} \times 40 \times 10^3\,\text{s}^{-1}} \geqslant 8.55\,\mu\text{H}; \quad \text{specified, } 8.6\,\mu\text{H}$$

δ_{Tmin} from Equation (5.18):

$$\delta_{\text{Tmin}} = 0.45\frac{210\,\text{V}}{340\,\text{V}} = 0.28$$

$$I_{\text{Lmax}} = I_{\text{0max}} + \Delta I_L/2 = 60\,\text{A} + 6\,\text{A} = 66\,\text{A}$$

If a ferrite core is chosen for the inductor, the following types are to be considered in accordance with Figure 2.5:

$$I^2 \,(A) \times L(mH) = 66^2 \,A^2 \times 8.6 \,\mu H = 37.5 \,A^2 \,mH \,(mW \,s)$$

The EC70 core would be generously dimensioned, the CC50 core more closely. A more suitable choice, because of its higher d.c. magnetization capability, would be a nickel-iron ring core No. 4 available from Krupp. (Krupp—Widia, D4300, Essen 1.)

The stored energy $LI^2/2$ is quoted here as a selection criterion.

$$\frac{LI^2}{2} = 18.75 \,mW \,s$$

The No. 4 core, with a stored energy of 20 mW s, is selected. From the data manual, maximum loss $= 4$ W; mean length of turn $= 9$ cm $= 0.09$ m. The maximum resistance of the winding may then be

$$R_L \leqslant \frac{4 \,W}{40^2 \,A^2} \leqslant 2.5 \,m\Omega$$

$$N = \sqrt{\frac{8.6 \times 10^{-6} \,H}{315 \times 10^{-9} \,H}} = 5.22$$

or approximately 5 turns ($L \approx 8 \,\mu H$). Wire length: $l = 5 \times 0.09$ m $= 0.45$ m. If litz wire, with an appropriate number of strands, is used, the copper space factor f_{Cu} can be taken as 0.42, from Figure 2.42. The outside diameter of the litz wire is then

$$d \geqslant \sqrt{\frac{4\,l}{\kappa R \pi f_{Cu}}} = \sqrt{\frac{0.45 \,m \times 4\,\Omega \,mm^2}{57 \,m \times 2.5 \times 10^{-3} \,\Omega \times 3.14 \times 0.42}}$$

$$\geqslant 3.1 \,mm$$

It is thus possible to use a suitable litz wire with an outside diameter of 4 to 5 mm. Given the 20 mm internal diameter of the core, a still thicker litz could be used if necessary.

The requirements for the two diodes D_2 and D_3 are as follows:

- very short reverse recovery time,
- small recovery current and stored charge,
- as far as possible no sharp cut-off of recovery current (soft recovery) in order to avoid very high-frequency interference,
- lowest possible forward voltage drop to minimize conduction loss.

Core no.	$1/2 \cdot L \cdot I^2 \cdot 2/N$	A_L	Internal dimensions in mm		
	(mWs)	(nH)	D_a	D_i	H
1	1.3	110	21	13	16
2	3.2	205	26	13	10
3	8.0	175	35	16	10
4	20.0	315	40	20	20
5	50.0	460	55	23	20

Fig. 5.20 Data and dimensions for nickel-iron ring cores with ground air gaps (Krupp)

These requirements are best met by Schottky power diodes, which are therefore used in most cases in circuits with low output voltages. (See also Figures 5.3 and 3.2.)

Diodes D_2 and D_3 must meet the following specifications: [7].

Reverse voltage

$$V_{Rmax} = \frac{V_0^*}{\delta_{Tmin}} + V_F \frac{(1 - \delta_{Tmin})}{\delta_{Tmin}} \quad (V) \tag{5.24}$$

$V_0^* V_0 +$ voltage drop across inductor and secondary winding (V).

Current loading

For I_{FAV} see Equation (2.34)

$$I_{FM} = 2I_0 \quad (A) \tag{5.25}$$

The diodes and the power transistor are subjected to an additional voltage stress as a result of the reverse recovery time of the two diodes (albeit very short in Schottky diodes) in conjunction with the leakage inductance of the transformer. To suppress this effect, RC networks should be connected in parallel with each diode. If the two diodes are mounted very close together, or if a double diode is used, a single RC network across the two outer terminals is sufficient.

The values of R and C are given by: [7]

$$R_s \leqslant \frac{V_{Rmax}}{I_{RM}} \quad (\Omega) \tag{5.26}$$

(R_{s3} and R_{s4} in Figure 5.18), where V_{Rmax} is the maximum applied reverse voltage (V) and I_{RM} is the maximum reverse current (A) (measured with current probe and oscilloscope or calculated from Equations (5.28) and (5.31))

$$C_s \geqslant \frac{L_s I_{RM}^2}{V_{Rmax}^2} \quad (F) \tag{5.27}$$

(C_{s3} and C_{s4} in Figure 5.18), where L_s is the secondary-side leakage inductance including leads (H).

Since it is difficult to measure the leakage inductance of the transformer including the leads, the values obtained for the suppression networks must be checked experimentally and corrected. For the present example (300 W forward converter with $V_0 = 5$ V) and an estimated total leakage inductance of 10 μH the values to be observed were as follows:

$$V_{Rmax} \geqslant \frac{5\,V + 0.5\,V}{0.28} + 0.6\,V \times \frac{0.72}{0.28} \geqslant 21.2\,V$$

Approximate calculation of V_{Rmax}: $V_{Rmax} \approx n V_{imax} + V_F = 21$ V

$$I_{FM} \geqslant 2 \times 60\,A \geqslant 120\,A;$$

Equation (2.34):

$$I_{FAV} = 60\,A\,(1 - 0.28) = 43\,A$$

These requirements are met by, for example, the TRW Schottky diode Type SD-51

($V_{Rmax} = 45$ V; $I_{FMmax} = 120$ A); further data are: $C_j = 1.8$ nF at 20 V. With these figures the values for the protective components R and C can be calculated.

For the reverse recovery charge, which is not quoted in the data sheet, it is acceptable for present purposes to calculate the charge Q stored in the junction capacitance.

$$Q = CV = 1.8 \times 10^{-9} \text{ F} \times 20 \text{ V} = 36 \text{ nA s}$$

The recovery time t_{rr} is quoted as 50 ns with $\theta_c = 125$ °C. The maximum reverse current I_{RM} can be calculated from the stored charge Q_{rr} and the recovery time t_{rr} as

$$I_{RM} = \frac{2Q_{rr}}{t_{rr}} \quad \text{(A)} \tag{5.28}$$

where Q_{rr} is the stored charge (A s) and t_{rr} is the recovery time (s).

In the present example,

$$I_{RM} = \frac{2 \times 36 \times 10^{-9} \text{ A s}}{50 \times 10^{-9} \text{ s}} = 1.44 \approx 1.5 \text{ A}$$

$$R_s = \frac{20 \text{ V}}{1.5 \text{ A}} = 13 \, \Omega; \quad \text{selected, } R_s = 4.7 \, \Omega/2 \text{ W}$$

$$C_s \geqslant \frac{10 \times 10^{-6} \text{ H} \times 1.5^2 \text{ A}^2}{20^2 \, V^2} \geqslant 0.06 \, \mu\text{F};$$

selected, $C_s = 0.1 \, \mu\text{F}/100$ V

$$P_{Rs} = f \frac{CV^2}{2} = 0.5 \times 0.1 \times 10^{-6} \text{ F} \times 20^2 \, V^2 \times 40 \times 10^3 \text{ s}^{-1}$$

$$= 0.8 \text{ W}$$

The switching loss is obtained from Equation (2.43) as

$$P_{Ds} = Q_{rr} V_R f$$

So for the example

$$P_{Ds} = 36 \times 10^{-9} \text{ A s} \times 20 \text{ V} \times 40 \times 10^3 \text{ s}^{-1} = 30 \text{ mW}$$

This low value is negligible in comparison with the other losses. Even though the calculation of the stored charge in the selected SD-51 Schottky diode can only be approximate, the optimization of the 300 W converter with $C_s = 0.1 \, \mu\text{F}$ and $R_s = 3.9 \, \Omega$ provided good confirmation of the estimate. Since the leakage inductance also can only be estimated approximately—although the figure of $10 \, \mu\text{H}$ is perfectly realistic—the circuit must be tested when the construction is completed. At the same time it is necessary to calculate the values of the protective elements at least approximately, to avoid the overloading of individual components.

The conduction losses in diodes D_2 and D_3 are as follows:

Rectifier diode D_2:

$$P_{Dc} = I_0 V_F \delta_{Tmax} \quad \text{(W)} \tag{5.29}$$

Free-wheel diode D_3, from Equation (2.42):

$$P_{Dc} = I_0 V_F (1 - \delta_{Tmin})$$

110

If D_2 and D_3 are combined in a double diode with a common cathode, or a common anode in the case of a negative output voltage, the conduction loss becomes

$$P_{Dc} = I_0 V_F \quad (W) \tag{5.30}$$

In the present example:

Rectifier diode D_2:

$$P_{Dc} = 60 \, A \times 0.6 \, V \times 0.45 = 16.2 \, W$$

Free-wheel diode D_3:

$$P_{Dc} = 60 \, A \times 0.6 \, V = 36 \, W$$

Both diodes on a common heatsink, or a double diode:

$$P_{Dc} = 60 \, A \times 0.6 \, V = 36 \, W$$

The sum of the losses in the individually cooled diodes (42.1 W) is not equal to the loss in a double diode (36 W). With separate cooling allowance must be made for the least favourable case, namely δ_{Tmax} for D_2 and $(1 - \delta_{Tmin})$ for D_3. With the internal thermal resistance of 1 kW for the SD-51 and a maximum permissible junction temperature of 150 °C, the external thermal resistance with a maximum ambient temperature of, say, 50 °C, is given by Equation (2.45) as

$$R_{thca} \leqslant \frac{150° - 50°}{36 \, W} - 1 \, kW \leqslant 1.8 \, kW; \quad \text{selected,} \ R_{th} = 1 \, kW$$

(both diodes on a single heatsink). Where the two diodes are mounted on a common heatsink, either they must both be insulated by means of mica washers, whose thermal resistance, of a few tenths kilowatt must of course be allowed for, and the heatsink

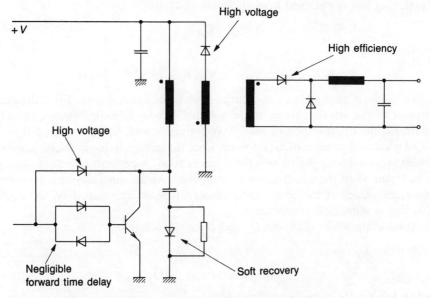

Fig. 5.21 Essential properties of diodes used in a switched-mode power supply as exemplified by a forward converter (Thomson CSF)

connected to the 0 V output line (although this is not essential), or the two diodes can be screwed without insulation directly into the heatsink, which is then live. The insulated heatsink can then be connected to the 0 V line from the point of view of h.f. through a capacitor. In the SD-51 the cathode is connected to the threaded stud. If it is required to have the anode connected to the stud, for example with a negative output voltage, the letter 'R' is added to the type number, as with many diodes.

If the output voltage V_0 is relatively high—this applies roughly to voltages above 15 V—Schottky diodes cannot be used. As shown by the numerical values in the example, the maximum reverse voltage that occurs is about four times the output voltage. Schottky diodes are available with reverse voltage ratings only up to 45 to 60 V; individual types attain 90 V. In these cases 'high-efficiency diodes' are used, e.g. Types BYW29 to BYW99 (Thomson, Unitrode and others). These diodes have reverse voltage ratings from 50 to 200 V, maximum permissible direct currents from 1 A to about 50 A and recovery times between 35 ns and about 60 ns. They are thus extremely fast. This means that the peak collector current when the transistor turns on is very low. The switching loss is also very small, because of the small stored charge. The high efficiency is, however, essentially a result of the fact that in these diodes (also referred to as 'epitaxial diodes') the forward voltage drop at maximum current is generally less than 1 V. Figure 5.21 illustrates the most important criteria in the selection of diode types for a switched-mode power supply.

Figure 5.22 shows the forward voltage drop V_F of Types BYW80-50 to BYW80-200 (the number following the hyphen indicates the maximum reverse voltage in volts) over the initial range of current (a) and over the full range (b).

Figure 5.23 shows the forward conduction loss as a function of the mean direct current I_{FAV} (for a single diode) with the duty cycle δ_T as parameter.

The stored charge Q_{rr}, required for tne calculation of the switching loss W_{Ds}, can be obtained from Figure 5.24 as a function of the rate of fall of current $(-)dI_F/dt$, and similarly the recovery time $t_{rr} = f(dI_F/dt)$ from Figure 5.25.

The rate of fall of current dI_F/dt can be calculated as follows:

$$\frac{dI_F}{dt} = \frac{(-)V_R}{L_s} \quad \text{(A/s)} \tag{5.31}$$

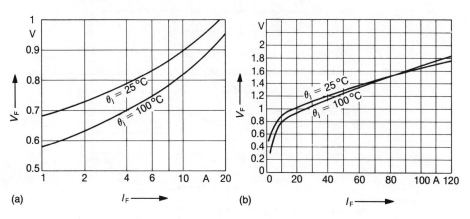

Fig. 5.22 Variation of forward voltage drop V_F of the BYW80 epitaxial diode with current: (a) initial region; (b) complete forward characteristic (Thomson CSF)

where V_R is the maximum applied voltage (calculated from (5.24)) (V) and L_s is the total leakage inductance of transformer and wiring (H).

Example

In a forward converter the maximum reverse voltage $V_{Rmax} = 90$ V; output current $I_{Omax} = 5$ A; $f = 40$ kHz; leakage inductance $L_s = 12\,\mu$H; $\theta_{jmax} = 75\,°$C (provisional assumption); $\theta_{amax} = 50\,°$C; $R_{thca} = 2.5$ kW. It is required to calculate the maximum

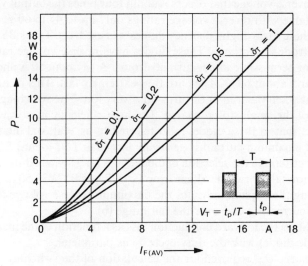

Fig. 5.23 Forward conduction loss P_{Dc} of the BYW80 diode as a function of mean direct current I_{FAV} with diode duty cycle δ_T as parameter (Thomson CSF)

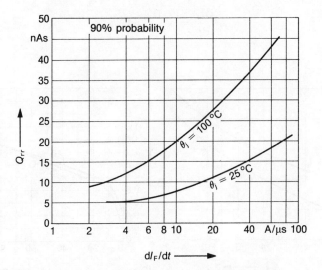

Fig. 5.24 Stored charge Q_{rr} (nA s) of the BYW80 diode as a function of the (negative) rate of rise of current dI_F/dt (A/μs) at two temperatures (Thomson CSF)

losses in the BYW80 rectifier and free-wheel diodes and to dimension the RC suppression network.

Conduction loss

With a maximum applied reverse voltage of 90 V the BYW80-150 version can be selected with some certainty. The BYW80-100 is probably too closely rated in view of the possible overshoot voltages. Since in the normal circuit of the forward converter the cathodes of the two diodes are connected together, they can be mounted on a common heatsink (the cathodes are connected to the diode cases); the heatsink must, of course, be adequately insulated. The appropriate calculation is then that for the rectifier and free-wheel diodes together, Equation (5.30).

From Figure 5.22a, $V_F = 0.75$ V at $I_F = 5$ A and $\theta_j = 75\,°C$ (interpolating between the two curves).

$$P_{Dc} = V_F I_0 = 0.75\ V \times 5\ A = 3.75\ W$$

From Figure 5.23, with $I_0 = I_{AV}$ and $\delta_T = 1$ (for a single diode it would be necessary to read off at approximately half the current and $\delta_T = 0.5$), $P_{Dc} = 3.5$ W—practically the same.

Switching loss

Equation (5.31):

$$dI_F/dt = \frac{90\ V}{12 \times 10^{-6}\ V\,s/A} = 7.5\ A/\mu s$$

From Figure 5.24, by interpolation between the two curves, $Q_{rr} = 12\,nA\,s$ with dI_F

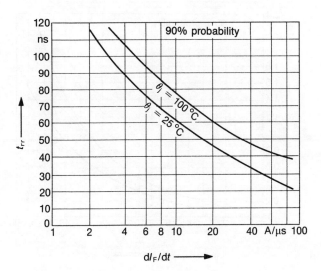

Fig. 5.25 Reverse recovery time t_{rr} (ns) of the BYW80 diode as a function of the (negative rate of rise of current dI_F/dt (A/μs) at two temperatures (Thomson CSF)

114

$dt = 7.5$ A/μs (calculation as before for the rectifier and free-wheel diodes):

$$P_{Ds} = 2Q_{rr}V_R f = 2 \times 12 \times 10^{-9}\,\text{A s} \times 90\,\text{V} \times 40 \times 10^3\,\text{s}^{-1}$$
$$= 86.5\,\text{mW}$$

This switching loss is so small as to be negligible in comparison with the conduction loss of 3.75 W.

From Figure 5.25, as before, $t_{rr} = 80$ ns

Equation (5.28):

$$I_{RM} = \frac{2Q_{rr}}{t_{rr}} = \frac{2 \times 12 \times 10^{-9}\,\text{A s}}{80 \times 10^{-9}\,\text{s}} = 0.3\,\text{A}$$

Figure 5.26: $I_{RM} = 0.3$ A—the same.

Equation (5.26):

$$R_s \leqslant \frac{90\,\text{V}}{0.3\,\text{A}} \leqslant 300\,\Omega; \text{ selected, } R_s = 220\,\Omega/0.5\,\text{W}$$

Equation (5.27):

$$C_s \geqslant \frac{12 \times 10^{-6}\,\text{H} \times 0.3^2\,\text{A}^2}{90^2\,\text{V}^2} \geqslant 0.133\,\text{nF}; \text{ selected, } C_s = 0.68\,\text{nF}/100\,\text{V}$$

$$P_{Rs} = \frac{fCV^2}{2} = 0.5 \times 0.68 \times 10^{-9}\,\text{F} \times 90^2\,\text{V}^2 \times 40 \times 10^3\,\text{s}^{-1}$$

$$= 0.11\,\text{W}; \text{ selected, } 0.5\,\text{W}$$

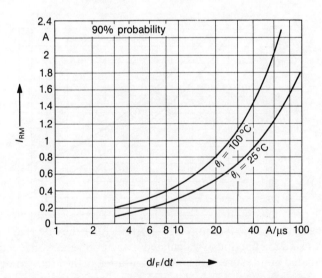

Fig. 5.26 Maximum reverse current I_{RM} (A) of the BYW80 diode as a function of the (negative) rate of rise of current dI_F/dt (A/μs) at two temperatures (Thomson CSF)

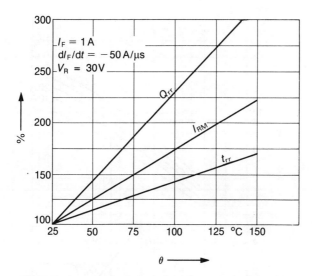

Fig. 5.27 Correction factors for the stored charge Q_{rr}, reverse current I_{RM} and recovery time t_{rr} of the BYW80 diode as functions of junction temperature (Thomson CSF)

$$R_{thca} \leqslant \frac{\theta_{jmax} - \theta_{amax}}{W} - R_{thjc} = \frac{150° - 50°}{3.8 \text{ W}} - 2.5 \text{ kW}$$

$$\leqslant 23 \text{ kW}$$

In view of the relatively large value of R_{thca}, a (small) heatsink with a thermal resistance R_{thca} of 10 kW is selected.
Maximum heatsink temperature:

$$\theta_{max} = \theta_{amax} + R_{thca}W = 50\,°\text{C} + 10 \text{ kW} \times 3.8 \text{ W} = 88\,°\text{C}$$
$$\theta_{jmax} = \theta_{max} + R_{thjc}W = 88\,°\text{C} + 2.5 \text{ kW} \times 3.8 \text{ W} = 97.5\,°\text{C}$$

It would have been more appropriate, therefore, to use the curves for $\theta_j = 100\,°$C for the determination of Q_{rr} and t_{rr}. A check (not detailed here) shows that in fact, due to the rounding-off involved, the same values of R_s and C_s can be retained.

For junction temperatures higher than 100 °C it is best to take the values for $\theta_j = 25\,°$C from the curves and correct them in accordance with Figure 5.27.

If the reverse voltages exceed about 200 V, it is necessary to use the so-called double-diffused fast silicon diodes. These are available, according to type, with voltage ratings up to 1000 V and current ratings up to 60 A or more. The stored charge in these diodes is, however, considerably greater than that of the epitaxial diodes, so that the switching losses are no longer negligible. The first requirement is to calculate the rate of fall of current from Equation (5.31), using the previously calculated value of reverse voltage and the leakage inductance (estimated or measured). From a presentation of the kind shown in Figure 5.28 the stored charge can then be read off at the intersection with the curve for the maximum diode current. Some manufacturers provide a graph such as that of Figure 5.29, which enables the switching loss to be determined quickly without calculation.

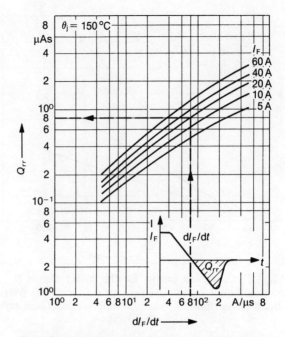

Fig. 5.28 Maximum stored charge Q_{rr} (μA s, μC) of
the BYX65 double-diffused fast diode as a function of
the (negative) rate of rise of current dI_F/dt (A/μs) with
various forward currents (Thomson CSF)

Example

The rate of fall of current has previously been calculated as 80 A/μs; the maximum reverse
voltage of the BYX65-400 rectifier diode is 400 V; $f = 20$ kHz. From Figure 5.28,

$$Q_{rr} = 0.8 \times 10^{-6} \text{A s at } I_F = 20 \text{A}.$$

Equation (2.43):

$$P_{Ds} = Q_{rr} V_R f = 0.8 \times 10^{-6} \text{A s} \times 400 \text{V} \times 20 \times 10^3 \text{s}^{-1}$$
$$= 6.4 \text{W}$$

The same can be read off Figure 5.29 (note the direction of the arrow). The principal
selection criteria and characteristics of the three types of diode discussed are illustrated in
Figures 5.30 and 5.31.

Calculation of the output capacitance C_0 follows exactly the procedure described in
Section 2.3. With very high output currents, and therefore with very high values of ΔI_L
(e.g. in the 300 W example considered above $\Delta I_L = 12$ A), it is not possible to connect ten
or more capacitors in parallel to obtain the very low values necessary for ESR and L_C. The
only solution is to use special electrolytic capacitors which are made specifically for
switched-mode power supplies.

Conversion from the triangular-wave alternating current given in peak-to-peak
amperes to the required r.m.s. value must be applied using Equation (2.27). Where
applicable, the correction factor from Figure 2.20 should be allowed, if the maximum
permissible ambient temperature is not reached. Correction for different frequencies, as

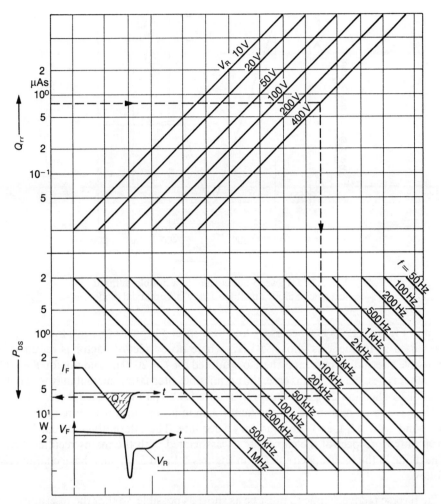

Fig. 5.29 Diagram for the determination of switching loss P_{Ds} of the BYX65 diode in terms of the stored charge Q_{rr} (μA s, μC), reverse voltage V_R (V) and switching frequency f (Hz) (Thomson CSF)

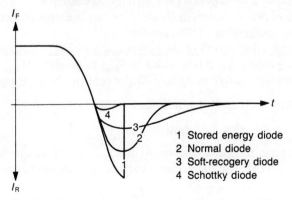

1 Stored energy diode
2 Normal diode
3 Soft-recogery diode
4 Schottky diode

Fig. 5.30 Reccovery characteristics of various rectifier diodes

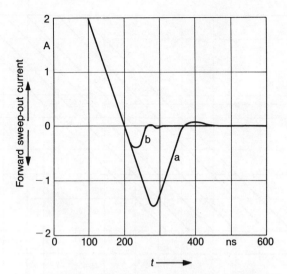

Fig. 5.31 (a) Sweep-out current in a double-diffused
fast silicon diode and (b) in a very fast epitaxial diode

shown in Figure 2.19, is not necessary for these special capacitors, since they are almost always used at frequencies above 20 kHz, where no correction factor is required. The permissible alternating current is therefore quoted for $f = 20$ kHz, and not for $f = 100$ Hz as for other types. In fact, in this case, the permissible alternating current may well not be a limitation (see Section 2.3), as is shown by Figure 5.32. The values listed there, of the order of 10 A, have to be multiplied by the factor 3.46 in accordance with Equation (2.27) (to correspond with ΔI_L), and are thus very adequate. The alternating-current loading is critical only in respect of the capacitor C_1 associated with the mains rectifier.

The diode in the resetting circuit (D_1 in Figures 5.1 and 5.18) must have characteristics as follows:

- The reverse recovery time t_{rr} must be as short as possible, so that the resetting current flows with minimum delay.
- The repetitive peak current I_{FRM} must be not less than I_{pmax}, since the collector current has to be carried for a short period ($< 1 \mu s$) until the current in D_3 reaches its maximum value. The rate of rise of current in D_3 is reduced by the stray inductance of the circuit and the leakage inductance of the transformer.
- During the conducting period of the power transistor T_s, a voltage equal to V_{imax} is induced in the winding N_2; this is added to V_{imax} to produce $2V_{imax}$. The calculations should also include a safety margin of 10 per cent to allow for possible overshoot.

Thus, for the resetting diode D_1 (as in Figure 5.18),

$$I_{FRM} \geqslant I_{pmax} \quad (A) \tag{5.32}$$

where I_{FRM} is the repetitive peak current (A) and I_{pmax} is the maximum primary current from Equation (5.21) (A). The mean direct current is given approximately by

$$I_{FAV} \approx \frac{I_M}{2} \quad (A) \tag{5.33}$$

where I_M is the maximum magnetizing current from Equation (5.22) (A).

C_N	V_N	R_{ESR} 20 kHz 20 °C	I_{Rmax} 5 min 20 °C	$I_{C\sim max}$ 20 kHz 85 °C	L_{ESL} ca.
(μF)	(V$-$)	(mΩ)	(mA)	(A)	(nH)
18000		5.8	0.18	9.8	20
32000	5	4.5	0.32	13.0	20
46000		3.8	0.46	15.8	20
15000		6.0	0.23	9.6	20
27000	7.5	4.6	0.41	12.8	20
39000		3.9	0.59	15.6	20
10000		6.4	0.32	9.3	20
18000	16	4.9	0.58	12.4	20
26000		4.0	0.84	15.4	20
8800		6.6	0.36	9.1	20
16000	20	5.0	0.64	12.3	20
22000		4.1	0.88	15.3	20
6300		7.1	0.36	8.8	20
11000	28	5.3	0.62	11.9	20
16000		4.3	0.90	14.9	20
4500		7.5	0.32	8.6	20
8100	35	5.5	0.57	11.7	20
12000		4.5	0.84	14.6	20
2800		8.7	0.31	8.0	20
5000	55	6.3	0.55	11.0	20
7300		5.0	0.81	13.8	20

Fig. 5.32 Data for special electrolytic capacitors for exacting requirements (Siemens)

The required reverse voltage rating (with allowance for repetitive transient voltages) is

$$V_{RRM} \geqslant 2 V_{imax} \times 1.1 \quad (V) \tag{5.34}$$

Thus for the 300 W power supply

$$I_{pmax} = 4.25 \, A; \quad I_{FAV} = 0.5 \times 0.264 \, A = 0.132 \, A;$$
$$V_{RRM} = 340 \, V \times 2.2 = 750 \, V$$

These requirements are met by, among others, the Valvo Type BYV96D ($I_{FM} = 10$ A; $I_{FAV} = 0.8$ A; $V_{RRM} = 800$ V, or 1000 V for the BYV96E), or the TRW Type DSR 3800 or DSR5800 ($I_{FAV} = 1$ A; $V_{RRM} = 800$ V). $t_{rr} \approx 300$ ns is quoted for the Valvo diode, and $t_{rr} \approx 75$ ns for the TRW diodes. Because high-voltage silicon diodes of this type are produced by double diffusion, they have forward voltage drops of around 1.6 to 1.8 V.

The forward conduction loss can be calculated by means of Equation (2.42). For the example (with I_{FAV} in place of I_0), $P_{Dd} I_{FAV} V_F (1 - \delta_{Tmin}) = 0.132 \, A \times 1.8 \, V \times 0.72 = 0.17$ W, which is insignificant.

There is no switching loss to be considered in connection with the resetting diode, because this diode does not have to be switched from full forward conduction to the blocking state; it is not required to block until the decreasing resetting current has already died away.

5.4 CALCULATIONS FOR THE BIPOLAR POWER TRANSISTOR

The requirements for the power transistor Tr_s are as follows:

- Rise and fall times of the collector current should be very short; this means either high initial overdrive and large sweep-out current (somewhat delayed because of recombination in the collector zone) or operation in the quasi-saturated mode in order to obtain short storage and fall times, and hence low turn-off loss at the expense of higher conduction loss.
- Highest possible transition frequency at I_{pmax} in order to obtain short switching times; $I_{pmax} < I_{Cmax}$.
- A blocking voltage V_{CEX} (V_{CEV}) (at $I_C = 0$ and with negative V_{BE}) of at least 800 V, and preferably 1000 V (feasible with RCD suppression); $V_{CE0} \geqslant 400$ V.

In the data sheets the values of t_r (rise time) and t_f (fall time) are usually quoted for conditions which are of little use in relation to operation in a switched-mode power supply. It is, moreover, an advantage to use a transistor which is over-dimensioned in terms of current, because the maximum transition frequency, and also the maximum current gain, occur at much less than the maximum permissible collector current I_{Cmax}. Since in fact $f_T = f(I_C)$ is not usually quoted, but $B = f(I_C)$ is, and these two values reach their maxima at approximately the same collector current, the shortest switching times are obtained with a dimensioning such that $I_{pmax} \approx I_C(B_{max})$. In the example of the 300 W forward converter considered here the BUV48A transistor ($I_{Cmax} = 15$ A; $V_{CEX} = 1000$ V) was operated at $I_{pmax} \approx 4$ A, which afforded the maximum transition frequency and hence the minimum switching times.

The losses in the transistor consist, as previously explained in Section 2.4 in connection with the inductor-coupled step-down converter, of conduction loss and switching losses. The most critical element here is the turn-on process, because the transistor in the forward converter is switched on to a high current. The turn-off loss can be kept very low by means of an RCD suppression network in parallel with the collector-emitter path of the transistor. The energy which is released when the transistor turns off can be harmlessly converted into heat in a resistor.

Since the switching times t_r and t_f are strongly dependent upon the drive circuit, the switching losses can only be calculated approximately. A final determination is possible only when the equipment has been assembled.

The losses are calculated in the same way as for the inductor-coupled step-down converter.

Conduction loss:

$$P_{Trc} = V_{CE}I_{pmax}\delta_{Tmax} \quad \text{(W)} \tag{5.35}$$

where V_{CE} is the collector-emitter voltage in the conducting state (V) (either V_{CEsat} if the transistor is saturated or about 1.8 to 2 V in the quasi-saturated mode) and I_{pmax} is the maximum primary collector current (A) (calculated from Equation (5.21)).

Turn-on loss:

$$P_{Trson} = V_{imax}I_{pmax}(t_r/2)f \quad \text{(W)} \tag{5.36}$$

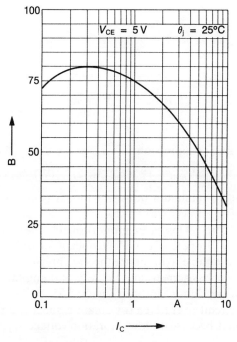

Fig. 5.33 Variation of the current gain B of a power switching transistor with I_C (Valvo)

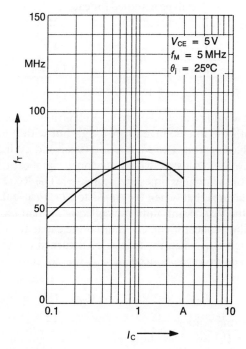

Fig. 5.34 Variation of the transition frequency f_T of the power switching transistor of Figure 5.33 with I_C (Valvo)

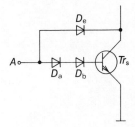

Fig. 5.35 Basic form of the Baker clamp anti-saturation circuit for bipolar transistors

Turn-off loss without RCD suppression:

$$P_{\text{Trsoff}} = V_{\text{imax}}I_{\text{pmax}}(t_f/2)f \quad (\text{W}) \tag{5.37}$$

To optimize the circuit for minimum transistor loss, it is necessary to consider all the components.

Since the turn-on and turn-off losses are proportional to the switching frequency f, and since it is advantageous to operate at high frequency in order to keep the inductive components small, it is desirable to minimize the switching losses, even at the expense of increasing the conduction loss somewhat. One way of doing this is to employ the anti-saturation circuit or Baker clamp shown in Figure 5.35.

Even with the base circuit overdriven (to obtain a short rise time) saturation of the transistor Tr_s is prevented, because the low saturation voltage V_{CEsat} cannot be attained in this circuit.

Operation:
A forward voltage drop of 0.7 V will be assumed for each of the three diodes D_a, D_b and D_e; for the transistor, $V_{\text{BE}} = 1$ V. If the current path from point A to earth is now considered,

$$V_{\text{A}-0} = 2 \times 0.7\,\text{V} + 1\,\text{V} = 2.4\,\text{V}$$

The same voltage appears in the circuit $D_e + V_{\text{CE}} = 0.7\,\text{V} + V_{\text{CE}}$. This means

$$V_{\text{CE}} = 2.4\,\text{V} - 0.7\,\text{V} = 1.7\,\text{V}$$

The saturation voltage would be around 0.3 to 0.5 V. It is thus impossible for the transistor to be in saturation with the low voltage drop V_{CEsat}, but with a large concentration of charge carriers in the base region. The conduction loss is indeed increased by a factor of about three, but the switching losses, which are more significant in this case, are very greatly reduced, since the switching times are decreased. The RCD suppression network shown in Figure 5.18 ensures that, when the transistor Tr_s turns off, the collector voltage V_{CEmax} cannot exceed the V_{CEO} rating until the collector current has fallen to zero. In the turn-off process the falling collector current charges the capacitor C_{s1} through D_{s1}; C_{s1} is only discharged, through R_{s1}, when the transistor is turned on again. Until resetting is complete, because of the voltage induced in N_1, the collector-emitter voltage, albeit delayed by C_{s1}, is

$$V_{\text{CEmax}} = 2 \times V_{\text{imax}} < V_{\text{CEX}} \quad (\text{V}) \tag{5.38}$$

C_{s1} is charged to the same voltage.

The requirements for D_{s1} are: very short recovery time; current rating equal to I_{pmax}; voltage rating equal to $2 \times V_{\text{imax}}$ + safety margin ≈ 800 to 1000 V.

$$I_{\text{FAV}} \approx I_{\text{pmax}}\frac{(2 \text{ to } 3)t_f}{T} \quad (\text{A}) \tag{5.39}$$

$$C_{s1} \geqslant \frac{I_{pmax}t_f}{2V_{CE0}} \quad (F) \tag{5.40}$$

Voltage rating: $V_{max} = 800$ to $1000\,V$. t_f is the collector current fall time (2) and V_{CE0} is the rated collector-emitter voltage (V) with $I_B = 0$ (data sheet)

$$R_{s1} \leqslant \frac{t_{on}}{4C_{s1}} \leqslant \frac{\delta_{Tmin}T}{4C_{s1}} \quad (\Omega) \tag{5.41}$$

and

$$R_{s1} \geqslant \frac{V_{imax}}{\Delta I_C} \quad (\Omega) \tag{5.42}$$

where I_C is the permissible increase in current above $I_{pmax}(A)$ ($\leqslant 0.1 I_{pmax}$).

R_{s1} should be so dimensioned, in accordance with Equation (5.41), that the capacitor C_{s1} is discharged, under the least favourable conditions, during the 'on' time of the transistor. Equation (5.42) determines R_{s1} so that the additional discharge current of C_{s1} does not significantly increase the collector current during the turn-on process, which is in any case critical.

The energy stored in C_{s1} must be converted into heat in the resistor R_{s1}. The loss in R_{s1} is

$$P_R = 0.5C_{s1}(2V_{imax})^2 f \quad (W) \tag{5.43}$$

The turn-off loss is considerably reduced by the RCD suppression network, because the rise in collector voltage is so delayed by the charging of C_{s1} that the collector current has already fallen almost to zero [15], [4], [19].

Turn-off loss with RCD suppression network:

$$P_{Trsoff} = \frac{I_{pmax}^2 t_f^2}{24C_{s1}} f \quad (W) \tag{5.44}$$

The suppression components may now be determined for the previous example of the 300 W forward converter.

Transistor BUV48A (Thomson): $V_{CEX} = 100\,V$; $V_{CE0} = 450\,V$; t_f and t_r (with suitable base drive) estimated at $0.2 \times 10^{-6}\,s$. Previous values: $I_{pmax} = 4.25\,A$; $\delta_{Tmin} = 0.28$; $V_{imax} = 340\,V$.

Equation (5.40):

$$C_{s1} \geqslant \frac{4.25\,A \times 0.2 \times 10^{-6}\,s}{2 \times 450\,V} \geqslant 0.95 \times 10^{-9}\,F$$

$$C_{s1} = 1.5\,nF/1000\,V \text{ (selected)}$$

$$R_{s1} \leqslant \frac{0.28 \times 25 \times 10^{-6}\,s}{4 \times 1.5 \times 10^{-9}\,F} \leqslant 1.17\,k\Omega$$

$$R_{s1} \geqslant \frac{340\,V}{0.42\,A} \geqslant 810\,\Omega; \text{ selected, } 1\,k\Omega$$

$$P_R = 0.5 \times 1.5 \times 10^{-9}\,F \times (2 \times 340\,V)^2 \times 40 \times 10^3\,s^{-1}$$

$$= 13.9\,W$$

t is thus necessary to use a resistor with a dissipation rating of 20 to 25 W.

Turn-on loss, from Equation (5.36):

$$P_{Trson} = 0.5 \times 340 \, V \times 4.25 \, A \times 0.2 \times 10^{-6} \, s \times 40 \times 10^3 \, s^{-1}$$
$$= 5.8 \, W$$

In [15] and [19] the authors indicate that in fact the turn-on loss tends to be more than twice the amount given by Equation (5.36). In all cases it is greater.

Conduction loss from Equation (5.35) with $V_{CE} = 1.8 \, V$:

$$P_{Trc} = 1.8 \, V \times 4.25 \, A \times 0.45 = 3.45 \, W$$

Equation (5.37): turn-off loss without RCD suppression network:

$$P_{Trsoff} = 0.5 \times 340 \, V \times 4.25 \, A \times 0.2 \times 10^{-6} \, s \times 40 \times 10^3 \, s^{-1}$$
$$= 5.8 \, W$$

which is the same as the turn-on loss.

Equation (5.44): turn-off loss with RCD suppression network:

$$P_{Trsoff} = \frac{4.25^2 \, A^2 \times (0.2 \times 10^{-6})^2 \times 40 \times 10^3}{24 \times 1.5 \times 10^{-9} \, F} = 0.80 \, W$$

Comments on the various losses:

- The conduction loss is of an acceptable order despite the relatively high collector voltage V_{CE}.
- The turn-off loss, as was to be expected, has become very small as a result of the RCD suppression. If the loss indicated by Equation (5.37) were doubled, in accordance with [5] and [19], the turn-off loss would be about 10 W. With the dimensioning chosen, however, the loss of about 14 W converted into heat in the resistor R_{s1} is somewhat greater than the loss of which the transistor is relieved. The overall efficiency with RCD suppression is therefore somewhat lower than without it. This could be offset by choosing a somewhat smaller capacitor (which calculation shows to be possible), albeit at the expense of the chosen safety margin for the transistor. At all events, it can be taken that an RCD suppression network is essential from the point of view of long transistor life.
- The high value obtained for the turn-on loss is not very satisfactory.

The total transistor loss is thus

$$P_{Trtot} = 3.45 \, W + (5.8 \text{ to } 11.6 \, W) + 0.8 \, W = 10.1 \text{ to } 15.85 \, W$$

The BUV48A transistor selected for the example has, according to the data sheet, a permissible total dissipation $W_{tot} = 150 \, W$. The definition 'W_{tot}' is, however, not very useful in practice, except in comparing different transistor types. 'W_{tot}' only means the loss which may be dissipated as heat between the junction and the case with an external thermal resistance $R_{thca} = 0$; i.e. the case must be kept exactly to the ambient temperature. This is completely unrealistic. To a first approximation it can be taken that a semiconductor device can operate at around $W_{tot}/2$. For the BUV48A this would be about 75 W in an ambient temperature $\theta_a = 25 \, °C$, or 38 W in an ambient temperature of 60 °C and the realistic thermal resistance of 2 kW for the heatsink. This is still twice the loss calculated for the least favourable conditions. The turn-on loss can be reduced by a facto

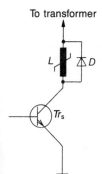

To transformer

L D

Tr_s

Fig. 5.36 Insertion of a saturable inductor L with a free-wheel diode D to reduce the turn-on loss in the switching transistor Tr

of 5 to 10 by inserting a small inductor, preferably saturable, in the collector lead. By this means the rise of collector current is somewhat delayed (in a manner analogous to the effect of the RCD suppression network on the rise of collector voltage at turn-off) [15], [19].

To enable the energy stored in the saturable inductor to be released from the core, and to prevent the generation of overvoltages, the inductor is shunted by a free-wheel diode D; this is dimensioned in exactly the same way as D_{s1}.

The most suitable core is a ring core without an air gap. A sufficient delay time t_d for the rise of collector current in this arrangement is about $1.5 t_r$.

$$t_d \geqslant 1.5 t_r \quad \text{(s)} \tag{5.45}$$

The number of turns is calculated as in Equation (2.20).

$$N = \frac{t_d V_{\text{imax}} \times 10^4}{A_e \hat{B}_s} \tag{5.46}$$

where t_r is the current rise time delay according to Equation (5.45) (s), A_e is the effective magnetic cross-sectional area (cm^2) and $\hat{B}_s$ is the saturation flux density (T) (about 0.3 to 0.5 T according to material).

The protective elements R_{s2}, C_{s2} and D_{s2} connected in parallel with N_1 (Figure 5.18) prevent voltage overshoot due to the unavoidable stray inductance when the transistor turns off. The parallel RC protective circuit operates in quite a different way from the RCD suppression network for the transistor, with its series connection of R and C and diode decoupling. The time constant of this protective network should be large in comparison with the cycle period T.

$$R_{s2} C_{s2} \geqslant 10 T \tag{5.47}$$

Thus, unlike C_{s1}, C_{s2} is not discharged in the course of one cycle. The voltage across C_{s2} remains practically constant, equal to the input voltage V_i. Since variations in the input voltage take place only slowly, they can be followed by the parallel circuit of C_{s2} and R_{s2}, so that C_{s2} remains charged to V_i. When an overvoltage occurs as the transistor Tr_s turns off, D_{s2} conducts and C_{s2} is charged to a slightly higher voltage. The energy associated with such overvoltages is usually relatively low, so that the additional charge in C_{s2} is small. The circuit behaves rather like a breakover diode in which the breakover voltage is equal to the input voltage V_i at any particular time.

For the present example,

$$R_{s2} C_{s2} \geqslant 250 \, \mu s; \text{ selected, 1 ms}$$

If R_{s2} is assumed to be $47\,\text{k}\Omega$, C_{s2} becomes

$$C_{s2} = \frac{1 \times 10^{-3}\,\text{s}}{47 \times 10^3\,\Omega} = 21.3\,\text{nF; selected, } 27\,\text{nF}/400\,\text{V}$$

A sufficient voltage rating for C_{s2} is $V_{imax} = 340\,\text{V}$, or rounded up, $400\,\text{V}$.

$$P_{Rs2} = \frac{V_{imax}^2}{R_{s2}} = \frac{(340\,\text{V})^2}{47 \times 10^3\,\Omega} = 2.5\,\text{W; selected, } R_{s2} = 47\,\text{k}\Omega/5\,\text{W}$$

The maximum voltage across D_{s2}, which should be very fast, is about $400\,\text{V}$. The mean current is very low.

5.5 DRIVE CIRCUITS

To obtain satisfactory switching behaviour in the transistor Tr_s, particular attention needs to be directed to the drive circuit. There is a choice to be made from various possibilities; the ultimate choice depends upon the requirements of the complete circuit.

To make clear the essentials of the circuits that follow, inessential details have been omitted. Thus, in all the drive circuits, the switching transistor Tr_s is shown without the necessary protective networks, and similarly the output transistor Tr_0 in the drive IC is depicted without the suppression network which is necessary for this component. Figure 5.37 shows a transformer-coupled drive circuit described in [4] for a 240 W forward-converter power supply with a flyback-converter drive transformer.

When transistor Tr_0 in the drive IC turns on, the BD230 transistor is turned off. The voltage induced in the secondary winding of the transformer is in such a direction, because of the opposite polarities of the windings (especially to be noted), that the switching transistor Tr_s is turned on. As a result of the constant-current source in the primary circuit, the secondary current is also forced, which is important from the point of view of fast turn-on. To avoid a gradual tailing-off of collector current, which would cause severe loading of

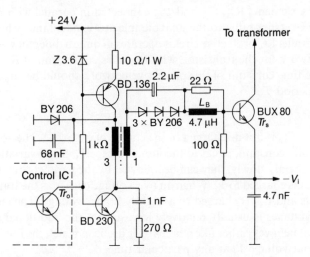

Fig. 5.37 Transformer-coupled drive circuit for a bipolar *n-p-n* power transistor with the transformer connected as for a flyback converter

the switching transistor at turn-off, the fall of base current during the turning-off process is delayed by a series inductance L_B. This matching of dI_B/dt must be exactly suited to the requirement. The optimum turn-off characteristic is obtained when the collector-base region is cleared of charge carriers before the emitter region. The negative base current has therefore to be somewhat delayed. The dimensioning is such that in the conducting state the switching transistor Tr_s is saturated. If Tr_s is now to be turned off, the BD230 transistor is turned on, and a negative voltage pulse is introduced into the base lead of Tr_s. This negative base voltage is augmented by about 2.1 V (three times the diode voltage drop during the 'on' period of Tr_s) across the 2.2 µF electrolytic capacitor and the voltage induced in L_B. This gives rise to a high negative (in the case of the n-p-n power transistor shown, and normally used) sweep-out current, which results in a short fall time and low turn-off loss. The storage time of Tr_s is relatively long, however, because of the saturation. In principle, the drive transformer could be operated in the forward-converter, as well as the flyback, mode. With the forward-converter drive system the control of the drive IC needs to be changed. The flyback version gives shorter turn-on and turn-off times, and is widely used in consequence. The diode D_e shown in Figure 5.18 is omitted when this drive circuit is used. At the same time, the drive transformer cannot be dimensioned in the way so far discussed (see Chapter 7, 'Flyback converters').

Significantly better switching performance is afforded by so-called self-regulating drive circuits. A circuit of this type with direct coupling [16] is shown in Figure 5.38.

In this case the switching transistor Tr_s operates in the 'quasi-saturated' mode; see also Figure 5.35.

Notable features of this arrangement are the one or two diodes in the base lead and the anti-saturation diode in front of these diodes, connected to the collector. Depending on the design of the circuit, the diodes in the base lead may be replaced by the base-emitter paths of transistors. In Figure 5.38, one diode is replaced by the base-emitter path of transistor Tr_2. In this type of circuit, the switching transistor Tr_s demands, as it were, under all conditions (varying according to the characteristic spreads of Tr_s and the

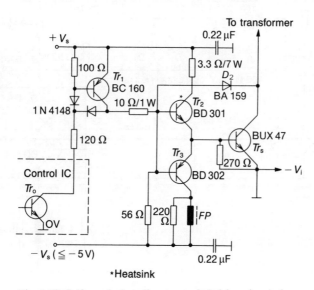

Fig. 5.38 Self-regulating direct-coupled drive circuit for a
bipolar power transistor

collector current flowing at any particular time) just so much base current that the collector voltage $V_{CE} = V_{BE/Trs} + V_{BE/Tr2} - V_{De} \approx 1\,V + 0.8\,V - 0.7\,V \approx 1.1\,V > V_{CEsat}$.

As a result of the automatic control of the base current of Tr_s, its switching times are rendered largely independent of the supply voltage (which does not therefore need to be stabilized), load variations in the output circuit and the differences in characteristics between individual switching transistors. Since saturation is avoided, the storage time t_s is short ($< 1\,\mu s$ according to information in [16]) and the current fall time very short (about 100 ns). This means a very small turn-off loss in Tr_s. The rate of rise of the negative base current is limited by a ferrite bead in the collector circuit of Tr_3 (damped by a parallel resistor) to about $2\,A/\mu s$, so that the relatively few charge carriers in the collector-base region have sufficient time to recombine.

The circuit operates in the following way.

When transistor Tr_0 in the drive IC turns on, the base of Tr_1 assumes 0 V potential. Since its emitter is connected to $+ V_s$, the p-n-p transistor Tr_1 turns on. Base current is supplied through its collector-emitter path and the $10\,\Omega$ resistor to Tr_2, and hence Tr_s also receives base current. Transistors Tr_2 and Tr_3 operate here in the linear mode. Transistor Tr_3 is initially still non-conducting, since its base, (at the V_{BE} of Tr_2) is more positive than its emitter. If transistor Tr_0 is now turned off, Tr_1 is turned off, and consequently so is Tr_2. The base of Tr_3 now receives a negative potential through the $56\,\Omega$ resistor, and Tr_3 turns on with a slight delay. The base of Tr_s can thus be cleared of charge carriers by a high sweep-out current. In addition the base of Tr_s is kept negatively biased, through Tr_3, until Tr_0 turns on again. A self-regulating drive circuit similar to that of Figure 5.38, but with an additional circuit to protect Tr_s against overload and short-circuit, is shown in Figure 5.39.

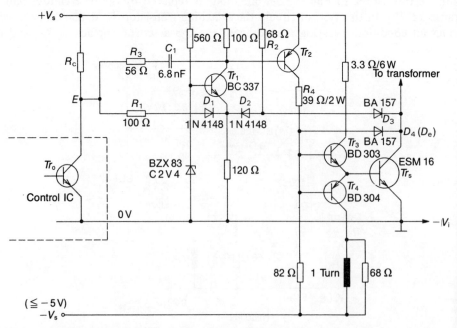

Fig. 5.39 Self-regulating direct-coupled drive circuit for a bipolar n-p-n power transistor with overload protection

So long as transistor Tr_0 in the drive IC has not turned on, the input I to the circuit is held at a positve potential $+V_s$ through R_C. Through R_C, R_1 and diode D_1, and also through R_2 and D_2, the emitter of Tr_1 is held at a higher positive potential than its base, whose potential is fixed at 2.4 V by the Zener diode. Tr_1 and Tr_2 are thus non-conducting. The drive transistor Tr_3 and the switching transistor Tr_s therefore receive no base current, and are also non-conducting. The conducting p-n-p transistor Tr_4 holds the base of Tr_s to a negative potential $-V_s$. If Tr_0 is now turned on, the potential of I is brought down to 0 V. Tr_2 is turned on for about 1.5 s through R_3 and C_1. The switching transistor Tr_s is turned on by base current received through Tr_2, R_4 and Tr_3. Simultaneously Tr_4 is turned off by the now positive potential of the bases of Tr_3 and Tr_4 relative to their common emitter connections. If the collector-emitter voltage Tr_s falls to less than 2.4 V within the 1.5 μs period, the current flowing in R_2 is diverted through D_3 and the collector-emitter path of Tr_s. Diode D_2 blocks, and the emitter of Tr_1 becomes less positive. Since D_1 is already blocking as a result of the turning on of Tr_0, Tr_1 is by now conducting and holds Tr_2 in the conducting state after the pulse through R_3 and C_1 has died away. If the collector voltage of Tr_s rises above 1.8 V—e.g. due to an overload or a short circuit on the output—D_3 blocks again, and the emitter of Tr_1 becomes more positive. Tr_1 turns off, and thereby switches the whole drive circuit to the 'off' state. The same happens when Tr_0 is turned off: the emitter of Tr_1 is then driven positive, and the drive circuit is turned off. The self-regulating properties of this circuit through Tr_3 and D_4 have already been discussed in connection with the circuit of Figure 5.38, and need not be explained again.

A further self-regulating drive circuit, with a pulse transformer T_1 in a forward-converter circuit to provide potential isolation between the drive and power stages, is shown in Figure 5.40 [17].

This drive circuit has been used to control the circuit of Figure 5.18. Unlike that of Figure 5.38, the base circuit of Tr_s in this case includes two diodes ($D_9 + V_{BE/Tr3}$), so that the collector-emitter voltage of Tr_s is higher by about 0.7 V. This of course means a higher conduction loss, but much shorter storage and fall times, and hence a lower turn-off loss,

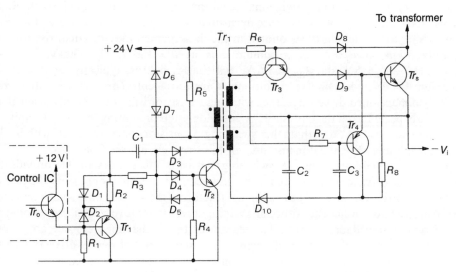

Fig. 5.40 Transformer-coupled self-regulating drive circuit for a bipolar n-p-n power transistor

particularly at high switching frequencies. The pulse transformer T_1 operates on the forward converter principle (see Figure 2.37). To improve the switching speed, the drive transistor Tr_2 for the pulse transformer is also operated with an anti-saturation circuit (D_3 to D_5).

The circuit operates in the following manner:

When transistor Tr_0 in the drive IC turns on (in this circuit the collector and emitter of this transistor must be accessible, which is not the case for all drive ICs), the control current flows through the collector-emitter path of Tr_0, diode D_2 (causing Tr_1 to be held 'off' by reason of the positive 0.7 V base-emitter voltage), resistor R_2, R_3 and C_1 in parallel (for faster turn-on) and D_4 to the base of Tr_2. A pulse of current thus flows in the primary winding of T_1. The resulting induced voltage is in such a direction that the ends of the windings shown dotted become positive with respect to the other ends. This means that the collector and base of Tr_3 assume a positive potential (about 6 to 8 V) relative to the emitter of Tr_s. Tr_3 turns on and supplies as much base current to Tr_s as is required by the loading in the collector circuit. In this case also the object is to maintain a collector voltage of about 1.8 V.

Capacitor C_2 (about 1 to 2.2 μF) is charged negatively (to about -0.7 V) by the turn-on pulse. When Tr_0 turns off again, Tr_1 is turned on via R_1 and rapidly sweeps out the base of Tr_2 through R_2, C_1 and R_3 in parallel and D_5. Transistor Tr_2, which was not in saturation by virtue of the anti-saturation circuit in its base circuit (D_3 and D_4), turns off very quickly. The transformer T_1 is reset through D_7 and D_6, R_5 and its primary winding (see also the calculation and discussion in connection with Figure 2.37). At the same time a reversed voltage pulse is generated in the secondary winding. A voltage is applied to the base of the p-n-p transistor Tr_4 through R_7 and C_3 with a slight delay (so that the charge carriers in the collector region can recombine before the high negative base sweep-out current is applied) and connects the base of Tr_s to the negative potential of C_2 (about -6 to -8 V). The base of Tr_s is now rapidly swept out. Because of the negative voltage on C_2, the base of Tr_s remains at a negative potential, through R_8, until it is turned on again. This drive circuit results both in a very short turn-on time and in a still shorter turn-off time, and hence in very low losses in the switching transistor Tr_s. According to [17], rise times of 100 ns and fall times of 50 ns were measured with the BUV48A transistor. For comparison, the switching times quoted in the data manual may be cited: rise time t_r, typically 3.5 μs at $\theta_j = 25\,^{\circ}$C, maximum 5 μs at $\theta_j = 100\,^{\circ}$C ($V_i = 300$ V, $I_C = 8$ A, $L_B = 3\,\mu$H, $-V_B = 5$ V and $I_{Bfin} = 1.6$ A); fall time $t_f = 80$ ns, typically, at $\theta_j = 25\,^{\circ}$C, maximum 0.4 μs under similar conditions of measurement. Thus, as a result of the carefully dimensioned drive stage, the switching transistor was considerably faster than was to be expected from the data sheet. An oscillogram representing the SOAR family of characteristics has shown that the switching transistor operates well within the permissible limits. (For the dimensioning of this drive circuit, see Chapter 6.)

While all drive circuits for bipolar switching transistors must be designed to provide the optimum base drive in order to obtain short switching times, this is not so in the case of MOSFET power transistors. The field-effect structure of these transistors is such that they do not need any drive current. Although their input resistance is very high, they should be driven by low-impedance circuits. The reason for this is that their considerable input capacitance has to be charged and discharged. From the point of view of drive, the size of the power transistor makes practically no difference. For high power requirements it is possible to drive a number of parallel-connected MOSFET power transistors with one

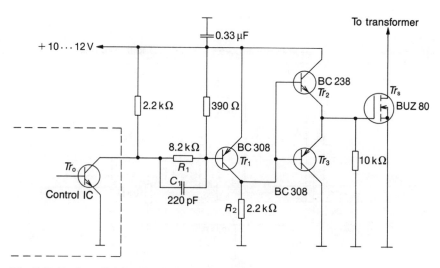

Fig. 5.41 Push–pull drive circuit with bipolar drive transistors for an *n*-channel power
MOSFET

drive circuit. Figure 5.41 shows a possible drive system in the form of a push–pull drive
circuit using bipolar transistors.

The operation of this drive circuit is as follows. When transistor Tr_0 in the drive IC
turns on, Tr_1 receives base current rapidly through R_1 in parallel with C_1 and is similarly
turned on. The bases of Tr_2 and Tr_3 are thereby connected to the positive line, so that Tr_2
can turn on. A positive voltage is applied to the gate of the MOSFET Tr_s through the low-
impedance output of Tr_2 (emitter follower) and the power transistor Tr_s turns on in about
70 ns. When Tr_0 turns off again, Tr_1 is turned off rapidly by the positive voltage now
stored on C_1, and base current is fed to Tr_3 through R_2. The gate is thus rapidly connected
to earth, and the power transistor switches typically in about 100 ns. If oscillations occur
during the switching process as a result of an unsatisfactory conductor layout, they can be
suppressed by inserting a low-value resistor or a ferrite bead directly in the gate lead. A
RCD suppression circuit is not usually necessary, because MOSFET transistors do not
exhibit second breakdown (associated with high current and high voltage simultaneously). Such a circuit may be provided, however, as a safety measure and to reduce the
dissipation in Tr_s. Since MOSFETs are available with off-state voltage ratings of 800 to
1000 V, the voltage in the denominator of Equation (5.40) can be set higher. A larger
proportion of the turn-off loss is thus dissipated in the transistor, and the capacitor C_{s1} is
in consequence smaller. The power to be dissipated in the associated resistor is also
reduced. A protective network against overvoltages due to the leakage and stray
inductance of the transformer, such as that in Figure 5.18 (C_{s2}, R_{s2} and D_{s2}) is, however,
recommended here also.

Calculation of the losses (conduction, turn-on and turn-off losses) follows the same
procedure as for the bipolar transistor, except that for the MOSFET the conduction loss is
calculated from Equation (2.61). The turn-on loss is obtained from Equation (5.36), turn-
off loss from Equation (5.37) or (5.44) depending on the circuit.

An alternative drive circuit for MOSFET power transistors, widely used, is shown in
Figure 5.42. In this circuit the gate charging current is supplied by the four parallel-

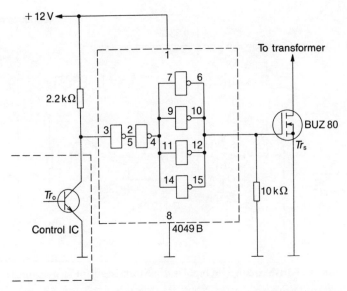

Fig. 5.42 Push–pull drive circuit with four MOSFET inverters
connected in parallel for an *n*-channel power MOSFET

connected inverters of the 4049B CMOS IC (each inverter stage is constituted by a complementary push–pull stage in a common-drain connection, and has a low-impedance output, like the complementary pair used with bipolar transistors; an inverter stage can supply about 40 mA). The circuit operates in the following manner.

When transistor Tr_0 in the drive IC turns on, a '0' signal is applied to Input 3 of the 4049B, and a '1' signal appears at its Output 2 (= Input 5 of the second inverter). Output 4 of this inverter therefore produces a '0' signal. A '0' signal is thus applied to all the inputs (6, 10, 12 and 15) of the four parallel-connected interters. A positive voltage is thus applied rapidly (because of the low output impedance of the four inverters) to the gate of the power MOSFET, and Tr_s turns on. In all other respects operation is similar to that of the circuit of Figure 5.41. The switching times $t_r \approx t_f$ attainable with this circuit are of the order of 50 ns.

If d.c. isolation is necessary, the circuit of Figure 5.43 can be used. This corresponds to that of Figure 2.43, but whereas in that case a low-voltage MOSFET was used, in this case it is necessary to use a high-voltage type. Since the mode of operation, including the calculations for the pulse transformer, have already been discussed, it need not be considered further here.

Since MOSFET power transistors have relatively large input capacitances to be charged and discharged, both the delay times (which represent dead times from the point of view of control and monitoring systems) and the switching times depend upon the rise times and magnitudes of the charging and discharging currents that can be supplied by the drive circuit used. The maximum possible peak gate currents are limited by the internal gate resistance (effectively about 10 Ω in series with the gate terminal). The following five drive circuits for *n*-channel MOSFET power transistors represent various degrees of complication, and give good, albeit various, results [20].

When the output transistor Tr_0 of the drive circuit of Figure 5.44 turns on, about 60 per cent of the supply voltage is applied to the base of Tr_1, while the emitter is at the full

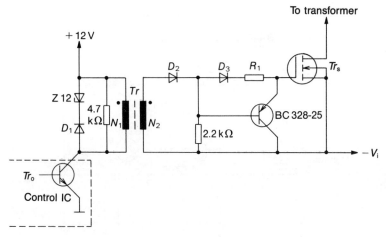

Fig. 5.43 Transformer-coupled drive circuit for an *n*-channel power MOSFET

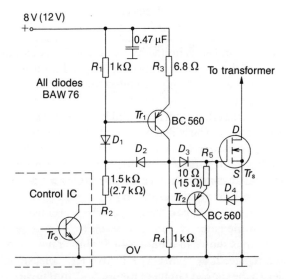

Fig. 5.44 Push–pull direct-coupled drive circuit with bipolar drive transistors for an *n*-channel power MOSFET (Siemens)

voltage. The base is thus more negative than the emitter, and the *p-n-p* transistor Tr_1 turns on, its collector current being limited by R_3. Diodes D_1 and D_2 prevent Tr_1 from saturating. D_3 ensures that Tr_2 is turned off when Tr_1 is conducting. Simultaneously, the gate of Tr_s is driven via D_3, and Tr_s turns on. When Tr_0 turns off again, Tr_1 turns off in consequence, and Tr_2 conducts. The gate capacitance is discharged through the collector-emitter path of Tr_2, the current being limited by R_5.

If higher gate currents than the circuit of Figure 5.44 provides are required, a Darlington circuit as shown in Figure 5.45 is more suitable. The Darlington circuit for the charging process consists of the two transistors Tr_1 and Tr_3: the discharging function is provided by Tr_2 and Tr_4. When the output transistor Tr_0 of the drive circuit turns on, Tr_1

134

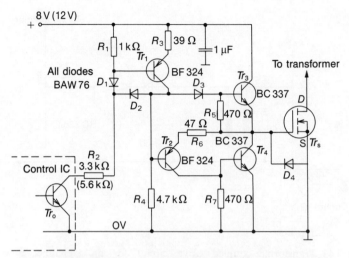

Fig. 5.45 Push–pull direct-coupled drive circuit with complementary Darlington transistors for an n-channel power MOSFET (Siemens)

conducts, its collector current being limited by R_3. A positive potential is applied through D_3 and Tr_3 to the gate of the switching transistor Tr_s, and it turns on. At the same time, Tr_2 and consequently Tr_4 are positively turned off by virtue of D_3. Saturation of Tr_1 is prevented by diodes D_1 and D_2.

When transistor Tr_0 turns off, Tr_1 turns off and hence also Tr_3. Tr_2 receives a negative input through R_4 and turns on. The collector current of Tr_2 is also the base current of Tr_4; the gate of the switching transistor Tr_s can now be rapidly discharged through Tr_4 and Tr_s turns off. This circuit has a low supply current demand in operation, but can nevertheless supply high charging and discharging currents to the MOSFET power transistor. The rise and fall times (see Figure 5.49) are therefore very short and the losses in Tr_s accordingly very low. Possible gate-source voltage overshoot can be prevented by a resistance of about 1 to $4.7\,\Omega$ in series with the gate.

Another circuit variant is shown in Figure 5.46. In this circuit the bipolar transistors of the input stage are replaced by a MOSFET inverter, consisting of three elements connected in parallel. In the simpler circuit of Figure 5.42, four inverters were connected in parallel, driving the MOSFET directly. Since a MOSFET inverter can supply a maximum of about 40 mA, the output of four in parallel is limited to 160 mA. In the circuit of Figure 5.46, however, the three inverters operate through a push–pull stage using bipolar transistors, so that considerable current amplification is obtained. As shown by the measurement results of Figure 5.49, the gate currents are considerably higher than would be possible with the inverters alone, and correspondingly the switching times are extremely short. The circuit operates as follows: when Tr_0 in the drive IC turns on, a low potential ('0' signal) is applied to the three parallel-connected inverters. The potential at the outputs is then high ('1' signal) and almost the whole supply voltage is applied to the base of Tr_1. Its emitter assumes a potential about 0.7 V lower, and the MOSFET power transistor Tr_s is turned on. Transistor Tr_2 is non-conducting, since its base is at a higher potential than its emitter. When Tr_0 turns off, there is a '1' signal at the inputs to the inverters and a '0' signal at their outputs. Transistor Tr_1 now turns off and Tr_2 turns on.

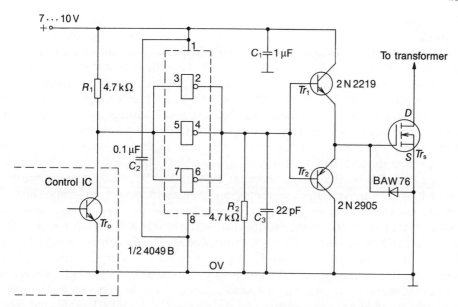

Fig. 5.46 Push–pull direct-coupled drive circuit with bipolar drive transistors controlled via three parallel-connected MOSFET inverters for an *n*-channel power MOSFET (Siemens)

The gate of Tr_s is discharged rapidly at a high current, and Tr_s turns off in a very short time. This circuit requires fewer components than that of Figure 5.45 and has roughly similar characteristics. The other half of the 4049B circuit can be used, for example, for a second similar circuit in push–pull operation. The resistor R_2 provides a continuous connection from the base of Tr_2 to earth. This prevents the occurrence of an undesirable potential at the gate of the MOSFET when Tr_2 turns on, which could cause the turning-on

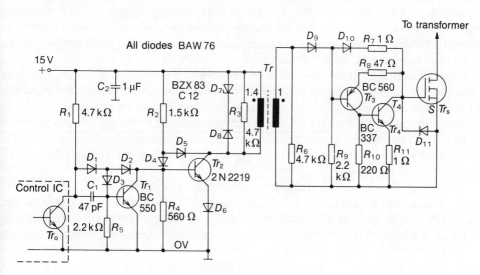

Fig. 5.47 High-power transformer-coupled drive circuit for an *n*-channel power MOSFET (Siemens)

of Tr_s. Capacitor C_3 shortens the time constant $R_2 C_{BE/Tr2}$, but can in many cases be omitted.

If potential isolation is necessary between the drive and switching stages, a transformer must be interposed (Figure 5.47). In the turn-on phase the gate of the MOSFET power transistor is driven from a low impedance through R_7, and is therefore charged rapidly. The discharge is effected when the driver turns off through Darlington transistor, in order to achieve a high discharge current and hence a short fall time t_f. Here again, anti-saturation diodes in both drive stages ensure that the switching times are very short. Operation is as follows: when Tr_0 in the drive IC turns on, the base current to Tr_1 through R_1 is diverted to earth, and Tr_1 turns off, assisted by C_1. Tr_2 now receives base current through R_2 and D_4, and turns on. Diodes D_2 and D_3, and D_4 and D_5, prevent transistors Tr_1 and Tr_2, respectively, from saturating, and hence switching more slowly. Diodes D_1 and D_6, by virtue of their threshold voltages, increase the input potentials of their associated transistors and ensure that they turn off positively. Resistors R_1 and R_5, and R_2 and R_4, are so dimensioned that Tr_2 is not turned on unintentionally as the supply voltage rises. When Tr_2 turns on, the current flowing in the primary winding generates an induced voltage in a direction opposite to that of the current, and a corresponding voltage, in the same direction, is produced in the secondary winding. Thus the upper (dotted) end of the secondary winding becomes positive and the gate of the MOSFET is charged rapidly through D_9, D_{10} and R_7. When Tr_a turns off, this causes Tr_1 to turn on and Tr_2 to turn off. A voltage in the opposite direction is now induced in the transformer windings, D_9 blocks and the magnetic energy stored in the transformer is discharged through D_8 and D_7. Resistors R_3 and R_6 damp possible oscillations during the resetting period. Since the positive voltage applied to the base of Tr_3 is removed with the blocking of D_9, Tr_3 now receives base current through R_9 and turns on, in consequence of which Tr_4 also turns on (Tr_3 and Tr_4 form a complementary Darlington circuit). The gate capacitance is rapidly discharged, the discharge current being limited only by R_{11} (1 Ω). The MOSFET is thus driven from a low impedance both in turning on and in turning off. This results in extremely short switching times. Because a certain amount of power has to be put into the transformer, this circuit consumes more power than those of Figures 5.44 to 5.46; on the other hand it is capable of supplying high drive currents, such that it is possible to operate several MOSFET power transistors in parallel if necessary. The same applies to the circuits of Figures 5.41, 5.43, 5.45 and 5.46.

If several MOSFET transistors are to be driven with d.c. isolation (if, for example, two transistors in series in the circuit of Figure 5.50 are switched simultaneously and in phase), further secondary windings, with appropriate suppression, can be provided.

The latter reason for the choice of a drive circuit where d.c. isolation is required points naturally to a transformer-coupled circuit. Where only a single power transistor or several in parallel are to be driven, the d.c. isolation can alternatively be provided by an optocoupler as shown in Figure 5.48.

The operation of this circuit is as follows. When the output transistor Tr_0 in the drive IC turns on, a current, set at 4 mA by the dimensioning of R_1, flows in the light-emitting diode of the CNY10 optocoupler. Owing to the inverter stage incorporated, the output is driven to 0 V. A '0' signal at the inputs to the following six parallel-connected inverters produces a '1' signal at their outputs, i.e. practically the full supply voltage of 7 V. Since the six inverters can supply about 240 mA, the gate capacitance is charged rapidly, and Tr_s turns on. When Tr_0 turns off, the LED is extinguished, and the output of the optocoupler returns to a '1' potential. This leads to a '0' signal at the inverter outputs, the gate

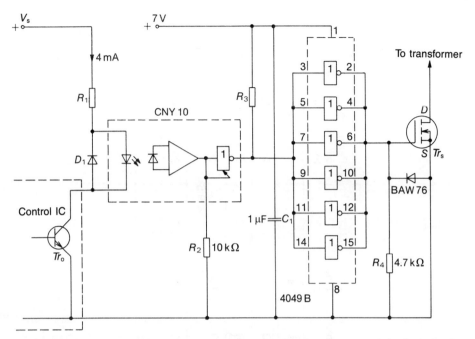

Fig. 5.48 Drive circuit with six parallel-connected MOSFET inverters and d.c. isolation by means of an optocoupler (Siemens)

capacitance is rapidly discharged, and Tr_s turns off. The 7 V supply for the receiver side of the optocoupler must be obtained from a special isolated power supply. This may be, for example, an additional winding on a small mains transformer with a rectifier and stabilizing circuit. The gate terminal of the optocoupler is connected to earth through R_2; the effect of this is that the output is open with a supply voltage of up to 4 V, i.e. both of the push–pull output transistors are non-conducting up to this point. This is necessary because the optocoupler gives a '0' signal until the supply voltage has risen to about 3 V, which would mean that the following inverters would turn on the MOSFET. During the period in which the optocoupler output is open, the inputs of the inverters are held at a '1' potential through R_3, so that a '0' signal appears at their outputs and the MOSFET remains non-conducting. R_4 holds the gate of the MOSFET at earth potential until the inverters are ready for operation. This circuit enables isolated operation to be combined with an 'on' time of any duration.

To facilitate comparison of the various alternative drive circuits, the principal characteristics of those shown in Figures 5.44 to 5.48—as far as they are known—are set out in the table of Figure 5.49.

5.6 SINGLE-ENDED FORWARD CONVERTER WITH TWO POWER TRANSISTORS CONNECTED IN SERIES (ASYMMETRICAL BRIDGE CIRCUIT)

This variant of the single-ended forward converter is shown in Figure 5.50 in a simplified representation without the mains rectification, drive circuit and protective networks.

The advantage of this arrangement lies in the fact that the transistors Tr_{s1} and Tr_{s2} are

138

Circuit from figure	Supply voltage	Maximum current consumption	Delay times		Fall time t_f	Rise time t_r	V_{as}	Peale gate current	
			t_{din}	t_{dout}				In	Out
	V	mA	ns	ns	ns	ns	V	mA	mA
5.44	8(12)	7	30	80	60	40	7	160	170
5.45	8(12)	5	40	40	30	10	6.5	500	500
5.46	7…10	5	80	80	30	10	6.5 (max. 9.5)	500	700
5.47	15	21	30	35	30	10	7.5	350	500
5.48	7	24	130	100	—	—	7	—	—

Fig. 5.49 Data for circuits of Figures 5.44 to 5.48 [20]

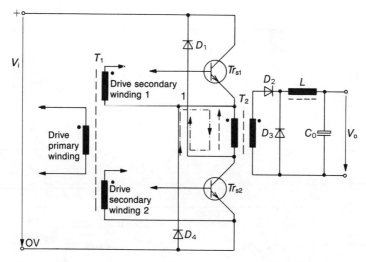

Fig. 5.50 Single-ended forward-converter power supply with two transistors connected in series (mains rectification, drive circuit and protective networks omitted)

subjected to only half the maximum voltage in comparison with the circuit of Figure 5.18, i.e. only V_{imax} instead of $2V_{imax}$. Also the power transformer T_2 require no resetting winding, and is thus simpler. The disadvantage lies in the cost of the two switching transistors (although they are cheaper, because of the lower voltage rating required), the two resetting diodes D_1 and D_4 and the more complicated drive transformer with its two secondary windings together with the components necessary on the secondary side. While Figure 5.50 shows n-p-n bipolar switching transistors, MOSFET power transistors can of course be used, it being necessary merely to use a different transformer-coupled drive circuit. Since many n-channel power MOSFETs have voltage ratings of only 400 to 500 V, this circuit represents a good compromise, permitting the use of such transistors. Alternatively, high-voltage transistors may be used, as in the simple circuit of Figure 5.18, and a greater safety margin is then afforded. Which circuit is ultimately preferred depends also on an economic comparison at the time of manufacture of an equipment.

The circuit operates as follows. When the drive transistor in the primary circuit of the drive transformer turns on, voltages are induced in the secondary windings such that the dotted ends are positive with respect to the undotted ends. The two switching transistors Tr_{s1} and Ts_{s2} turn on simultaneously and primary current flows from $+V_i$ through Tr_{s1}, the primary winding of T_2 and Tr_{s2} to 0 V. In this period a back voltage in induced (dashed) in the primary winding, and—since a forward converter is in question—simultaneously in the same direction on the secondary side. The rectifier diode D_2 conducts, and current is delivered to the output V_0 (see also Figure 5.18). When the drive transistor turns off, voltages are induced in the secondary windings of the drive transformer in the opposite direction, and both switching transistors are turned off. Since an abrupt interruption of the primary current of T_2 would produce a high induced voltage, a path is provided for the decay of this current, and for the magnetization of T_2, through diodes D_1 and D_4 (chain-dotted). The resetting winding and resetting diode can be dispensed with, their function being assumed automatically by the primary winding. See Section 5.3 in connection with the dimensioning of diodes D_1 and D_4; the same applies for

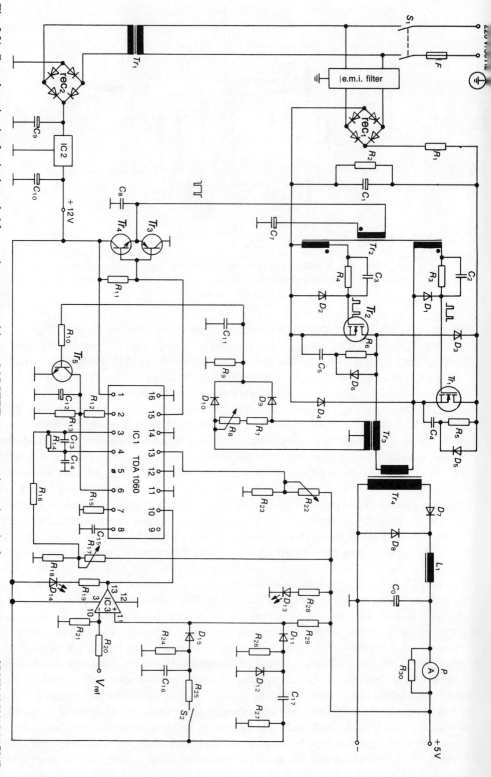

Fig. 5.51 Complete circuit of a single-ended forward converter with two MOSFET power transistors in series for an output power of 75 W (5 V/15 A) [26]

the whole secondary circuit of T_2 including the suppression networks for diodes D_2 and D_3. For diodes D_1 and D_4, the voltage value to be inserted in Equation (5.34) is halved; similarly, as already mentioned, the voltage is to be halved for the purposes of the voltage rating of the switching transistors in Equation (5.38). Everything else remains as discussed in connection with the circuit of Figure 5.18. Drive circuits to be considered (with duplicated suppression networks for the two secondary windings) include those of Figures 5.37 and 5.40 for bipolar switching transistors and those of Figure 5.43 and 5.47 for MOSFET power transistors. It will be evident that RCD suppression networks and the protective network for the primary winding shown in Figure 5.18 are also necessary here.

Figure 5.51 shows the complete circuit of a single-ended forward converter with two power transistors connected in series (in this case MOSFET power transistors, because of the high switching frequency $f = 90\,\text{kHz}$) for an output power of 75 W (5 V at $I_{\text{max}} = 15\,\text{A}$) [26].

As the control IC (IC1) the Valvo Type TDA1060 is used (see also Chapter 8). The supply to the drive IC and the driver stage is provided by a special transformer with a linear regulator (see also Figure 5.59). A driver stage (Tr_3/Tr_4) is introduced, since the two power transistors have to be driven simultaneously. To obtain the correct phase relationship on the secondary side of T_2, the connections of the secondary windings were reversed (see dots). C_7 is provided for the purpose of d.c. blocking for the primary winding of T_2. Since this winding is energized in alternate directions by the drive circuit, no resetting circuit is necessary. Overcurrent monitoring is effected by T_3, so that the voltage on Terminal 6 of IC1 is reduced via Tr_5, causing the duty cycle to the reduced. Switching on the equipment with S_1 would cause Output 13 of IC3 to assume a low potential, which would correspond to a response of the overcurrent indicator diode D_{14}. To prevent this, 'output voltage' is applied briefly through C_{17} (positive potential on Terminal 11 of IC3), so that a high potential is applied to Terminal 10 of IC1, enabling the circuit to start up with a delay owing to C_{12}. When the output voltage has built up, this is indicated by D_{13}. In the event of a short circuit Terminal 6 of IC1 is connected to earth and the output is thereby turned off. A reset pushbutton S_2 is provided to avoid the necessity of switching on again with the mains switch S_1. By this means a voltage input to Terminal 11 of IC3 is simulated afresh. If the output becomes too high as a result of a fault in the control lead, switch-off is effected at Terminal 13 of IC1 by virtue of a response threshold of 0.6 V (for further details see Sections 5.7 and 8.1).

5.7 OVERCURRENT AND SHORT-CIRCUIT DETECTION

Since the switching transistors in a switched-mode power supply are rated only for the maximum current for which the equipment is designed, protective measures must be provided against excessive loading or inadvertent short circuits. Ideally the current sensor which is necessary for this purpose should be located in the secondary circuit. However, this leads to problems, because of the magnitude of the secondary current, which is often very high; it would also be necessary to convey the signal proportional to current to the control IC with d.c. isolation (e.g. through an optocoupler with an amplifier). Since in a forward converter the primary current (apart from the small magnetizing current) is directly proportional to the secondary current (see Equation (5.21)), the current monitoring can more simply be effected on the primary side, where the current is much lower. There is an additional advantage in this, namely that the collector current of the

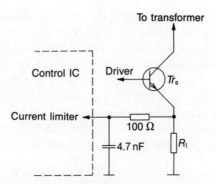

Fig. 5.52 Current limiting circuit on the primary side and direct coupling to the control IC

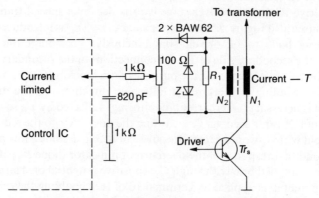

Fig. 5.53 Primary-side current limiting circuit with a current transformer for a single-ended forward converter

switching transistor is monitored directly, and not just the secondary current through the necessary control loop. If no d.c. isolation from the control IC is necessary (this is a question of the way in which the control IC is placed in the circuit), the current monitoring can be effected at little cost as shown in Figure 5.52.

The current-sensing resistor R_1 should be selected so that at 30 to 40 per cent above the maximum nominal current I_{pmax} a voltage is produced across it which causes the current-limiting circuit in the control IC to operate. The response threshold lies between 0.2 V and 5 V, depending on the type.

If d.c. isolation is required, a current transformer must be provided. This usually consists of a small ring core with a single primary turn (the primary conductor is simply passed through the ring) and around 50 to 100 secondary turns. The burden on the secondary side (R_1) is either fixed and predetermined—in which case the secondary voltage is made adjustable—or the burden resistor is itself an adjustable resistor. In single-ended circuits a simple rectifier is sufficient, as shown in Figure 5.53; push–pull circuits require a double-way rectifier as in Figure 5.54.

The RC networks in the monitoring circuit are intended to prevent the circuit from responding to short-duration current spikes, which would otherwise cause the switching transistor to be turned off (time constant about 0.5 to 1 μs). The resetting of the current

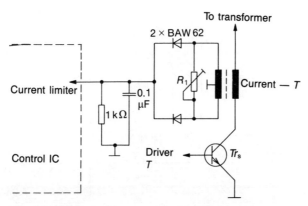

Fig. 5.54 Primary-side current limiting circuit with a current
transformer for a push–pull forward converter

transformer which is necessary in single-ended circuits is effected in Figure 5.53 by means
of the Zener diode Z, with a Zener voltage about 50 per cent above the maximum
secondary voltage, in series with the BAW62 diode. Since the choice of control IC cannot
be presumed, the Zener diode could for example be, for low response voltages, a ZTE1.5
(1.35 to 1.55 V) or a ZPD 6.8 to ZPD 7.5 (6.4/7.2 V to 7.0/7.9 V).

5.8 POWER SUPPLIES FOR CONTROL ICS AND DRIVE CIRCUITS

Since the converter circuits must come into operation before they can produce output
power, in principle the control ICs and the drive circuits cannot initially be supplied from
the converter itself. At least in the starting phase, a special low-voltage power supply must
be available. Whether the supply is subsequently switched over automatically to the
converter output is another question.

In all circuit variants in which the power transistor Tr_s is driven directly, i.e. without
transformer coupling, the control IC and associated drivers can be supplied directly from
the mains. The simplest way of producing a low voltage is to use a Zener diode with the
appropriate Zener voltage. The cheapest way of supplying the Zener diode from the d.c.
input voltage V_i is to connect it through a series resistor; this method does not call for
consideration here, since it entails an excessive power loss. With a maximum direct voltage
of 340 V (see the previously discussed example of a single-ended forward converter), and a
total Zener current of, for example, 30 mA (including the load current that may be drawn),
the series resistance would need to be about 10 kΩ: in the worst case this would lead to a
loss of over 10 W. In a small converter this would result in an unacceptable efficiency,
while for a larger converter a better solution could be afforded. If a capacitive impedance is
used instead of a resistor, however, no power loss arises; Figures 5.55 and 5.56 show two
circuits of this kind.

In these two circuits the always necessary r.f.i. filter has been drawn in, but this has no
special significance here. In Figure 5.55, two similar capacitors $C_1 = C_2 = 1$ to $2\,\mu\text{F}/400\,\text{V}$
are provided. The two resistors $R_1 = R_2 = 180\,\Omega/0.5\,\text{W}$ are only for protective purposes,
and have practically no effect on the normal current. If the capacitors are each $1\,\mu\text{F}$, a
direct current of 25 mA flows in the Zener diode with a mains voltage of 180 V; with the
maximum voltage of 242 V the current rises to 36 mA. The same values are arrived at
theoretically, by calculating the current at a particular voltage based on the reactance of

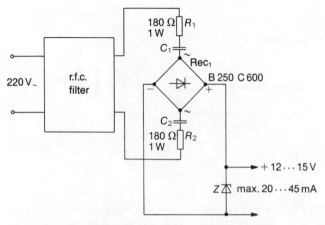

Fig. 5.55 Power supply for the control IC, derived from the mains via two series capacitors with protective resistors and stabilized by a Zener diode

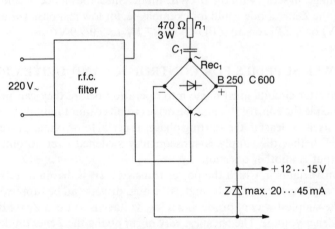

Fig. 5.56 Power supply for the control IC, derived from the mains via one series capacitor with a protective resistor and stabilized by a Zener diode

the 1 μF capacitors at 50 Hz (3.18 kΩ). Since a current of at least 5 mA must be maintained in the Zener diode, this means that there is a current of up to 20 mA available in the worst case to supply the control IC and the drive circuit. If the capacitances are increased to 2 μF, the current is also doubled; the load current can then be taken as 50 mA − 5 mA = 45 mA maximum. Since the mains input voltage is high in comparison with the Zener voltage, the current varies little with Zener voltages between 12 V and 24 V. If the ZX12 Zener diode is selected (for a 12 V output), no external cooling is necessary: the maximum permissible Zener current in an ambient temperature of 45 °C is 86 mA, and the current that flows with the maximum mains voltage of 242 V using 2 μF capacitors is only 70 mA. With a Zener voltage of 24 V the permissible current is only 45 mA (sufficient with 1 μF capacitors) without additional cooling, but as high as 345 mA with an aluminium cooling fin 12.5 × 12.5 cm, 2 mm thick. The use of still larger capacitors is possible in principle, but not

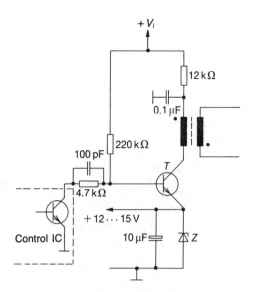

Fig. 5.57 Power supply for the control IC obtained from the emitter of the switching transistor in a flyback converter by means of a Zener diode

worthwhile. The series connection of two capacitors reduces the likelihood of a possible breakdown. If one capacitor is short-circuited, the result is merely to double the current— otherwise there is no effect. If it is desirable to save cost, one capacitor and one resistor can be omitted, as in Figure 5.56. To ensure high reliability in this case also, a self-healing type of capacitor should be used. The capacitor C_1 needs to be of half the capacitance of those in Figure 5.55. Each of these circuits draws practically the same current loaded and unloaded, since any current not taken by the load flows in the Zener diode. Another simple power supply is shown in Figure 5.57.

This circuit is a modification of the drive circuit of Figure 5.37. The secondary side of the drive transformer, not shown, can be dimensioned in the same way as in Figure 5.37, but whereas the supply voltage in Figure 5.37 is $+24$ V (it could be lower, with a different transformation ratio) the supply for the driver stage in this case is the input V_i. The current supplied cannot be very large if the overall efficiency is not to be reduced to much, but a large current is quite unnecessary, since this power supply does not have to supply the driver stage. As in Figure 5.37, the drive circuit is here designed as a flyback converter. As previously explained, this results in shorter switching times, but in this case a flyback converter is the only possible choice, since V_i builds up immediately when the converter is switched on. The transistor Tr is supplied with base current through the 220 kΩ base resistor, and a collector current flows. This is turn produces an emitter voltage, corresponding to the Zener voltage of the Zener diode (about 12 to 15 V). The supply obtained here can then be applied to the control circuit. The available current is equal to the mean transistor current, less a few mA for the Zener diode.

The three power-supply circuits for control ICs and drive circuits so far discussed have all incorporated Zener diodes. In such circuits, the current consumed is practically independent of the output current, any current not required flowing in the Zener diode. In power-supply circuits which draw much less current on no load than on full load, it may be worthwhile to take the required current from the converter itself, once this has started up. This has the advantage that there is always some load on the converter, even when no load is present on the secondary side. Figure 5.58 shows a simple stabilizing circuit using an amplifier transistor.

According to [22], resistor R_1 must be so determined that at $V_{imin} = 212$ V sufficient current flows in the Zener diode ($I_{zmin} = 5$ mA) and the base current I_B is adequate with the maximum collector current I_{Cmax}. Since this circuit produces excessive losses with too high a load current, the maximum load current should be limited to, say, 20 mA. A transistor with a voltage rating of at least V_{imax}, i.e. about 340 V, is required at a collector current of about 30 to 50 mA; suitable types are BD232, BUX86 etc. R_2 is chosen so that with the maximum load current of, for example 20 mA, collector-emitter voltage of Tr is only a few volts. R_2 is thus given by

$$R_2 \leqslant \frac{V_{imin} - V_z}{I_{Cmax}} \quad (\Omega) \tag{5.48}$$

For $V_{imin} = 212$ V, $V_z = 12$ V and $I_{Cmax} = 20$ mA,

$$R_2 \leqslant \frac{212\,\text{V} - 12\,\text{V}}{20 \times 10^{-3}\,\text{A}} \leqslant 10\,\text{k}\Omega/6\,\text{W}$$

The maximum power loss at V_{imin} is

$$P_{R2} = \frac{V^2}{R_2} = \frac{200^2\,V^2}{10 \times 10^3\,\Omega} = 4\,\text{W}$$

There is hardly any loss in the transistor at maximum current and V_{imin}, because the collector-emitter voltage is then very low. Neither is there any loss on no-load, since there is then no collector current. The maximum loss results from maximum current and maximum input voltage V_{imax}:

$$P_{Trmax} = (V_{imax} - I_{Cmax}R_2)\,I_{Cmax} \quad (\text{W}) \tag{5.49}$$
$$P_{Trmax} = (330\,\text{V} - 20 \times 10^{-3}\,\text{A} \times 10 \times 10^3\,\Omega) \times 20 \times 10^{-3}\,\text{A}$$
$$= 2.6\,\text{W}$$

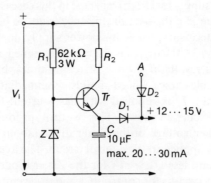

Fig. 5.58 Power supply for the control IC obtained from the input voltage via a stabilizing circuit (Zener diode and transistor)

The thermal resistance required of the cooling fin can be calculated from Equation (2.45). Equation (2.45):

$$R_{thca} \leqslant \frac{\theta_{jmax} - \theta_{amax}}{W} - R_{thjc}$$

With $\theta_{jmax} = 125\,°C$ (data sheet), $\theta_{amax} = $ e.g. $50\,°C$ and $R_{thjc} = 4.5\,kW$ (data sheet),

$$R_{thca} \leqslant \frac{125° - 50°}{2.6\,W} - 4.5\,kW \leqslant 24\,kW$$

This high thermal resistance can be obtained with, for example, an aluminium sheet $40 \times 40 \times 0.5\,mm$, finished matt black [22].

The Zener diode can be a low-power type, e.g. ZPY13, since the transistor functions as an amplifier. To allow for the voltage drop of about $0.7\,V$ in the base-emitter path of Tr, the Zener voltage must be increased by this amount. Thus for an output of about $12\,V$ a Zener voltage of $13\,V$ is suitable. The series resistance R_2 for the Zener diode is calculated, according to [22], from

$$R \leqslant \frac{V_{imin} - V_{zmax}}{I_{Bmax} + I_{zmin}} \quad (\Omega) \tag{5.50}$$

and

$$R \geqslant \frac{V_{imax} - V_{zmin}}{I_{Bmin} + I_{zmax}} \quad (\Omega) \tag{5.51}$$

where V_{imin} is the minimum input voltage (V), V_{imax} is the maximum input voltage (V), I_{Bmax} is the maximum base current (A), I_{Bmin} is the minimum base current (A) (in many cases zero) and I_{zmin} is the minimum required Zener current (A) (usually 2 to 5 mA).

For the circuit of Figure 5.58, with $V_{imin} = 212\,V$, $V_{imax} = 340\,V$; $V_{zmax} = 14.1\,V$, $V_{zmin} = 12.4\,V$ (from the data sheet for the ZPY13) at $I_{zmin} = 5\,mA$ and $I_{Bmax} = 1\,mA$ ($I_{Bmin} = 0$) from Equations (5.50) and (5.51); $I_{zmax} = 78\,mA$,

$$R = R_1 \leqslant \frac{212\,V - 14.1\,V}{1\,mA + 2\,mA} \leqslant 66\,k\Omega$$

$$R = R_1 \geqslant \frac{340\,V - 12.4\,V}{0 + 78\,mA} \geqslant 4.2\,k\Omega$$

selected, $R_1 = 47\,k\Omega/4\,W$

Maximum loss in R_1:

$$P_R = \frac{(340\,V - 13\,V)^2}{47 \times 10^3\,\Omega} = 2.3\,W$$

If the power supply for the control circuit and the driver is to be taken from the converter itself once the converter output has built up, a slightly higher voltage must be applied to Terminal A in Figure 5.58 (about 1 to 2 V above the output voltage of the start-up circuit). D_2 then conducts, while D_1 blocks, so that the current is drawn from the converter output. The converter is thus provided with a desirable minimum residual load and the start-up circuit draws only the low current in the Zener diode.

If the transistor Tr in Figure 5.58 is replaced by a Darlington transistor, output currents higher by a factor of 2 to 4 can be drawn without requiring a significant base current

148

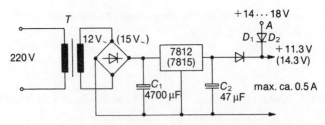

Fig. 5.59 Power supply for the control IC and driver stage using an auxiliary transformer, rectification and stabilization by means of a linear regulator

(considerably less than 1 mA). The minimum Zener current can also be reduced to about 1 mA. This would increase R_1 to around 100 to 150 kΩ and correspondingly reduce the (continuous) loss in R_1. An especially high degree of stabilization is not a primary consideration here. The Zener diode must merely limit the output voltage to approximately V_z. In any case the start-up circuit is disconnected after a very short period in operation. Neither is it necessary to dimension R_2 strictly in accordance with Equation (5.48); if it is made smaller, the loss in R_2 is lower, and that in the transistor Tr higher, but since the circuit operates for only a few seconds, these considerations are not critical.

If somewhat higher currents are required, in particular for a more powerful drive circuit, the most satisfactory solution is to use a start-up circuit such as that of Figure 5.59.

Since the transformer T produces a low voltage directly, this is one of the best circuits for efficiency. From this point of view it is satisfactory if the nominal alternating voltage is approximately equal to the d.c. output voltage of the 78XX linear regulator. Since these regulators require only a small difference between the input and output voltages, the necessary minimum difference is obtained even with low mains voltage, provided that a relatively large (electrolytic) input capacitor is used, so that the input voltage contains very little ripple. If a smaller value is chosen for C_1, the a.c. input voltage must be increased by about 1 V. The maximum secondary current rating of the transformer should be about twice the maximum direct current to be drawn at the output [22]. This circuit, like the previous ones, may be decoupled by means of the two diodes D_1 and D_2 when the converter has started up. No special requirements attach to these diodes: they should be able to block a reverse voltage of about 50 V and should be rated for a maximum forward current of a few hundred mA. Compared with the circuits previously discussed, that of Figure 5.59 has the great advantage that mains isolation is afforded by the transformer T. The reference earth for the control IC and the drive circuit can therefore be chosen without restriction. With the power supply circuits tied to the mains as in Figures 5.55 to 5.58, the control signal has to be transmitted from the output to the control IC through an optocoupler. With the circuit of Figure 5.59, on the other hand, the output voltage can be applied directly to the control circuit; it is of course necessary to provide mains isolation in the drive circuit by using transformer coupling to the switching transistor. In the case of a *flyback* converter, the control IC, together with the drive circuit, can be supplied directly from the converter output, assuming the voltage to be sufficient (shown dashed in Figure 5.60). Otherwise a separate winding with a rectifier and reservoir capacitor must be provided for this supply. Direct coupling to the output, which must be isolated from the mains, is only possible, however, if the start-up circuit is also isolated from the mains; the choice is then restricted to the circuit of Figure 5.59. With a separate winding (N_3 in

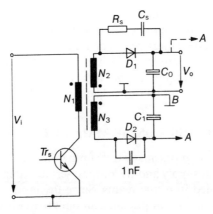

Fig. 5.60 Power supply for the control
IC and driver stage obtained from an
additional winding on the power transformer in a flyback converter

Figure 5.60), any of the start-up circuits discussed may be used, since the required isolation
is provided by the extra winding. Which solution is appropriate in a given complete circuit
is a question of cost, and must be assessed individually for each requirement.

In a *forward converter*, as in the flyback converter, the power supply can be taken
directly from the output if the output voltage V_0 is high enough. If it is too high, it can be
reduced to the required level by means of a linear regulator (chain-dotted in Figure 5.61).
Here again the question of potential relationships has to be considered, since the output
must be isolated from the mains. An elegant solution, which affords the necessary
isolation, is to obtain the power supply from a second winding on the inductor L. Since
this inductor carries not only the direct current I_0 but also a superimposed alternating
current (of triangular waveform) (see Chapter 2), a voltage of suitable magnitude can be

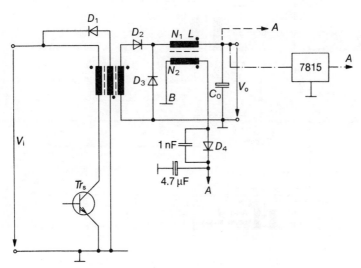

Fig. 5.61 Power supply for the control IC and driver stage obtained
from an additional winding on the inductor in a forward converter

induced in a secondary winding, and this can be used, after rectification, to supply the control IC together with the drive circuit. The diode D_4 used to rectify the auxiliary supply must be a fast-switching type (short recovery time), since it operates at the switching frequency f. The voltage rating needs to be about twice the magnitude of the auxiliary voltage; 100 V should meet all requirements. High currents are not required, and a current rating I_{FAV} of 0.5 A is sufficient in all cases. There is an adequate choice of such diodes.

5.9 PROBLEMS OF MAINS ISOLATION

While in general there are no problems in transformer-coupled converters with regard to isolation between the input supply and the output, serious attention must be paid to this point in converters supplied from the mains. Since the insulation voltage between the mains (usually 220 V a.c.) and the output must amount to at least 2.5 kV, a high degree of isolation has to be provided. An arrangement frequently used is represented in Figure 5.62.

Here the control IC is connected directly to the output, which represents the most convenient arrangement for the control of the output. This means that the control IC, together with the driver stage, must be supplied through a special auxiliary power supply incorporating mains isolation (see Figure 5.59); in addition, the driver stage must be isolated from the switching transistor Tr_s by a pulse transformer.

Alternatively, there may be direct coupling between the control IC, the driver stage and the switching transistor, as in Figure 5.63. In this case the auxiliary power supply does not

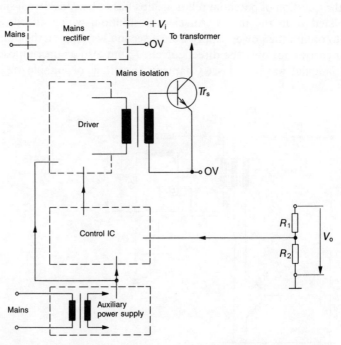

Fig. 5.62 Control IC connected directly to the output; mains isolation is necessary in the power supply to the IC and in the driver stage

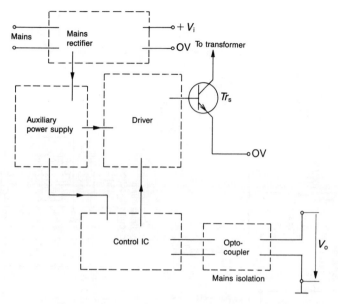

Fig. 5.63 Direct-coupled control and drive circuit and auxiliary power supply without mains isolation; isolation is provided by a optocoupler

have to be isolated from the mains (see Figures 5.55 to 5.58). The required mains isolation is then provided by an optocoupler with an adequate insulation voltage. An accurately stabilized output voltage can be obtained by either of the methods shown in Figures 5.62 and 5.63.

A further method of mains isolation, frequently used in flyback converters, is by means of an auxiliary winding on the mains transformer (e.g. N_3 in Figure 5.60). The necessary isolation is automatically provided by the main transformer. This solution is the cheapest, but it introduces larger control errors due to the unavoidable leakage reactance of the transformer.

5.10 CIRCUIT EXAMPLES

220 W forward converter, Figure 5.64 [28]

This is a single-ended forward converter for mains operation with bipolar switching transistor Type BUX82 and flyback-converter driver stage providing matched base current; power transistor operated in saturated mode; RCD protective network. Short-circuit protection with current transformer on primary side. Auxiliary power supply 24 V with mains isolation as in Figure 5.59. Switching frequency $f = 50\,\text{kHz}$. Control IC connected to output. The maximum duty cycle δ_{Tmax} is limited to 0.37 by the network connected to Terminal 2 of the TDA1060 control IC. The gain of the error voltage amplifier (R_{16}/R_{17}) is about 100. C_{15} limits the initial rate of increase of the duty cycle. To provide an overshoot of base current at turn-on, C_1 is charged to 24 V during the non-conducting phase of Tr_1; the collector current of this transistor is limited by R_2. C_4 also effects a powerful base-current overshoot. The output diodes D_4 and D_7 are Schottky diodes, or a double Schottky diode. Overvoltages owing to reverse blocking of the diodes

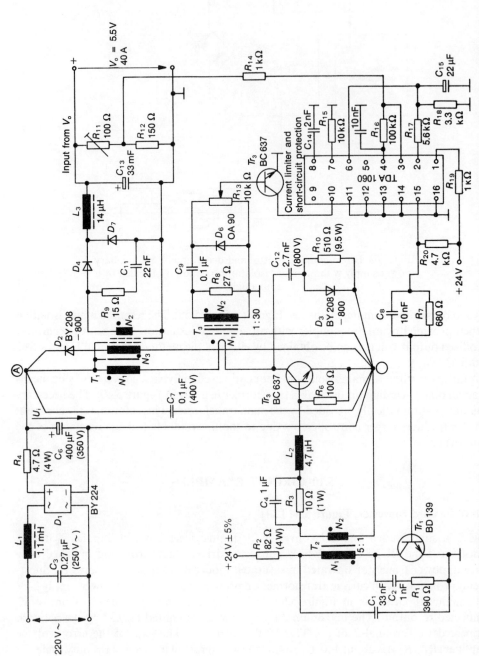

Fig. 5.64 Mains-supplied single-ended forward converter for $V_o = 5.5$ V at $I_{0max} = 40$ A using the BUX82 power transistor and the TDA1060 control IC (Valvo)

are prevented by an RC suppression network. EMI suppression is effected by the input filter C_3/L_1, capacitor C_7 and a double screen in the transformer T_1. Other measures against interference are: shortest possible connections between points A and B; BUX82 power transistor insulated from the heatsink; core of T_1 connected to $+V_1$.

Transformer T_1:   Core EC52/24/14 (2x)-3C8 (N27) without air gap.
                        $N_1 = N_{1a} + N_{1b}$: 72 turns 0.5 mm CuL (2 layers, $L_1 = 12$ mH)
                        N_2: 5 turns 23.5 × 0.2 mm copper strip
                        N_3: 72 turns 0.25 mm CuL (1 layer); screen 23 × 0.05 mm copper
                        Core U30/50/16-3C8 (total gap 5.5 mm) (2x) Winding 13 turns 25 × 0.5 mm copper strip

Drive transformer T_2:   Core U15/22/6-3C8 without air gap (2x)
                        N_1: 55 turns 0.18 mm CuL
                        N_2: 11 turns 0.35 mm CuL

Current transformer T_3: Core U15/22/6-3C8
                        N_1: 1 turn 1 mm CuL
                        N_2: 20 turns 0.22 mm CuL

Performance figures for the circuit of Figure 5.64: Mains voltage 220 V~ + 10 per cent, − 15 per cent; $V_0 = 5.5$ V at $I_{0max} = 40$ A Variation of V_0 over stated mains voltage range at I_{0max}:

$$\Delta V_0 \leqslant 30 \text{ mV (0.6 per cent)}$$

Variation of V_0 at nominal mains voltage with $I_0 = 2$ to 40 A:

$$\Delta V_0 \leqslant 20 \text{ mV (0.4 per cent)}$$

Variation of V_0 with a step-load change $\Delta I_0 = 36$ A:

$$\Delta V_0 \leqslant 0.4 \text{ V (8 per cent)}$$

Recovery time following a step-load change $\Delta I_0 = 36$ A:

$$t_{tr} \leqslant 250 \ \mu s$$

Open-circuit and short-circuit protection; efficiency ≈ 75 per cent.

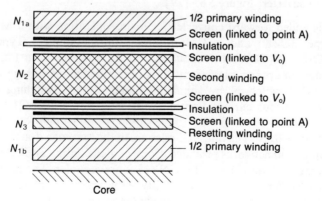

Fig. 5.65 Winding arrangement of the transformer T_1 in the circuit of Figure 5.64 (Valvo)

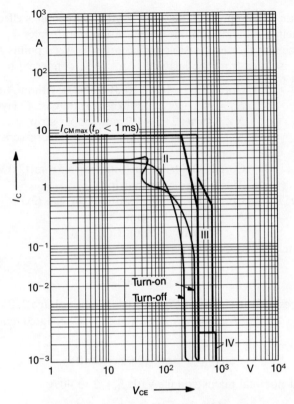

Fig. 5.66 Turn-on and turn-off loci for the circuit of
Figure 5.64 superimposed on the SOAR diagram for the
BUX82 switching transistor (Valvo)

Figure 5.66 shows the SOAR diagram for the BUX82 power transistor with the switching locus marked on it. It can be seen clearly that the turn-on loss (with a higher collector voltage) is larger than the turn-off loss.

100 W forward converter, Figure 5.67 [29]

This is a single-ended forward converter for mains operation with a MOSFET power transistor Type BUZ80 and a CMOS driver (direct-coupled) Type 4049 B (see Figure 5.42). Drive IC TDA4718 supplied direct from mains, as in Figure 5.55; mains isolation through an optocoupler; monitoring of rectified input voltage V_i for over- and undervoltage, with switch-off by simulated undervoltage. Switching frequency $f = 50\,\text{kHz}$; dynamic output current limiting with set point at $I_0 = 21\,\text{A}$ and $I_k = 25\,\text{A}$ (Figure 5.68). Feed-forward for mains ripple suppression through R_R; soft start through C_A (start-up delay $\approx 0.3\,\text{s}$). The output diodes are in the form of a double Schottky diode; protective suppression networks provided. Control of the output is effected by means of a TAA761A operational amplifier and a CNY17-2 optocoupler. The reference voltage is obtained from a 3 V Zener diode. If the stability of this reference voltage is not acceptable a band-gap reference must be used to provide better constancy.

Interference suppression is provided by an e.m.i. filter in the mains lead and a screen in the transformer.

Single-ended SNT 220 V~ − 5 V/20 A/50 kHz
(for 117V ~/220V~ mains input from Figure 8)

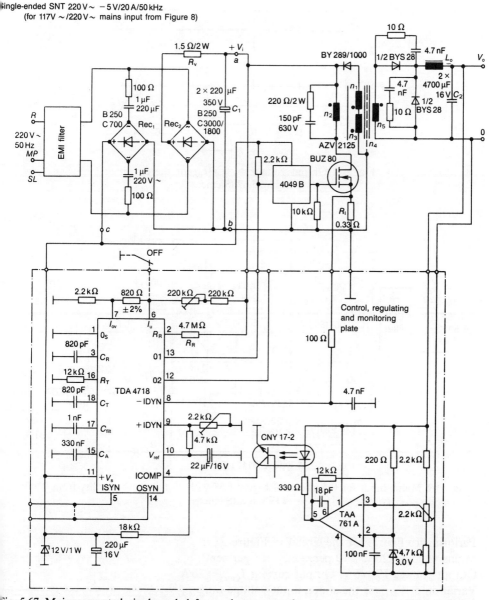

Fig. 5.67 Mains-operated single-ended forward converter for $V_0 = 5$ V at $I_{0max} = 20$ A using the BUZ80 MOSFET power transistor and the TDA4718 control IC (Siemens)

Transformer: Type AZV2125 (Siemens): Core ER42/15
$$N_1 + N_3 = N_2$$

Inductor L_0: Pot core CC36-N27; $A_L = 150$ nH with $s = 4$ mm.
Winding: 12 turns $3 \times 120 \times 0.1$ mm CuL

If a facility for an alternative 117 V mains input is required, the input circuit is provided with a split input capacitor C_1 (becoming C_1 and C_2) to form a voltage doubler, as shown in Figure 5.69. If necessary the $1\,\mu$F capacitors may be doubled in value in accordance with rectifier Gl.2.

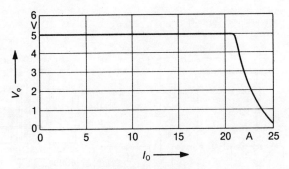

Fig. 5.68 Output voltage $V_0 = f(I_0)$ for the circuit of Figure 5.67 (Siemens)

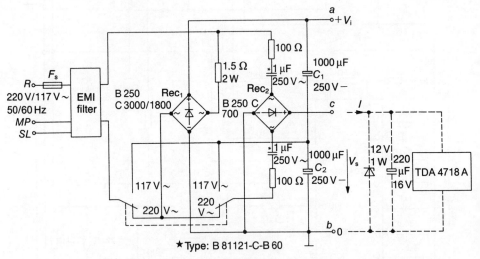

Fig. 5.69 Mains input for the circuit of Figure 5.57 with a changeover facility from 220 V to 117 V a.c. (Siemens)

Performance figures for the circuit of Figure 5.67:
Mains voltage 220 V + 10 per cent, − 15 per cent
Output voltage $V_0 = 5$ V; Output current $I_{0max} = 20$ A
Variation of V_0 over the stated input voltage range:

$$\Delta V_0 \leqslant 5 \text{ mV (0.1 per cent)}$$

Variation of V_0 with $I_0 = 0$ to 20 A:

$$\Delta V_0 \leqslant 20 \text{ mV (0.4 per cent)}$$

Variation of V_0 with a step-load change $\Delta I_0 = 18$ A:

$$\Delta V_0 \leqslant 0.2 \text{ V (4 per cent)}$$

Recovery time following a step-load change $\Delta I_0 = 18$ A:

$$t_{tr} \leqslant 3 \text{ ms}$$

Open-circuit and short-circuit protection; efficiency ≈ 80 per cent. Output voltage ripple (50 kHz):

$$\Delta V_0 \leqslant 40\,\mathrm{mV_{pp}}$$

100 W forward converter with multiple outputs, Figure 5.70 [30]

Apart from the multiple windings on the transformer and the inductor, this circuit is identical with that of Figure 5.67, and needs no further discussion.

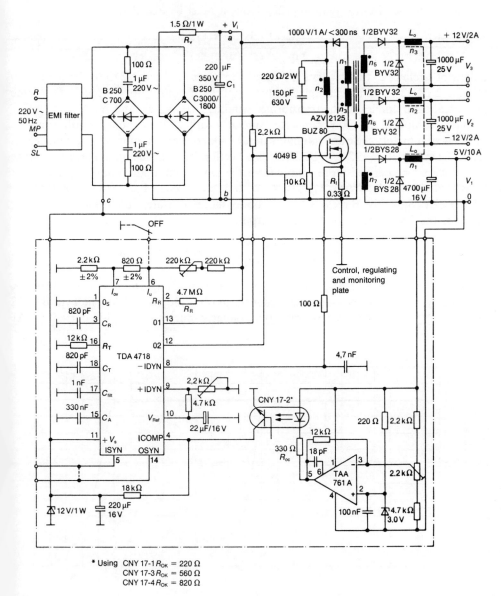

* Using CNY 17-1 $R_{OK} = 220\,\Omega$
CNY 17-3 $R_{OK} = 560\,\Omega$
CNY 17-4 $R_{OK} = 820\,\Omega$

Fig. 5.70 Mains-operated single-ended forward conberter for $V_1 = 5\,\mathrm{V}$ at $I_{1max} = 10\,\mathrm{A}$, $V_2 = -12\,\mathrm{V}$ at $I_{2max} = -2\,\mathrm{A}$ and $V_3 = 12\,\mathrm{V}$ at $I_{3max} = 2\,\mathrm{A}$ using the BUZ80 MOSFET power transistor and the TDA4718 control IC (Siemens)

While the total output power of a flyback converter can be divided between any number of outputs, in the case of a forward converter this is possible only by means of a trick, because of the dependence of the duty cycle upon the inductor voltage. Here again, in principle, only one output can be regulated; usually it is the 5 V output. The characteristics in regard to this output correspond to those of the circuit of Figure 5.67, except that the maximum current is limited to 10 A, since the other two outputs account for the same output power. The transformation ratio for the 5 V output is calculated in the usual way according to Equation (5.19). The number of secondary turns for the regulated output (N_7) is determined from Equation (5.20). Now, it must be true that

$$\frac{N_{2L}}{N_{1L}} = \frac{V_2 + V_{D2} + V_{L2}}{V_1 + V_{D1} + V_{L1}} \tag{5.52}$$

Subscript 1 relates to the (regulated) Output 1 (5 V), and subscript 2 to the (unregulated) Output 2 (-12 V). A similar relationship applies, of course, to Output 3. The same numerical ratio applies to the transformer windings, i.e. N_6/N_7 and N_5/N_7.

In the example of the 300 W converter, similarly with a 5 V secondary winding and the same maximum duty cycle and minimum input voltage, the maximum reverse voltage was found, from Equation (5.24), to be 20.1 V (see Section 5.3). The maximum reverse voltage can, however, be derived more simply from the maximum input voltage V_{imax} and the transformation ratio. With $V_{imax} = 340$ V and $n = 0.0603$, $V_{Rmax} = 340$ V $\times$ 0.0603 + 0.5 V = 21 V. Since the BYS28 Schottky diode (more accurately BYS28-45), this is suitable for the 5 V output. For the 12 V output,

$$V_{R12} = V_{R5} \times \frac{12\,V + 0.5\,V + 0.2\,V}{5\,V + 0.5\,V + 0.2\,V} = 21\,V \times 2.23 = 46.8\,V$$

In this case the BYS28-45 would be too closely rated in voltage; the BYS28-90 would be satisfactory, but greatly over-dimensioned in terms of current (2×20 A). The Type BYV32 (BYV32/100 with a 100 V reverse rating) is similarly much over-rated in current (20 A).

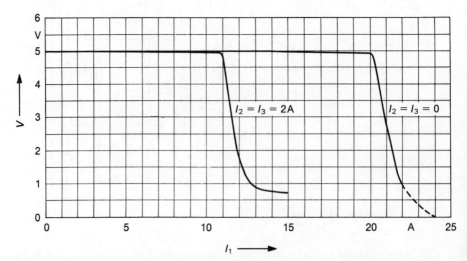

Fig. 5.71 Output voltage $V_1 = f(I_1)$ with $I_2 = I_3 = 2\,\text{A}$ and $I_2 = I_3 = 0$ for the circuit of Figure 5.70 (Siemens)

The BYS24-90 Schottky diode is therefore suggested as a better solution, with $I_{FAV} = 2 \times 5\,A$ and $V_{Rmax} = 90\,V$.

Since the short-circuit current limiting is effected on the primary side (at the $0.33\,\Omega$ resistor), the maximum current in the 5 V output is limited to about 11 A if both the other outputs are fully loaded (the current limiting must be so adjusted by means of the $2.2\,k\Omega$ potentiometer on Terminal 9 of the TDA4718). If the other two outputs are unloaded, however, the limiting current in the 5 V output is approximately doubled, as shown in Figure 5.71.

While the regulated 5 V output maintains very good constancy from 0 to 10 A, this cannot be true of the other two outputs. With increasing load on Output 1, the duty cycle becomes somewhat higher to compensate for the voltage drops in the secondary circuit. Hence the voltages of the other two outputs increase, as shown in Figure 5.72. Connecting in parallel a (cooled) Zener diode Type ZX13 holds the two 12 V outputs to a maximum of 12.4 to 14.1 V according to the Zener voltage tolerance. With the circuit arranged as in Figure 5.70 ($R_T = 12\,k\Omega$) the maximum duty cycle $\delta_{Tmax} = 0.43$ (see Chapter 8). In the worst case (with light load on the two 12 V outputs, when current limiting does not occur until $I_1 = 20\,A$) it can rise to nearly 0.5. This would result in some cases in an excessive voltage on the 12 V outputs. Transformer Type AZV2138 (Siemens):

Inductor L: Pot core CC36-N27: $A_L = 250\,nH$ with $s = 2\,mm$

                Winding $N_1 = 14$ turns $2 \times 1\,mm$ CuL, insulation $1 \times 0.05\,mm$

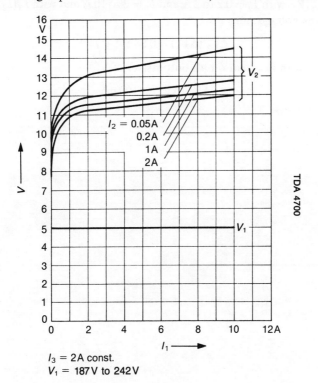

$I_3 = 2A$ const.
$V_1 = 187\,V$ to $242\,V$

Fig. 5.72 Variation of output voltages V_1 and V_2 with output current I_1, with I_2 as parameter and $I_3 = 2\,A$, for the circuit of Figure 5.70 over the full mains voltage range (Siemens)

Winding $N_2 = N_3 = 32$ turns $2 \times 0.5\,\text{mm}$ CuL, insulation each $1 \times 0.05\,\text{mm}$

Performance figures for the circuit of Figure 5.70:
Mains voltage $220\,\text{V} + 10$ per cent, -15 per cent
Output voltages:

$$V_1 = 5\,\text{V at } I_{1\text{max}} = 10\,\text{A}; \quad V_3 = 12\,\text{V at } I_{3\text{max}} = 2\,\text{A};$$
$$V_2 = -12\,\text{V at } I_{2\text{max}} = -2\,\text{A}$$

Variation of V_1 over the stated mains voltage range with full load on all outputs:

$$\Delta V_1 = 20\,\text{mV (60 per cent)}$$

Variation of V_1 with $I_1 = 6\,\text{A}$ and $I_2 = I_3 = 0$ to $2\,\text{A}$:

$$\Delta V_1 = 70\,\text{mV (1.4 per cent)}$$

Ripple (50 kHz) with full load on all outputs:

$$\Delta V_1 = 40\,\text{mV}_{\text{pp}}$$

Output current $I_1 = 2$ to $10\,\text{A}$ (0 to $10\,\text{A}$); $I_{\text{sc}} = 24\,\text{A}$ with $I_2 = I_3 = 0$

Output voltage $V_2\,(V_3) = 11.9\,\text{V}$ with $I_1 = 6\,\text{A}$ and $I_2\,(I_3) = 2\,\text{A}$
Variation of $V_2\,(V_3)$ with $I_2 = 0.2$ to $2\,\text{A}$ and $I_1 = 2$ to $10\,\text{A}$ and with $I_3\,(I_2) = 0$ to $2\,\text{A}$ over the stated mains voltage range:

$$\Delta V_2\,(\Delta V_3) = 0.84\,\text{V}\ (\pm 7\%)$$

Ripple (50 kHz) with full load on all outputs:

$$\Delta V_2\,(\Delta V_3) = 100\,\text{mV}$$

6 Push–pull and bridge converters for mains operation

The simple push–pull converter of Figure 1.9, the half-bridge converter of Figure 1.10 and the full-bridge converter of Figure 1.11 all operate with the power transistors conducting alternately. All these converters are in principle forward converters (see also Figure 1.1).

In all push–pull converters precautions must be taken to ensure symmetry of operation, so that the power transformer is not subject to d.c. magnetization and possible saturation (see Section 8.2).

Since it would exceed the scope of this book to discuss every push–pull circuit, consideration will be restricted to the half-bridge push–pull forward converter, albeit in bipolar and MOSFET versions.

Half-bridge push–pull forward converter

A block diagram of the half-bridge forward converter is shown in Figure 6.1.

The mains supply is fed to a bridge rectifier via an e.m.i. filter in the mains lead. Half the rectified input voltage appears across each of the two series-connected high-voltage electrolytic capacitors. The input voltage is switched alternately by the transistors Tr_1 and Tr_2 and applied to the primary winding of T_1. A dynamic current-limiting signal is produced by a current transformer in series with the primary winding. On the secondary side of T_1, bi-phase rectification is effected by two diodes which function both as rectifiers and as free-wheel diodes. As a result of the push–pull operation the frequency of the inductor current is twice the switching frequency. In addition to the dynamic current limiting on the primary side, static current limiting is provided in the output circuit. The auxiliary power supply to the TDA4700 control circuit, including the driver stages, which is necessary for starting, is obtained initially from an auxiliary transformer T_2, a changeover being effected to a supply obtained from the inductor once the converter has started up. This provides a small residual load at all times, and the problematical no-load condition is avoided.

6.1 MAINS RECTIFICATION

The first step, as in Section 5.1, is to determine the smoothing capacitor C_i (Figure 6.2: C_1 as in Figure 5.1). Since the converter is of the half-bridge type, the smoothing capacitor has to consist of at least two units in series; in most cases a number of capacitors are connected in series-parallel. The complete power supply to the converter is shown in Figure 6.2.

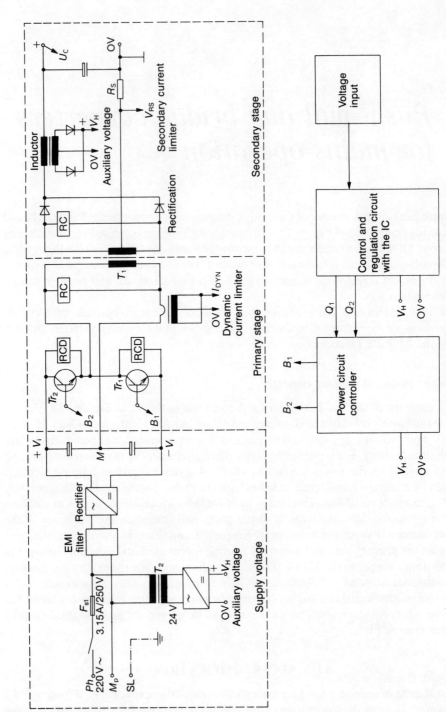

Fig. 6.1 Block diagram of a half-bridge push–pull converter

In order to determine the magnitude of C_i it is necessary to know the input power. This in turn is derived from the maximum output power divided by the efficiency η (for a preliminary estimate the efficiency is usually assumed to be 75 per cent).

Example [27]

A push–pull forward-converter power supply is to meet the following requirements:

$V_i = 220\,V_\sim + 10$ per cent, -15 per cent; $a = 1$ (loss of one mains half-cycle)

$V_0 = 12\,V \pm 10$ per cent; adjustable to $12\,V \pm 10$ per cent $= 10.8$ to $13.2\,V$

$I_{0max} = 15\,A$; $I_{0min} = 1.5\,A$; max. step change $\Delta I_{0max} = 0.5 I_{0max}$

Maximum output voltage variation due to ΔI_{0max}: $0.1\,V_0 = 1.2\,V$

Maximum permissible output ripple: $\Delta V_0 = 100\,mV$

Switching frequency (bipolar version) $f = 40\,kHz$

(MOSFET version) $f = 60\,kHz$

Maximum ambient temperature: $\theta_{amax} = 45\,°C$

(It would of course have been possible to conceive this converter, with $P_{0max} = 13.2\,V \times 15\,A \approx 200\,W$, without any further consideration as a single-ended forward converter in accordance with Figure 1.1, but neither the components nor the measuring equipment for a much larger push–pull converter were available, and the special features of a push–pull half-bridge converter can be realized even at this power level.)

The input power is given by

$$P_i = \frac{13.2\,V \times 15\,A}{0.75} = 264\,W \approx 265\,W$$

The customary dimensioning of $C_i / P_i = 1.5$ to $2\,\mu F/W$ gives

$$C_i = (1.5 \text{ to } 2)\,\mu F/W \times 265\,W = 398 \text{ to } 530\,\mu F$$

The maximum possible direct input voltage with the mains 10 per cent high is

$$V_{imax} = 220\,V \times 1.1 \times \sqrt{2} - 2V_F = 342\,V - 2\,V = 340\,V$$

Since the electrolytic capacitors in Figure 6.2 are subjected to only half the input voltage, the maximum voltage applied is $170\,V$. The requirement is thus for units of 400 to $500\,\mu F/200$ to $250\,V$.

If only two capacitors are used, they must be of twice the capacitance, because of the series connection, i.e. 800 to $1000\,\mu F$. These, however, are of an inconvenient overall height from the constructional point of view. If four units are chosen from Figure 5.10, the only type to be considered is $430\,\mu F/250\,V$, but since $220\,\mu F/250\,V$ capacitors (from another series) were readily available (B43 564 series from Siemens), eight of these were used, giving $C_i = 440\,\mu F$. The data for these units is: $C = 220\,\mu F/250\,V$; $I_{C\sim max}$ at $\theta_a = 85\,°C = 0.77\,A$; correction factor for $I_{C\sim}$ (as in Figure 5.11) at $\theta_a = 45\,°C = 1.75$; I_{Rmax} at $\theta_a = 20\,°C = 0.11\,mA$; correction factor for I_R at $\theta_a = 45\,°C$ from Figure 6.3 $= 2.6$; $ESR = 0.8\,\Omega$ per capacitor; $ESR_{tot} = (0.8\,\Omega \times 2)/4 = 0.4\,\Omega$.

The four parallel-connected capacitors will thus withstand an alternating current $I_{C\sim} = 4 \times 0.77\,A = 5.4\,A$.

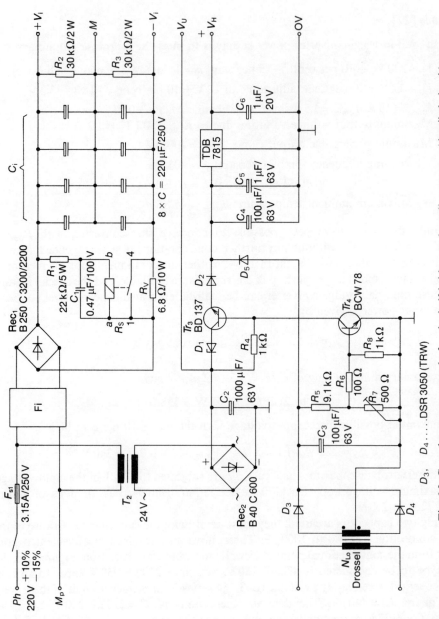

Fig. 6.2 Complete power supply for the half-bridge push–pull converter (detailed circuit)

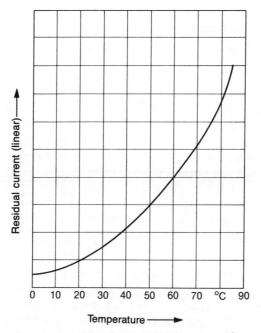

Fig. 6.3 Variation of the residual current of an electrolytic capacitor with temperature (relative to $\theta_a = 20\,°C$) (Siemens)

The maximum possible residual current at $\theta_a = 45\,°C$ is

$$I_{Rmax} = 0.11 \times 10^{-3}\,A \times 4 \times 2.6 = 1.14\,mA$$

Since a direct voltage cannot be divided by series-connected (ideal) capacitors, the division of voltage must be forced by parallel-connected resistors of a suitable value. The current in the resistors should on the one hand be as small as possible from the point of view of losses, but on the other should be large in comparison with the maximum possible residual current of the capacitors. If a factor of five is chosen as a compromise, then the current through $R_2 = R_3$ must be $5 \times 1.14\,mA = 5.7\,mA$. This gives, for the two resistors,

$$R_2 = R_3 = \frac{170\,V}{5.7 \times 10^{-3}\,A} = 29.8\,k\Omega$$

selected, $R_2 = R_3 = 30\,k\Omega/2\,W$

$$\text{Loss in } R_2/R_3 : P = \frac{(170\,V)^2}{30 \times 10^3\,\Omega} = 0.96\,W;$$

selected, $W = 2\,W$.

The determined value of $C_i = 440\,\mu F$ and the input power $P_i = 265\,W$ give a factor $C_i/P_i = 1.66\,\mu F/W$. For this factor, Figure 5.5 gives, for $a = 1$,

$$V_{ir} = 45\,V_{pp} \quad \text{and} \quad V_{imin} + 2V_F = 222\,V$$

Thus the limits of the input voltage are $V_{imin} = 220\,V$ and $V_{imax} = 340\,V$, or, at the junction of the capacitors, half these values, i.e. 110 V and 170 V.

It is now necessary to check that the maximum alternating current I_{Crms} is less than the maximum permissible current $I_{C\sim} = 5.4$ A.

Equation (5.5):

$$I_{iAV} = \frac{265\ W}{187\ V \times 1.41 - 0.5 \times 45\ V} = 1.10\ A$$

The maximum current I_{Crms} flowing in the capacitor C_i can be calculated from Equation (5.7); first the form factor k_{fD} must be determined from Equation (5.6) and Figure 5.9. With $V_{ir}/\hat{V}_a = 45\ V/265\ V = 0.17$, k_{fD} is given by Figure 5.9 as 3.8.

With $k_{fi} = 1.5$ (mean value from Equation (5.8)), I_{Crms} is given by Equation (5.7) as

$$I_{Crms} = 1.1\ A \times \sqrt{0.5 \times 3.8^2 + 1.5^2 - 2} = 3\ A$$
$$P_{Ci} = 3^2\ A^2 \times 0.4\ \Omega = 3.6\ W$$

Since I_{Crms} (3 A) $< I_{C\sim}$ (5.4 A), the electrolytic capacitors selected are suitable in regard to alternating current loading. For the rectifier Rec_1, the Type B250C1400 would be suitable according to the table of Figure 5.12, since the maximum direct input current amounts only to 1.1 A. To give a greater safety margin, however, Type B250C3200/2200 was preferred; no heatsink is necessary. The mains r.f.i. filter is dimensioned according to the requirements regarding interference levels and need not be discussed here. Suitable components for any requirements can be located in the relevant manufacturers' publications. The input current must in any case be known.

Equation (5.9):

$$I_{irms} = 1.1\ A \times \frac{3.8}{\sqrt{2}} = 2.96\ A \approx 3\ A$$

An r.f.i. filter from the Schiller company rated at 250 V/4 A was selected. To calculate the series resistor, only Equation (5.16) is needed, since it gives the largest value. As previously calculated, R_i should be about 3.5 to 7 Ω. With $I_{irms} = 3$ A, however, the power loss $W = 3^2\ A^2 \times (3.5\ \text{to}\ 7\ \Omega) = 31.5$ to 63 W, which is much too high for this converter. Either the resistor must be bypassed after a few milliseconds (when the electrolytic capacitors are charged) by an electronic circuit (see Figure 5.17) or a time-delay relay (as here or in Figure 5.16), or R_i must be in the form of a thermistor. A suitable type would be the SG220 (Thomatronik) with $R_{25} = 10\ \Omega$ and R_{220} (at $I = 3$ A) $= 0.2\ \Omega$. The loss at the full-load current of 3 A would then be less than 2 W, which is tolerable. Using a thermistor has the advantage of minimum cost, but the cooling time constant is up to a minute; following the switching-off of the power supply, therefore, it cannot be switched on again within this period. The data for the miniature relay used in the example are as follows (Type V23016-A13-A101): switching voltage 250 V; maximum switching current 15 A; operating voltage 42 to 110 V for a nominal voltage of 60 V; coil resistance 5 kΩ $\pm$ 0.75 kΩ. The series resistance should be so chosen that at least the operating voltage of 42 V is attained with low input voltage V_{imin} while with high input voltage V_{imax} the maximum permissible voltage of 110 V is not exceeded. Hence

$$R_1 \leqslant \frac{(220\ V - 42\ V)}{42\ V} \times 5.75\ k\Omega \leqslant 24.4\ k\Omega;\ \text{selected, } 22\ k\Omega/5\ W$$

$$I_{\text{Re1max}} = \frac{340\ \text{V}}{26.25\ \text{k}\Omega} = 13\ \text{mA}$$

$$P_{\text{R1max}} = (13 \times 10^{-3}\ \text{A})^2 \times 22 \times 10^3\ \Omega = 3.7\ \text{W} \approx 5\ \text{W}$$

$$V_{\text{Re1max}} = 13 \times 10^{-3}\ \text{A} \times 5.75 \times 10^3\ \Omega = 75\ \text{V} < 110\ \text{V}$$

$$P_{\text{Re1}} = (13 \times 10^{-3}\ \text{A})^2 \times 5 \times 10^3\ \Omega = 0.85\ \text{W}$$

The capacitor C_1 in parallel with the relay winding is best determined by experiment. The charging time constant of C_1 through R_1 with maximum capacitance ($+ 50$ per cent at the upper limit of tolerance) is $\tau = 6.8\ \Omega \times 660\ \mu\text{F} = 4.5\ \text{ms}$. The total operating delay of the relay should be at least 10 to 20 ms, so that the high charging current has time to decay. The operating delay depends on the mechanical characteristics of the relay as well as on the inductance of the winding. Neither is quoted in the data sheet, and cannot therefore be put into the (complicated) calculation. To be sure that the delay of about 20 ms will be attained with the selected relay type, it is necessary to measure it with an oscilloscope.

The 15 V auxiliary supply required for the control IC and the driver stages is provided initially by a miniature transformer T_2 with a secondary voltage of 22 to 24 V and a small rectifier Rec_2 with a smoothing capacitor C_2. The voltage developed on C_2 is fed to the changeover transistor Tr_3. The voltage drop of 4.7 V across the Zener diode D_1 (ZPY4.7) ensures that a sufficient voltage is applied to the base resistor R_4 to drive Tr_3 into saturation. To permit further calculation, however, the auxiliary current must first be estimated. Reference is made for this purpose to the part diagram of Figure 6.18, which shows the drive circuit. In the drive circuit used, with an anti-saturation feature in the output stage, the output transistor draws just so much base current as is necessary. The output stage thus operates in, as it were, a linear mode at a low collector-emitter voltage.

The duty cycle in a push–pull output stage must evidently not exceed 0.5, to avoid overlap and the consequent destruction of both transistors. δ_{Tmax} will thus lie in the region of 0.4 to 0.45 at V_{imin}. It may be assumed that $\delta_{\text{Tmax}} = 0.4$. The input power $P_i = 265\ \text{W}$ already calculated must be delivered alternately by the two transistors. The worst case applies with V_{imin}. As previously calculated, this is $220\ \text{V}/2 = 110\ \text{V}$ on the primary winding. The maximum collector current is therefore $I_{\text{Cmax}} = 265\ \text{W}/110\ \text{V} = 2.41\ \text{A}$. To this is added $\Delta I_L/2$ reflected from the secondary side ($+ 10$ per cent) and a proportion of about 20 per cent for the magnetizing current. A small air gap should be provided, in order to reduce the switching times, making the magnetizing current higher than it would otherwise be.

In total the correction factor amounts to 1.3, so that the maximum collector current I_{Cmax} becomes $2.41\ \text{A} \times 1.3 = 3.13\ \text{A} \approx 3.2\ \text{A}$. This current has to be multiplied by the duty cycle, i.e. 0.4, since the collector current, and similarly the mean base current, are not continuous. The current finally required is $3.2\ \text{A} \times 0.4 = 1.3\ \text{A}$. According to the data sheet the current gain for the BUV47A power transistor used at $I_C = 3\ \text{A}$ is about ten. The mean base current for the output stage is thus approximately $1.3\ \text{A}/10 = 0.13\ \text{A}$. Taking into account the transformation ratio of the drive transformer, $N_p/N_s = 2$, the drive current drawn from the 15 V supply on the primary side is about 65 mA. To this must be added 20 per cent magnetizing current in the drive transformer, a maximum of 20 mA for the TDB7815 linear regulator and a maximum of 8 mA residual current for the TDA4700 switching regulator, so that the total current is 88 mA, or in round figures 100 mA. This estimation is confirmed by measurement.

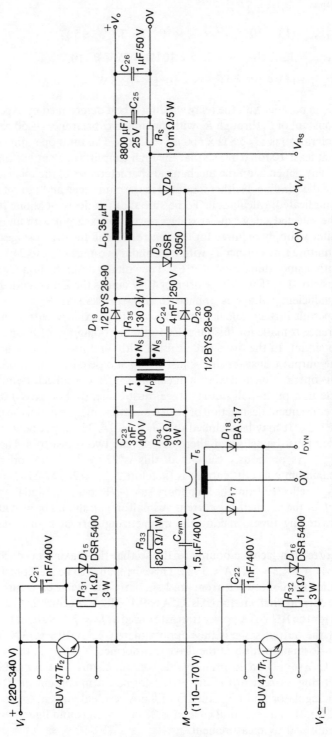

Fig. 6.4 Power stage of the bipolar version with output circuit

Calculation can now proceed in regard to Tr_3. With $B \approx 100$ at $I_C = 0.1$ A, the required base current is just 1 mA. In saturated operation $V_{CB} \approx 0$, and the current in R_4 is $I_{R4} = 4.7$ V/1 k$\Omega = 4.7$ mA. The base current which flows is clearly in excess of the minimum necessary, and Tr_3 is therefore saturated. Since the saturated voltage drop with a current ratio of 20 (which obtains here), according to the data manual, is 0.3 V maximum, the maximum loss in Tr_3 is $W_{Tr3} = 0.3$ V $\times 0.1$ A $= 30$ mW. This is so small that no heatsink is necessary. The maximum loss in R_4 occurs with Tr_3 turned off, since the base of Tr_3 is driven to earth. According to [22] the maximum voltage on C_2 can be calculated as $V_{C2max} = 24$ V $\times 1.41 \times 1.1 \times 0.85 = 31.6$ V ≈ 32 V. The dissipation in R_4 is then $W_{R4} = (32$ V$)^2/1$ k$\Omega = 1$ W; 1 kΩ/2 W is selected. Since the r.m.s. current in the secondary winding of T_2 is higher than the direct current of 0.1 V by a factor of 1.7 to 2 [22], the rating of T_2 becomes 24 V $\times (0.17$ to $2) = 4$ to 4.8 VA. A 5 VA transformer is thus adequate, particularly as it operates only in starting up and in the event of equivalent low mains voltage.

The supply voltage to the TDB7815 linear regulator must be between 16.5 V minimum and 35 V maximum.

$$V_{min} = 24 \text{ V} \times 0.85 \times 1.41 \times 0.85 - (4.7 \text{ V} + 0.7 \text{ V} + 0.3 \text{ V})$$
$$= 18.8 \text{ V} > 16.5 \text{ V}$$

$$V_{max} = 24 \text{ V} \times 1.1 \times 1.41 \times 0.85 - (4.7 \text{ V} + 0.7 \text{ V} + 0.3 \text{ V})$$
$$= 26 \text{ V} < 35 \text{ V}$$

Diode D_2 prevents feedback from the supply derived from the inductor through the base-emitter path of Tr_3, which breaks down at voltages above about 10 V. There are no special requirements for this diode; Type 1N4002 is suggested ($V_{Rmax} = 200$ V, $I_{Fmax} = 1$ A). The same applies to D_5.

When the converter has started up, which takes some seconds, because of the soft start, a voltage is generated in the secondary winding of the inductor and rectified by diodes D_3 and D_5. Since the frequency is around 100 kHz, very fast diodes must be used for this purpose, e.g. Type DSR3050 (TRW). When the voltage on C_3 reaches 18 to 19 V (adjustable by means of R_7), Tr_4 turns on and drives the base of Tr_3 to earth potential. Tr_3 is thus turned off, and the TDB7815 linear regulator is supplied from the inductor winding. For the dimensioning of the inductor secondary winding, see Section 6.3. This completes the calculation and discussion for the whole of the power supply.

The voltage V_u on C_2, which is proportional to the mains voltage, is used for the undervoltage trip in the TDA4700 control circuit.

6.2 CALCULATIONS FOR THE MAIN TRANSFORMER T_1 AND THE BALANCING CAPACITOR C_{bal}

The basic equation relating V_0 and V_p, according to [7], is

$$V_0 = 2\delta_T n V_p \quad \text{(V)} \tag{6.1}$$

where V_p is the voltage on the primary winding (V).

If V_p is represented, in the half-bridge converter, by $V_i/2$, this leads to the same relationship as for the single-ended forward converter:

$$V_0 = 2\delta_T n V_i/2 = \delta_T V_i n \quad \text{(V)} \tag{6.2}$$

where V_i is the total input voltage across the two capacitor groups in series.

Hence the same equation can be used as in Section 2.3:

$$n = \frac{N_s}{N_p} = \frac{V_{0max} + V_F + V_{Ls}}{\delta_{Tmax} V^*_{imin}} \tag{6.3}$$

where V^*_{imin} is the effective minimum input voltage $= V_{imin} - V_{CE} - V_{Lp}$, V_F is the forward voltage drop of rectifier diode (0.5 to 1 V), V_{Ls} is the voltage drop due to secondary winding and inductor (about 0.5 V) and V_{Lp} is the voltage drop due to primary winding (about 2 V). δ_{Tmax} is usually set in the forward converter at 0.45. To increase the safety margin somewhat under the most unfavourable operating conditions, $\delta_{Tmax} = 0.4$ [0.39] should be assumed for the push–pull converter. It may be observed here that the minimum value of δ_{Tmax} on the TDA4700 used is 0.35 (see also Chapter 8). n can now be calculated for the example.

$$V_{0max} = 13.2\,V; \quad V_F = 0.8\,V \text{ (estimated)}; \quad V_{Ls} = 0.5\,V \text{ (estimated)}$$

V_{CE} is relatively high, because of the anti-saturation drive circuit; it may be taken as 2 V.

$$V^*_{imin} = 220\,V - 2\,V - 2\,V = 216\,V$$
$$V^*_{imax} = 340\,V - 2\,V - 2\,V = 336\,V$$
$$n = \frac{13.2\,V + 0.8\,V + 0.5\,V}{0.4 \times 216\,V} = 0.1678,$$

or $1/n = 5.959$, rounded up to 6. Thus n becomes $1/6 = 0.1667$.

If the output voltage were constant, as is usually the case, δ_{Tmin} could be calculated from Equation (5.18). Since, however, V_0 has to be adjustable by ± 10 per cent to conform to the requirements, i.e. from 10.8 V to 13.2 V, δ_{Tmin} must be calculated from Equation (6.3), putting $V_0 = 10.8\,V$.

$$\delta_{Tmin} = \frac{10.8\,V + 0.8\,V + 0.5\,V}{1/6 \times 336\,V} = 0.216\,[0.22]$$

It can now be checked whether the minimum 'on' time t_1 is large in comparison with the switching times of the transistor. For the BUV47A switching transistor, which is used here, switching times of about $0.3\,\mu s$ are quoted (for resistive load, which corresponds approximately to the conditions in a forward converter). With the very effective drive circuit used here, the switching times can be taken as around 0.1 to $0.2\,\mu s$. With MOSFET power transistors with reasonably good drive, the switching times may be assumed to be $0.1\,\mu s$ or less. The cycle period for the bipolar version is

$$T = \frac{1}{40 \times 10^3\,Hz} = 25\,\mu s;$$

for the MOSFET version,

$$T = \frac{1}{60 \times 10^3\,Hz} = 16.7\,\mu s$$

$$t_1 = 0.216 \times 25\,\mu s = 5.4\,\mu s \text{ or } 0.216 \times 16.7\,\mu s = 3.6\,\mu s$$

Since the switching times amount to less than 10 per cent of the minimum 'on' time, the switching frequencies chosen are not too high. To prevent at all times the d.c. magnetization of the transformer T_1 which might occur in the starting-up process and

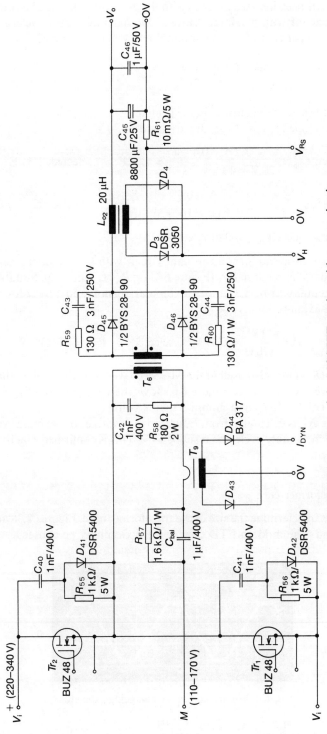

Fig. 6.5 Power stage of the MOSFET version with output circuit

with step changes in load, due to inevitable asymmetry in the switching transistors, capacitor C_{bal} has been inserted in series with the primary winding. The voltage developed across this capacitor with maximum current and maximum duty cycle should be about 20 V, or about 10 per cent of the maximum direct voltage. From [27], a value for C_{bal} is given by

$$C_{bal} \leqslant \frac{I_{pmax}\delta_{Tmax}}{\Delta V f} \quad \text{(F)} \tag{6.4}$$

where I_{pmax} is the maximum current in the primary winding (estimated at 3.2 A) and ΔV is the voltage developed across C_{bal}; already fixed at 20 V.

Bipolar version: $C_{bal} \leqslant \dfrac{3.2 \, \text{A} \times 0.4}{20 \, \text{V} \times 40 \times 10^3 \, \text{s}^{-1}} = 1.6 \, \mu\text{F}$;   selected, $1.5 \, \mu\text{F}$

MOSFET version: $C_{bal} \leqslant \dfrac{3.2 \, \text{A} \times 0.4}{20 \, \text{V} \times 60 \times 10^3 \, \text{s}^{-1}} = 1.1 \, \mu\text{F}$;   selected, $1 \, \mu\text{F}$

Voltage rating: at least $V_{imax} = 340 \, \text{V}$; e.g. 400 V

Since this capacitor can form a resonant circuit with stray inductances, it should be damped by a parallel resistor R_{33} (Figure 6.4) or R_{57} (Figure 6.5). Suitable values were found experimentally to be $820 \, \Omega$ for the bipolar version and $1.6 \, \text{k}\Omega$ for the MOSFET version. The maximum loss is

$$P = \frac{(20 \, \text{V})^2}{820 \, \Omega} = 0.49 \, \text{W}; \quad \text{selected,} \quad R_{33} = 820 \, \Omega/1 \, \text{W}$$

A 1 W resistor (R_{57}) was also used in the MOSFET version, although in this case a 0.5 W type would have been adequate. Figure 6.6 shows the voltage across C_{bal} with maximum output power (13.2 V/15 A) and nominal mains voltage.

If the voltage ΔV is calculated from Equation (6.4) with nominal mains voltage and the corresponding mean duty cycle, the result is 14 V. This is confirmed exactly by Figure 6.6 (28 V/2 = 14 V).

Choice of transformer core

The next step is to determine the size of the transformer core. Figures 2.38 and 2.39 can be used to this end. The modern ETD core construction may be chosen.

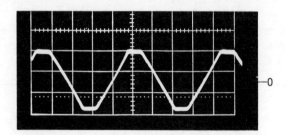

Fig. 6.6 Voltage across the balancing capacitor C_{bal} at P_{Omax} ($V_0 = 13.2 \, \text{V}$, $I_0 = 15 \, \text{A}$) and $V_a = 220 \, \text{V}$. Horizontal scale: $5 \, \mu\text{s}/\text{division}$. Vertical scale: 10 V/division

1. Trial with ETD39 core: Figure 2.38 gives about 210 W at $f = 20\,\text{kHz}$. According to Equation (2.50) the power that can be handled is proportional to frequency, but the losses increase at a more-than-proportional rate (see Figure 2.12); the power/frequency dependence can be taken approximately as $\sqrt{f}$ (see Figure 2.39). This would indicate a power-handling capability $P = 210\,\text{W} \times 1.4 = 294\,\text{W}$, which on this basis is sufficient. However, if exceeding the maximum permissible flux density of $2 \times 200\,\text{mT} = 400\,\text{mT}$ is to be avoided with certainty (starting, step load changes), the maximum flux density may be determined as $\hat{B} = 2 \times 150\,\text{mT} = 300\,\text{mT}$. The maximum power according to Equation (2.50) is then reduced by a factor of $300/400 = 0.75$, so that $P = 294\,\text{W} \times 0.75 = 220\,\text{W}$. This is too low, since the power to be transmitted is about 265 W. The next larger core, Type ETD44, must therefore be used. This is to be recommended for another reason, namely that because of the push–pull operation two secondary windings are required, which takes up more winding space. Figure 2.39 indicates a power-handling capability for the ETD44 core of 430 W at 50 kHz and 400 mT. At the lower frequency $f = 40\,\text{kHz}$ and the reduced flux density $\hat{B} = 300\,\text{mT}$ this is reduced to about 285 W, which is sufficient. According to the data manual, the manufacturer quotes the 'power-handling capability with insulation to IEC 435 without resin casting' as approximately two-thirds of the maximum possible power. With resin casting this becomes threequarters. The values in Figures 2.38 to 2.40, however, are only guidance values based on a core temperature rise of 30 K, and they can be increased or reduced according to the design. A small air gap of 0.2 mm should be introduced for the reasons previously mentioned. This can be done either by using one half-core with a 0.2 mm gap and one-half with zero gap or by interposing two strips of insulating material each of 0.1 mm thickness between the outer limbs. The sprung clamping device will accommodate this slight increase in dimensions without difficulty, particularly as the core tolerances are considerably greater.

Data for the ETD44 core:

Effective magnetic cross-sectional area:	$A_e = 173\,\text{mm}^2 = 1.73\,\text{cm}^2$
Minimum cross-sectional area:	$A_{min} = 172\,\text{mm}^2 = 1.72\,\text{cm}^2$
Effective magnetic path length:	$l_e = 103\,\text{mm} = 10.3\,\text{cm}$
Piece weight:	47 g
Usable winding cross-sectional area:	$A_N = 210\,\text{mm}^2$
Mean length of turn:	$l_N = 77.7\,\text{mm}$
Maximum winding width:	29.5 mm
Maximum winding depth:	7.25 mm

The core losses can be determined from Figure 2.12(a) for $f = 40\,\text{kHz}$ and $B = 150\,\text{mT}$ as 40 mW/g. Then, with the core weight of $2 \times 47\,\text{g} = 94\,\text{g}$, $W_{\text{core}} = 40 \times 10^{-3}\,\text{W/g} \times 94\,\text{g} = 3.76\,\text{W}$ or, from Figure 6.7, $W_{\text{core}} \approx 3.5\,\text{W}$. Because of the inclusion of the air gap the loss is actually significantly lower. Figure 6.8 indicates an inductance factor of 850 nH with a 0.2 mm gap.

Equation (2.16):

$$\mu_e = \frac{A_L \sum l_e / A_{min}}{\mu_0} = \frac{850 \times 10^{-9}\,\text{H} \times \dfrac{10.3\,\text{cm}}{1.72\,\text{cm}^2}}{12.57 \times 10^{-9}\,\text{H/cm}} = 405$$

174

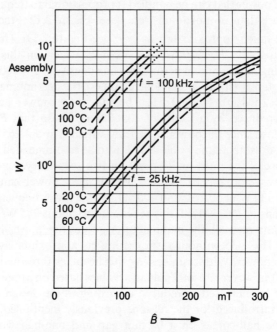

ETD 44 (B66365-G-X127)

Fig. 6.7 Variation of core losses (W/assembly) in the
ETD44 core (without air gap) with alternating flux
density $\hat{B}$ at $f = 25\,\text{kHz}$ and $f = 100\,\text{kHz}$, with
temperature as parameter (Siemens)

The core losses with an air gap in the magnetic circuit are given approximately by

$$P'_{\text{core}} = W_{\text{core}}\frac{\mu_e}{\mu_i} \quad \text{(W)} \tag{6.5}$$

$$P_{\text{core}} = 3.76\,\text{W} \times \frac{405}{2000} = 0.76\,\text{W}$$

The incorporation of a small air gap thus reduces the core losses to an insignificant level.

As with the single-ended forward converter, the determination of the numbers of turns is conveniently begun with N_s.

$$N_s \geqslant \frac{\delta_{\text{Tmax}}V_{\text{pmax}}n \times 10^4}{f\,A_{\text{min}}2\hat{B}} \tag{6.6}$$

The factor 2 in the denominator comes from the flux-density excursion $\Delta B = 2\hat{B}$ (i.e. from $+\hat{B}$ to $-\hat{B}$). To ensure that the maximum permissible flux density $\hat{B}_{\text{max}} = 150\,\text{mT}$ is not exceeded with step changes of load, the combination of δ_{Tmax} and V_{imax} or V_{pmax} is used here rather than the usual δ_{Tmax} and V_{imin} or V_{pmin}.

The maximum voltage V_{pmax} is taken as $170\,\text{V} + 20\,\text{V} = 190\,\text{V}$ (or if greater accuracy is required, $170\,\text{V} - V_{\text{CE}} - V_{\text{Lp}} + 20\,\text{V} = 186\,\text{V}$). The sign of the 20 V across the balancing capacitor is positive because at the changeover from one transistor to the other the

Variation of A_L values with total air gap s
for an assembly:
1 core B66365-G-X127   ($s \approx 0$) and
1 core B66365-G...   ($s > 0$)
    or
2 cores B66365-G...   ($s > 0$)

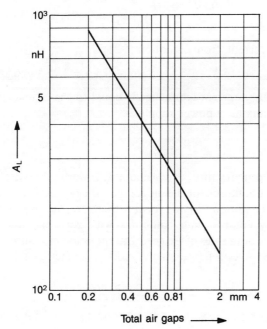

Fig. 6.8 Variation of inductance factor A_L (nH) of a
core assembly with total air gap s (mm) (Siemens)

capacitor is charged in opposition. In the example (bipolar version),

$$N_s \geqslant \frac{0.4 \times 190\,\text{V} \times (1/6) \times 10^4}{40 \times 10^3\,\text{s}^{-1} \times 2 \times 150 \times 10^{-3}\,\text{T} \times 1.72\,\text{cm}^2} \geqslant 6.14;$$

selected, $N_s = 8$.

For the MOSFET version it is only necessary to replace $f = 40\,\text{kHz}$ by $f = 60\,\text{kHz}$, the data otherwise remaining the same. The result is $N_s \geqslant 4.1$; selected, $N_s = 5$.

Since the losses rise more than proportionally to frequency, Figure 6.7 gives the core loss without an air gap as about 6 W. Here again the small gap reduces the loss to the acceptable value of 2.15 W. This value is non-critical, since once the circuit is started up, in the continuous operation which determines the temperature rise, it is the combination of δ_{Tmin} and V_{imax} or V_{pmax} that is significant. Hence the flux density in Equation (6.6) is halved, and the core losses much lower.

For the primary winding, from Equation (6.3),

Bipolar version: $N_p = N_s/n = 8 \times 6 = 48$ turns

MOSFET version: $N_p = 5 \times 6 = 30$ turns

The primary inductance can be calculated from Equation (2.17):

$$L_p = N^2 A_L$$

Bipolar version: $L_p = 48^2 \times 850 \times 10^{-9}\,\text{H} = 1.96\,\text{mH}\,[2.1\,\text{mH}]$

Magnetizing current in steady-state operation (with a step change in load in brackets), from Equation (5.22):

$$I_M = \frac{190\,\text{V} \times 0.216}{40 \times 10^3\,\text{s}^{-1} \times 2.1 \times 10^{-3}\,\text{H}} = 0.49\,\text{A} \approx 0.5\,\text{A}\,(0.9\,\text{A})$$

Maximum primary current, from Equation (5.21):

$$I_{pmax} = (15\,\text{A} + 1.5\,\text{A}) \times (1/6) + 0.5\,\text{A} = 3.25\,\text{A}\,(3.65\,\text{A})$$

This maximum primary current is practically equal to the estimated figure used in the calculations for the auxiliary power supply.

MOSFET version

Since the number of primary turns is considerably lower in this case, and the number of turns is squared in the equation for the inductance, the magnetizing current would be too high under the same conditions. In this version, therefore, the air gap should be halved, to 0.1 mm. This merely entails using half-cores with zero gap and inserting 0.05 mm insulating material on both sides of the core. Interpolation in Figure 6.8 (if necessary with the aid of Figure 2.8) gives a figure for the inductance factor of 1500 nH. The primary inductance is then

$$L_p = 30^2 \times 1500 \times 10^{-9}\,\text{H} = 1.35\,\text{mH}$$

I_M then becomes 0.76 A (1.4 A) and I_{pmax} is 3.5 A (3.9 A)

The conditions are thus similar to those in the bipolar version. Determination of the wire thickness in the windings is based on the available space, the permissible current density and, if litz wire is not used, the skin effect.

A current density of 3 to 5 A/mm^2 should be acceptable; for the skin effect, an increase in resistance of up to 5 per cent is tolerable.

Bipolar version

With $f = 40\,\text{kHz}$ Figure 2.11 gives $d_{max} = 0.78\,\text{mm}$; the standard thickness of 0.71 mm (outside diameter 0.77 mm) is selected from Figure 2.10. To obtain the necessary cross-sectional area, three wires are wound on *simultaneously*. The winding width of 29.5 mm just accommodates $29.5/0.77 = 38$ turns. Since three wires are wound on in parallel, only 36 turns (3×12) can be applied. A total of $3 \times 48/36 = 4$ layers is required. Allowing for a thin layer of insulating material of 0.05 mm between layers, the total primary winding depth is $4 \times 0.77\,\text{mm} + 4 \times 0.05\,\text{mm} = 3.28\,\text{mm}$. Allowing further for a protective screen, e.g. 0.1 mm thick between the primary and secondary windings, the winding depth becomes about 3.4 mm. The secondary winding may consist of litz wire of 2.5 mm outside diameter, with $2 \times 8 = 16$ turns. The winding width of 29.5 mm will only accommodate $29.5/2.5 = 11.8$, or 11 turns; since a winding space of $7.25\,\text{mm} - (3.4\,\text{mm} + 0.6\,\text{mm}$ insulation between primary and secondary) $= 3.25\,\text{mm}$ is all that remains, it is not possible to apply two secondary layers. If, however, a layer of 2×5 turns is wound on and the remaining 2×3 turns laid in the intervening spaces, the total winding space is just sufficient.

The cross-sectional areas of the windings are

$$A_p = \frac{3d_p^2 \pi}{4} = \frac{0.71^2 \, \text{mm}^2 \times 3 \times \pi}{4} = 1.19 \, \text{mm}^2$$

$$A_s = \frac{d_s^2 \pi f_{Cu}}{4} = \frac{2.5^2 \, \text{mm}^2 \times \pi \times 0.42}{4} = 2.06 \, \text{mm}^2$$

with $f_{Cu} = 0.42$ from Figure 2.42.

In relation to the maximum primary current which is significant from the point of view of heating, ΔI_L is not considered. The value of I'_{pmax} becomes $15 \, \text{A}/6 + 0.5 \, \text{A} = 3 \, \text{A}$.

The r.m.s. primary current which determines the temperature rise is given by

$$I_{prms} = I_{pmax} \sqrt{2 \delta_{Tmax}} \quad (A) \tag{6.7}$$

The relationship for the r.m.s. current in the secondary winding is more complicated, because the two rectifier diodes operate partly as rectifiers and partly as free-wheel diodes, with different current amplitudes (see also Section 6.3).

$$I_{srms} = I_{0max} \sqrt{0.25 + 0.75 \delta_{Tmax}} \quad (A) \tag{6.8}$$

Figures for the example are

$$I_{prms} = 3 \, \text{A} \times \sqrt{0.8} = 2.68 \, \text{A}$$

(here the value of 3 A designated as I'_{pmax} should be used)

$$I_{srms} = 15 \, \text{A} \times \sqrt{0.25 + 0.75 \times 0.4} = 11.1 \, \text{A}$$

With the cross-sectional areas already calculated the current densities are

$$J_p = \frac{2.68 \, \text{A}}{1.19 \, \text{mm}^2} = 2.25 \, \text{A/mm}^2 \quad \text{and} \quad J_s = \frac{11.1 \, \text{A}}{2.06 \, \text{mm}^2} = 5.38 \, \text{A/mm}^2$$

In view of the relatively low current density J_p, the transformer could alternatively be wound with only two parallel wires of 0.71 mm diameter. With a primary winding of three layers there is about 2.5 mm left at the sides for insulation. There is also a gain of 0.77 mm in the winding depth for further insulation. A new transformer calculation will, however, not be offered here. Additional insulation is afforded by the sheath of the litz wire. With $R' = 0.04355 \, \Omega/\text{m}$ (from Figure 2.10) and $l_N = 77.7 \times 10^{-3} \, \text{m}$,

Primary winding:

$$R_p = 0.04355 \, \Omega/\text{m} \times 48 \times 77.7 \times 10^{-3} \, \text{m}/3 = 54 \, \text{m}\Omega \, [70 \, \text{m}\Omega]$$

(measurement after a long period of operation and increase in resistance due to temperature rise)

$$P_p = 2.68^2 \, \text{A}^2 \times 70 \times 10^{-3} \, \Omega = 0.5 \, \text{W}$$

Secondary winding:

$$R_s = \frac{77.7 \times 10^{-3} \, \text{m} \times 16}{56 \, \text{S m/mm}^2 \times 2.06 \, \text{mm}^2} = 10.8 \, \text{m}\Omega$$

$$P_s = 11.1^2 \, \text{A}^2 \times 10.8 \times 10^{-3} \, \Omega = 1.33 \, \text{W}$$

Total loss in transformer:

$$P_T = P'_{core} + P_p + P_s = 0.76\,\text{W} + 0.5\,\text{W} + 1.33\,\text{W} = 2.6\,\text{W}$$

MOSFET version

Because of the much smaller number of turns ($N_p = 30$) on the same core, a litz wire, of 1.5 mm diameter, can also be used on the primary side. $29.5/1.5 = 19$ turns can be fitted into the winding width of 29.5 mm, so that two layers of 5 turns each are sufficient. The required winding width of $15 \times 1.5 = 22.5$ mm can be located in the middle of the winding space and generous room for insulation thereby gained at the sides. The winding depth then becomes 2×1.5 mm = 3 mm—the same as in the bipolar version. The secondary winding can likewise be designed with 2×5 turns, leaving still more room for insulation.

$$A_p = \frac{1.5^2\,\text{mm}^2 \times \pi \times 0.42}{4} = 0.74\,\text{mm}^2$$

With the maximum primary current $I'_{pmax} = 15\,\text{A}/6 + 0.76\,\text{A} = 3.26\,\text{A}$ the r.m.s. primary current and the current density become

$$I_{prms} = 3.26\,\text{A} \times \sqrt{0.8} = 2.92\,\text{A}$$

$$J_p = \frac{2.92\,\text{A}}{0.74\,\text{mm}^2} = 3.95\,\text{A/mm}^2$$

$$R_p = \frac{77.7 \times 10^{-3}\,\text{m} \times 30}{56\,\text{S m/mm}^2 \times 0.74\,\text{mm}^2} = 56\,\text{m}\Omega$$

$$P_p = 2.92^2\,\text{A}^2 \times 56 \times 10^{-3}\,\Omega = 0.48\,\text{W} \approx 0.5\,\text{W}$$

The secondary current density is the same as that in the bipolar version, i.e. $J_s = 5.38\,\text{A/cm}^2$.

$$R_s = \frac{77.7 \times 10^{-3}\,\text{m} \times 10}{56\,\text{S m/mm}^2 \times 2.06\,\text{mm}^2} = 6.7\,\text{m}\Omega$$

$$P_s = 11.1^2\,\text{A}^2 \times 6.7 \times 10^{-3}\,\Omega = 0.83\,\text{W} \approx 0.9\,\text{W}$$

With the core loss previously calculated the total transformer loss becomes

$$P_T = 2.15\,\text{W} + 0.5\,\text{W} + 0.9\,\text{W} = 3.6\,\text{W}$$

As was to be expected, the transformer loss is higher in total because of the higher frequency, but it is nevertheless quite acceptable.

To avoid voltage spikes arising from voltage overshoot owing to the inevitable stray inductances, a capacitor and resistor in series across the primary winding are recommended. This protective circuit is designed in the same way as for fast rectifier diodes (see Equation (5.27)):

$$C_s \geqslant L_s \frac{I_p}{\Delta V} \quad \text{(F)} \qquad (6.9)$$

where ΔV is the permissible overvoltage (V).

The capacitor should discharge to some extent during the fall time, so

$$R_s \leqslant \frac{t_f}{C_s} \quad (\Omega) \tag{6.10}$$

The values of these protective components can only be predicted approximately, because of the inexact estimation of L_s and ΔV, and should be optimized in the course of the final construction. In the first place an overshoot of 150 V should be assumed; L_s is estimated at $10 \, \mu H$.

Bipolar version: t_f estimated at $0.5 \, \mu s$

$$C_s \geqslant \frac{10 \times 10^{-6} \, H \times 3.25^2 \, A^2}{150^2 \, V^2} \geqslant 4.7 \, nF$$

$$R_s \leqslant \frac{0.5 \times 10^{-6} \, s}{4.7 \times 10^{-9} \, F} \leqslant 106$$

Final values: $C_s = 3 \, nF/400 \, V$ (C_{23} in Figure 6.4)

$$P_{Rs} = 0.5 \times 3 \times 10^{-9} \, F \times 170^2 \, V^2 \times 40 \times 10^3 \, s^{-1} = 1.75 \, W$$

$$R_s = 90 \, \Omega/3 \, W \, (R_{34} \text{ in Figure 6.4})$$

MOSFET version: t_f estimated at $0.2 \, \mu s$

$$C_s \geqslant 10 \times 10^{-6} \, H \times \frac{3.5^2 \, A^2}{150^2 \, V^2} \geqslant 5.4 \, nF$$

$$C_s = 4.7 \, nF \text{ is assumed initially.}$$

Final values: $C_s = 1 \, nF/400 \, V$ (C_{42} in Figure 6.5)

$$P_{Rs} = 0.5 \times 1 \times 10^{-9} \, F \times 170^2 \, V^2 \times 60 \times 10^3 \, s^{-1} = 0.9 \, W$$

$$R_s \leqslant \frac{0.2 \times 10^{-6} \, s}{1 \times 10^{-9} \, F} \leqslant 200 \, \Omega$$

selected, $R_s = 180 \, \Omega/2 \, W$ (R_{58} in Figure 6.5).

This completes the design of the transformer, including suppression.

6.3 DESIGN OF THE SECONDARY CIRCUIT

The inductance of the inductor is calculated in the same way as for the inductor-coupled step-down converter or the single-ended forward converter. For the push–pull forward converter it is necessary merely to introduce a factor of two in the duty cycle.

$$L \geqslant \frac{V_0^*(1 - 2\delta_{Tmin})}{\Delta I_L \times 2f} \quad (H) \tag{6.11}$$

$V_0^* = V_{0max} + V_F + V_L$ (V).

The doubled switching frequency $2f$ is required here on account of the double-way rectification (f = oscillator frequency in the control circuit). A pot core should be used, because it contains the air gap only in the enclosed centre core, and there is hardly any external stray flux. The selection criterion, as explained in Sections 2.2 and 5.3, is the stored energy in W s, or in this case in A^2 mH. The choice is made in accordance with Figure 2.5(c).

Bipolar version

$$L \geqslant \frac{14.5 \text{ V} \times (1 - 2 \times 0.216)}{3 \text{ A} \times 2 \times 40 \times 10^3 \text{ s}^{-1}} = 34.3 \, \mu\text{H} \approx 35 \, \mu\text{H}$$

$$= 35 \times 10^{-3} \text{ mH}$$

$$I^2 L = 16.5^2 \text{ A}^2 \times 35 \times 10^{-3} \text{ mH} = 9.53 \text{ A}^2 \text{ mH}$$

According to Figure 2.5(c) the CC36 core might be considered, but since a second winding has to be wound on, and a litz wire of optimum diameter is not obtainable, the CC50 core was selected. This is actually too large, but the calculations may be carried out with it initially.

The effective permeability can be estimated as about 70 to 80. From Figure 2.7, with $l_e = 91$ mm (data manual), the air gap is determined as approximately 1.5 mm. This need not be determined accurately, since cores are only available with gaps in steps of 1 mm, i.e. 1 mm or 2 mm. The larger value of 2 mm must therefore be chosen; $A_L = 440$ nH. The number of turns, from Equation (2.17), is

$$N = \sqrt{\frac{35 \times 10^{-6} \text{ H}}{440 \times 10^{-9} \text{ H}}} = 8.9, \text{ rounded up to } 9$$

Further data for the CC50 core:

Winding width:	24 mm
Winding depth:	15.7 mm
Mean length of turn:	$l_N = 111$ mm

A litz wire of 2.5 mm outside diameter may be used (as in the transformer).

24 mm/2.5 mm = 9.6 turns per layer, rounded down to 9 turns, can be accommodated per layer. If two litz wires are wound on simultaneously, to be connected in parallel, there will be two layers, giving a depth of 2×2.5 mm = 5 mm. The total cross-sectional area is

$$A = 2 \times 2.06 \text{ mm}^2 = 4.12 \text{ mm}^2$$

$$\text{Current density } J = \frac{15 \text{ A}}{4.12 \text{ mm}^2} = 3.64 \text{ A/mm}^2$$

The current density is thus not excessive. Further parallel connection of windings is not necessary.

Winding resistance:

$$R_L = \frac{0.111 \text{ m} \times 9}{56 \text{ S m/mm}^2 \times 4.12 \text{ mm}^2} = 4.29 \text{ m}\Omega$$

$$P_L = 15^2 \text{ A}^2 \times 4.29 \times 10^{-3} \, \Omega = 0.97 \text{ W} \approx 1 \text{ W}$$

The CC36 core may now be investigated again. The data for this core are:

Winding width:	19.6 mm
Mean length of turn:	$l_N = 77$ mm
Winding depth:	10 mm; $l_e = 69$ mm

From Figure 2.5c, $\mu_e = 45$, and from Figure 2.7, with $l_e = 69$ mm, an air gap of about 2.2 mm is required. Since the core is available only with $s = 1$, 2 or 3 mm, a design with $s = 2$ mm would be somewhat close; with $s = 3$ mm the inductance factor would be relatively low, and hence the number of turns would be high. Calculation may proceed on the basis of a 2 mm gap. With the inductance factor of 250 nH, the number of turns required is 12. Using 2.5 mm diameter litz wire, the winding width of 19.6 mm will accommodate only seven turns per layer; the two layers required, with a depth of 5 mm, fit easily into the winding space, but the current density, at 7.3 A/mm² is much too high. The parallel connection of two 2.5 mm litz wires makes the current density acceptable, but the winding depth then becomes 10 mm, leaving no room for the secondary winding. A litz wire of 2 mm diameter would seem fairly suitable, but is not available. It would be possible to wind on 4 litz wires of 1.5 mm diameter together in four layers, and connect these in parallel; this would give an acceptable current density of 5.1 A/mm² and a winding depth of 6 mm, leaving sufficient room for the secondary winding. With some restrictions it would thus be possible to use the CC36 core, but the CC50 core was selected on account of its better electrical characteristics. The secondary winding may now be designed, but a few reflections are necessary first.

Since the voltages across the secondary winding V_s and the primary winding V_L are in the ratio of the numbers of turns N_{AL}/N_L, the tolerances on V_L should first be calculated. V_{0max}^* is approximately 14.5 V at full load and V_{0min}^* is approximately 11.5 V at minimum load.

$$V_{Lmin} = 110 \text{ V}/6 - 14.5 \text{ V} = 3.8 \text{ V}$$
$$V_{Lmax} = 170 \text{ V}/6 - 11.5 \text{ V} = 16.8 \text{ V}$$

To dimension the secondary winding for this wide voltage range is not practicable. The voltage of 170 V applies, however, only with high mains voltage and absolutely no load, which never occurs, because of the inevitable losses. From the point of view of stresses with maximum input voltage, this figure of 170 V can be used, but in this case the maximum input voltage must be considered somewhat more exactly. With full load and high mains voltage $V_{imax} = 220 \text{ V} \times 1.1/0.85 = 284 \text{ V}$ or $V_{imax}/2 = 142$ V. With minimum load—no load on the output—the maximum primary voltage may be in the region of 155 V, because the losses have the effect of a small load on the reservoir capacitors. Based on this figure of 155 V,

$$V_{Lmax} = 170 \text{ V}/6 - 11.5 \text{ V} = 14.3 \text{ V}$$

The 7815 regulator requires, at the low maximum current of about 0.1 A (as previously calculated), a minimum input voltage of 16.5 V; the maximum input voltage is 35 V. However, with the calculated realistic value of 14.3 V for V_{Lmax} the range between V_{Lmin} and V_{Lmax}—1:3.8—is still much too large and impracticable. The nominal voltage V_{Lnom} with the nominal mains voltage of 220 V is 110 V/0.85/6 − (14.5 to 12.1 V) = 7.07 to 9.5 V. Allowing for a voltage drop of 0.5 V in the rectifier diodes,

$$N_{AL} = \frac{17 \text{ V} \times 9}{7.07} = 21.6 \text{ turns, rounded up to 22 turns.}$$

The secondary winding is thus designed with 2×22 turns (0.45 mm diameter). This means that with full load on the output and the output voltage adjusted to the maximum of $12\,V + 10$ per cent the auxiliary supply is provided by the inductor winding only from nominal mains voltage upwards; below nominal mains voltage it is supplied by the auxiliary transformer. With the output voltage set to $12\,V$, however, a mains voltage 5 per cent below nominal is sufficient at full load, and this is within the normal range of supply voltage.

The maximum secondary voltage is then

$$V_{smax} = \frac{22 \times 14.3\ V}{9} = 35\ V, \text{ which is permissible.}$$

MOSFET version

$$L \geqslant \frac{14.5\ V \times (1 - 2 \times 0.216)}{3\ A \times 2 \times 60 \times 10^3\ s^{-1}} = 22.9\ \mu H, \quad \text{or}$$

approximately $23\,\mu H = 23 \times 10^{-3}\,mH$.

$$I^2 L_{max} = 16.5^2\ A^2 \times 23 \times 10^{-3}\ H = 6.26\ A^2\ mH$$

From Figure 2.5(c) the CC36 core is quite adequate for this; the smaller CC26 cannot be used.

$\mu_e = 60$, and from Figure 2.7 the air gap s with $l_e = 69\,mm$ is about 1.6 mm. Since $s = 2\,mm$ is the only possibility, this core must be accepted.

$$A_L = 250$$

and

$$N = \sqrt{\frac{23 \times 10^{-6}\ H}{250 \times 10^{-9}\ H}} = 9.6\ \text{turns, rounded up to 10 turns}$$

Litz wire of 1.5 mm diameter may be used. To obtain an acceptable current density, six wires are wound on together and connected in parallel. The cross-sectional area of a 1.5 mm litz wire is

$$A = \frac{(1.5\ mm)^2 \times \pi \times 0.42}{4} = 0.74\ mm^2$$

Total cross-sectional area $A_{tot} = 6 \times 0.74\ mm^2 = 4.45\ mm^2$.

$$\text{Current density } J = \frac{15\ A}{4.45\ mm^2} = 3.37\ A/mm^2,$$

sufficiently low.

Since one layer will accommodate $19.6\,mm/1.5\,mm = 13\,$turns, $2 \times 6 = 12\,$turns per layer can be wound on. For the total of ten turns the winding depth is $5 \times 1.5\,mm = 7.5\,mm$. There thus remains sufficient space for the secondary winding, as will be seen.

$$R_L = \frac{0.111\ m \times 10}{56\ S\,m/mm^2 \times 4.45\ mm^2} = 4.45\ m\Omega$$

and

$$P_L = 15^2\ A^2 \times 4.45 \times 10^{-3}\ \Omega = 1\ W$$

For the secondary winding, on the same basis as before,

$$N_{AL} = \frac{17\,V \times 10}{7.07\,V} = 24 \quad \text{and} \quad V_{smax} = \frac{24 \times 14.3\,V}{10} = 34.3\,V$$

which is satisfactory. The wire used, of 0.45 mm diameter (maximum outside diameter 0.495 mm) permits 19.6 mm/0.495 mm = 39.6 turns per layer; since $2 \times 24 = 48$ turns are required, two layers must be applied. This leads to a winding depth of $2 \times 0.5\,mm = 1\,mm$. Together with the primary winding a total of $5 \times 1.5\,mm + 1\,mm = 8.5\,mm < 10\,mm$ is used up. There is thus some space remaining for insulation.

Since in both versions a relatively large air gap was provided, a check on the maximum flux density can be dispensed with. The core losses are very low, because of the air gap, and certainly do not exceed 0.5 W. With the copper losses of 1 W the maximum total loss in the inductor is $P_{Lmax} = 1.5\,W$.

Rectifier diodes

The maximum reverse voltage is calculated in accordance with Equation (5.24) ($V_0^* = 12.1\,V$ with $\delta_{Tmax} = 0.216$):

$$V_{Rmax} = \frac{12.1\,V}{0.216} + 0.7\,V \times \frac{(1 - 0.216)}{0.216} = 58.6\,V \approx 59\,V$$

V_F is estimated at 0.7 V.

$$V_{Rmax} \approx V_{imax}\,n = \frac{340\,V}{6} = 57\,V$$

The assessment of the forward current rating must be based either on the maximum permissible repetitive current I_{FM} from Equation (5.25), not to be confused with the much higher surge current I_{FSM}, or on the maximum mean current I_{FAV} from Equation (2.34). Different data are quoted by various manufacturers.

Equation (5.25):

$$I_{FM} = 2I_0 = 2 \times 15\,A = 30\,A$$

Equation (2.34):

$$I_{FAV} = 15\,A \times (1 - 0.216) = 11.8\,A \approx 12\,A$$

Since the two diodes operate alternately as rectifier and free-wheel diodes, and in part simultaneously, the maximum loading is very difficult to determine. If worst-case operation in the free-wheel mode is assumed, however, there is no risk of overloading.

Since primary-switched forward converters invariably incorporate two diodes, it is convenient to use a double diode. In the arrangement of Figure 6.1, or in that of Figure 5.18, with the inductor in the positive line, the two cathodes are connected together, and are connected to the mounting flange.

The maximum forward current I_{FM} indicated in the data sheets applies to a duty cycle of 0.5; that is, the maximum mean current $I_{FAV} = 0.5I_{FM}$, even though it is not always quoted. In accordance with Equation (2.34), the maximum current I_{FAV} flows in the free-wheel diode when the duty cycle approaches zero. This only occurs when short-circuit

184

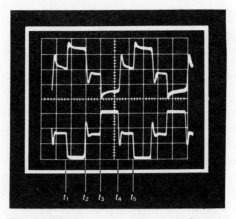

Fig. 6.9 Upper trace: current waveform in an output diode (measured with a current probe) at half maximum load ($V_0 = 12$ V and $I_0 = 7.5$ A). t_1–t_2, total current amplitude I_0; t_2–t_3, half current amplitude $I_0/2$; t_3–t_4, no current in the diode (only transient effect visible); t_4–t_5, half current amplitude $I_0/2$. Horizontal scale: 5 µs/division. Vertical scale: 2.5 A/division. Lower trace: voltage waveform across transistor Tr_1 under the same conditions. t_1–t_2, Tr_1 'on'; t_2–t_3, Tr_1 and Tr_2 'off', t_1–t_4, Tr_2 'on'; t_4–t_5, Tr_1 and Tr_2 'off'. Horizontal scale: 5 µs/division. Vertical scale: 100 V/division. Switching frequency $f = 40$ kHz (bipolar version)

current limiting is incorporated, because with $V_0 = 0$, in accordance with Equations (2.4) or (6.3) and (5.19), the duty cycle δ_T is also equal to zero. $(1 - \delta_T)$ then approaches 1, and the free-wheel diode has to carry almost the full output current. Since, however, in the push–pull converter the two diodes operate alternately and are both conducting for part of the time, this condition is not so critical as in the single-ended converter.

$$I_{FAVmax} = 0.5 I_{FM} \quad (A) \tag{6.12}$$

where I_{FAVmax} is the maximum permissible continuous current (A) (arithmetic mean) and I_{FM} is the maximum repetitive forward current (A).

Since a Schottky diode has in principle a lower voltage drop than an epitaxial diode, and is also extremely fast, an attempt should first be made to find a suitable diode of this type. Most Schottky diodes have a maximum reverse voltage rating of around 45 to 50 V, but the BYS28-90 (Siemens) has a maximum rating of 90 V. This is adequate in the present case.

Data for the BYS28-90:

$V_{RRM} = 90$ V (> 60 V); $I_{FAV} = 12.5$ A (> 12 A) at $\theta_c = 90\,°$C; $V_F = 0.9$ V at $I_F = 15$ A and $\theta_c = 25\,°$C; $I_{Rmax} = 20$ mA at $V_R = 90$ V; $C_j = 150$ pF at $V_R = 60$ V; $\theta_{jmax} = 125\,°$C $R_{thjc} = 1.2$ kW.

Maximum conduction loss, from Equation (5.30), for a double diode:

$$P_{Dc} = 15\,A \times 0.9\,V = 13.5\,W$$

Since the reverse current in a Schottky diode is relatively high (albeit low in absolute terms), the reverse loss may here be calculated from Equation (3.18):

$$P_{DR} = 20 \times 10^{-3}\,A \times 60\,V \times 0.4 = 0.5\,W$$

The total diode loss is thus

$$P_D = 13.5\,W + 0.5\,W = 14\,W$$

Heatsink:

$$R_{thca} \leqslant \frac{125\,°C - 45\,°C}{14\,W} - 1.2\,kW \leqslant 4.5\,kW$$

selected, $R_{thca} = 2.5\,kW$

The overvoltage protective components for the diodes can be determined only approximately here (as in Section 5.3), because various parameters can only be estimated.

$$Q_{rr} \approx 150 \times 10^{-12}\,F \times 60\,V = 9 \times 10^{-9}\,A\,s$$

Equation (5.28):

$$I_{RM} = \frac{2 \times 9 \times 10^{-9}\,A\,s}{30 \times 10^{-9}\,s} = 0.6\,A$$

For the reverse recovery time, which is not quoted for the BYS28-90 diode, the value of t_{rr} for a similar TRW diode was taken. This is quoted as 50 ns at $\theta_j = 125\,°C$, but, since the junction temperature is not so high, 30 ns would be a more realistic value. Having regard to the less-than-optimal construction of the circuit [27], a figure of 15 μH is assumed for the stray inductance.

Equation (5.27):

$$C_s \geqslant \frac{15 \times 10^{-6}\,H \times 0.6^2\,A^2}{60^2\,V^2} \geqslant 1.5\,nF$$

Equation (5.26):

$$R_s \leqslant \frac{60\,V}{0.6\,A} \leqslant 100$$

The optimized values were 4 nF for $C_s = C_{24}$ in Figure 6.4 for the bipolar version and 3 nF for $C_s = C_{43}$ and C_{44} in Figure 6.5 for the MOSFET version. For $R_s = R_{35}$ (bipolar version) and R_{59} and R_{60} (MOSFET version), 130 Ω/1 W resistors were used.

Loss in R_s:

$$P_{Rs} = 0.5 \times 4 \times 10^{-9}\,F \times 60^2\,V^2 \times 80 \times 10^3\,s^{-1} = 0.6\,W \approx 1\,W,$$

or

$$P_{Rs} = 0.5 \times 3 \times 10^{-9}\,F \times 60^2\,V^2 \times 120 \times 10^3\,s^{-1} = 0.7\,W \approx 1\,W$$

Output capacitors

As shown in detail in Chapter 2, the size of the output capacitors is determined by the equivalent series resistance (ESR). Their permissible operating voltage must of course be at least somewhat above V_{0max}.

At the high operating frequency of 80 or 120 kHz it is not to be expected that the relationship $\omega L_C < \text{ESR}$ is satisfied.

Equation (2.22):

$$\text{ESR} \leqslant \frac{100 \times 10^{-3}\,\text{V}}{2 \times 3\,\text{A}} \leqslant 16.7\,\text{m}\Omega$$

$$L_C \leqslant \frac{100 \times 10^{-3}\,\text{V}}{2 \times 3\,\text{A}} \times \frac{0.216 \times (1 - 0.216)}{(80 \text{ to } 120) \times 10^3\,\text{s}^{-1}} \leqslant 23.5 \text{ to } 35\,\text{nH}$$

$$\omega L_C = 2\pi \times (80 \text{ to } 120) \times 10^3\,\text{s}^{-1} \times 20 \times 10^{-9}\,\text{H} = 10 \text{ to } 15\,\text{m}\Omega$$

Based on these figures, it is possible to use either, from Figure 2.16, three or four capacitors of 4700 μF/16 V or three or four of 2200 μF/25 V or special capacitors from Figure 5.32, namely one unit of 10 000 μF/16 V or, better, one unit of 8800 μF/20 V. In the present example the latter unit was used. The values of ESR and L_C are considerably better than required. A check on the alternating current loading is unnecessary, since these electrolytic capacitors have very high ratings.

Figure 6.10 shows the output ripple at 80 kHz; the measurement at 120 kHz was the same. A ripple voltage of about 100 mV$_{pp}$ can be distinguished; the spikes could be due to the measuring equipment. The thickening of the trace arises from a small proportion of the 100 Hz input ripple, which is always present in the output. A ripple voltage at the input has the effect of a periodically varying input voltage, which is reflected in a varying output voltage. The capacitors in parallel with the output capacitor, C_{26} and C_{46}, are types with extremely low self-inductance, e.g. multi-layer Hi-K capacitors, which effectively suppress the spikes at the output.

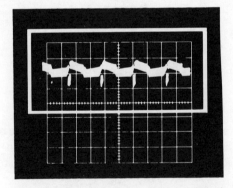

Fig. 6.10 Output ripple ΔV_0, measured across C_{25}/C_{26}, in the bipolar version with $f = 40$ kHz (the frequency of the inductor voltage is $2f = 80$ kHz because of the double-way rectification). Horizontal scale: 5 μs/division. Vertical scale: 100 mV/division

6.4 CALCULATIONS FOR THE OUTPUT TRANSISTORS

Bipolar version

The voltage stress amounts to $V_{imax}/2$ with collector current flowing and $2V_{imax}/2 = V_{imax}$ with $I_C = 0$ and negative base voltage; the current loading is about 3.5 A. A transistor with $V_{CE0} = 200\,V$ and $V_{CEX} = 400\,V$ would thus be adequate. The type selected was the BUV47, with $V_{CE0} = 400\,V$, $V_{CEX} = 800\,V$ and $I_{Cmax} = 9\,A$; the maximum current gain occurs at about 2 A. This transistor thus affords generous safety margins, and could exhibit its maximum transition frequency at about 3 A (see Figures 5.33 and 5.34). This signifies very short switching times. According to the data sheet, t_f can be taken as about $0.15\,\mu s$ and t_r as about $0.25\,\mu s$.

Equation (5.40): calculation of C_{s1} (from Figure 6.4, $C_{21} = C_{22}$)

$$C_{s1} \geqslant \frac{3.25\,A \times 0.15 \times 10^{-6}\,s}{2 \times 400\,V} \geqslant 0.6\,nF$$

selected, $C_{s1} = C_{21} = C_{22} = 1\,nF/400\,V$

Equations (5.41) and (5.42): calculation of R_{s1} ($R_{31} = R_{32}$)

$$R_{s1} \leqslant \frac{0.216 \times 25 \times 10^{-6}\,s}{4 \times 1 \times 10^{-9}\,F} \leqslant 1.35\,k\Omega$$

$$R_{s1} \leqslant \frac{170\,V}{0.32\,A} \geqslant 0.53\,k\Omega; \quad \text{selected, } R_{s1} = R_{31} = R_{32} = 1\,k\Omega/3\,W$$

$$P_{Rs1} = 0.5 \times 1 \times 10^{-9}\,F \times (2 \times 170\,V)^2 \times 40 \times 10^3\,s^{-1} = 2.3\,W$$

Diode $D_{s1} = D_{15} = D_{16}$: selected Type DSR5400 ($V_R = 400\,V$; $I_F = 4\,A$; $t_{rr} = 30\,ns$).

Calculation of losses:

Equation (5.35) with V_{CE} from Figure 6.18:

$$V_{CE} \approx 1\,V + 0.7\,V + 0.7\,V - 0.7\,V = 1.7\,V$$

$$P_{Trc} = 1.7\,V \times 3.25\,A \times 0.4 = 2.2\,W$$

Equation (5.36):

$$P_{Trson} = 170\,V \times 3.25\,A \times 0.5 \times 0.25 \times 10^{-6}\,s \times 40 \times 10^3\,s^{-1}$$
$$= 2.8\,W$$

Equation (5.44) (turn-off loss with RCD suppression):

$$P_{Trsoff} = \frac{3.25^2\,A^2 \times (0.15 \times 10^{-6}\,s)^2 \times 40 \times 10^3\,s^{-1}}{24 \times 1 \times 10^{-9}\,F}$$

$$= 0.4\,W$$

The total transistor loss is $P_{Tr} = 2.2\,W + 2.8\,W + 0.4\,W = 5.4\,W$. A heatsink Type KS65-50 (Austerlitz Elektronik) was chosen, with a thermal resistance of 5 kW. This results in a temperature rise of about 25 °C.

MOSFET version

The MOSFET transistor must meet the following requirements:

$$V_{DS} \geqslant 400\,V; \quad I_D \geqslant 3.5\,A$$

Because the input voltage is halved in the half-bridge circuit, the transistor does not have to handle a very high voltage. In selecting a suitable type the reduction in permissible drain current and the increase in $R_{DS(on)}$ with increasing temperature must be taken into account. With $\theta_{amax} = 45\,°C$ and $\Delta\theta = 25$ to $30\,°C$, due to losses, a case temperature of 70 to $80\,°C$ must be expected.

The current I_D is the mean current I_{DAV}, not the much higher permissible pulse current. As in the calculation of conduction loss, the mean current is given by

$$I_{DAV} = I_D \delta_{Tmax} \quad (A) \tag{6.13}$$

With the maximum current $I_D = 3.5\,A$ and $\delta_{Tmax} = 0.4$, the MOSFET must thus be capable of sustaining a mean current $I_{DAV} = 3.5\,A \times 0.4 = 1.4\,A$.

The choice of a suitable type may be made on various criteria. The off-state voltage rating must of course be at least 400 V and the permissible current $I_{DAV}(I_D)$ must be at least 1.4 A at $80\,°C$. If the lowest possible cost is the prime consideration, the BUZ40 ($I_{DAV} = 2\,A$ at $\theta_c = 80\,°C$; $V_{DS} = 500\,V$) could be chosen. The on-state resistance, which is significant in relation to the conduction loss, is about $6.8\,\Omega$ at the resonable junction temperature of $100\,°C$. This, however, would result, in accordance with Equation (2.56), in a loss of about 33 W, which is unacceptable. Quite apart from the cooling problems, this amount of loss would in itself reduce the efficiency by about 16 per cent. A transistor with a much lower $R_{DS(on)}$ must therefore be selected. The BUZ41 would be much more suitable, but since the lowest possible losses were sought the choice fell upon the BUZ48 (Siemens), with a maximum on-state resistance of $1\,\Omega$ at $\theta_j = 100\,°C$. The maximum conduction loss then becomes, from Equation (2.56), $W_{Trc} = 1\,\Omega \times 3.5^2\,A^2 \times 0.4 = 4.9\,W$. (For the BUZ41 with $R_{DS(on)max} = 1.75\,\Omega$, the figure would also have been tolerable at 8.6 W.)

Calculation for the suppression circuit as in Figure 6.5: Equation (5.40): (according to the data manual, $t_r = t_f = 0.1\,\mu s$; $V_{DS} = 500\,V$; approximately half values to be used in Equation (5.40))

$$C_{s1} \geqslant \frac{3.5\,A \times 0.1 \times 10^{-6}\,s}{2 \times 250\,V} \geqslant 0.7\,nF;$$

selected, $C_{s1} = C_{40} = C_{41} = 1\,nF/400\,V$

Equations (5.41) and (5.42):

$$R_{s1} \leqslant \frac{0.216 \times 17 \times 10^{-6}\,s}{4 \times 1 \times 10^{-9}\,F} \leqslant 918\,\Omega$$

$$R_{s1} > \frac{170\,V}{0.35\,A} > 485\,\Omega;$$

selected, $R_{s1} = R_{55} = R_{56} = 1\,k\Omega/5\,W$

$$P_{Rs1} = 0.5 \times 1 \times 10^{-9}\,F \times (2 \times 170\,V)^2 \times 60 \times 10^3\,s^{-1} = 3.5\,W$$

Diode $D_{s1} = D_{41} = D_{42}$: selected Type DSR5400 as in the bipolar version.

Calculation of losses:

Equation (5.36):

$$P_{\text{Trson}} = 170\,\text{V} \times 35\,\text{A} \times 0.5 \times 0.1 \times 10^{-6}\,\text{s} \times 60 \times 10^3\,\text{s}^{-1}$$
$$= 1.8\,\text{W}$$

Equation (5.44):

$$P_{\text{Trsoff}} = \frac{3.5^2\,\text{A}^2 \times (0.1 \times 10^{-6}\,\text{s})^2 \times 60 \times 10^3\,\text{s}^{-1}}{24 \times 1 \times 10^{-9}\,\text{F}} = 0.3\,\text{W}$$

Total transistor loss: $P_{\text{Tr}} = 4.9\,\text{W} + 1.8\,\text{W} + 0.3\,\text{W} = 7\,\text{W}$

The efficiency can now be determined by adding all the losses. This will be done here for the MOSFET version.

$$P_{\text{tot}} = P_T + P(R_{\text{ba1}}) + P_{R58} + P_L(\text{inductor}) + P_A(\text{auxiliary power supply})$$
$$+ P_D + P_{R56/55} + P_{59/60} + P_{Rs}(R_{61}) + P_{R2/R3}$$
$$+ P_{Ci} + P_{\text{rel}} + P_{\text{Tr}} + P_{R1} + P(\text{filter}) = 3.6\,\text{W} + 0.5\,\text{W}$$
$$+ 0.9\,\text{W} + 1\,\text{W} + 4.8\,\text{W} + 14\,\text{W} + 7\,\text{W}\,(2 \times 3.5\,\text{W})$$
$$+ 2\,\text{W}\,(2 \times 1\,\text{W}) + 22.5\,\text{W} + 2\,\text{W}\,(2 \times 1\,\text{W}) + 3.6\,\text{W} + 0.85\,\text{W}$$
$$+ 14\,\text{W}\,(2 \times 7\,\text{W}) + 3.7\,\text{W} + 1.1\,\text{W} = 61.3\,\text{W}$$

This leads to an efficiency with the maximum output power $P_{\text{0max}} = 13.2\,\text{V} \times 15\,\text{A} = 198\,\text{W}$

$$\eta = \frac{198\,\text{W}}{198\,\text{W} + 61.3\,\text{W}} = 76.4\%\,[83\%]$$

Here again the theoretical efficiency is somewhat lower than that calculated. The reason for this is that the worst case was assumed in considering the loading of each individual component. It is most unlikely, however, that in the model that was constructed the tolerances would all be unfavourable. It should also be said that the measurement of the input power placed great demands upon the measuring equipment (see the distorted capacitor current in Figure 5.7). It was not possible to establish the magnitude of the possible measurement errors. The agreement between the theory and the measured results is nevertheless quite good.

Figure 6.11 shows the variation with load current I_0 of the efficiency calculated from the measured input power.

Similar values are obtained for the bipolar version, since a large proportion of the losses are either the same or only slightly different.

Figures 6.12 and 6.13 show the constancy of the output voltage (relative to the nominal output voltage of 12 V) as a function of the mains voltage with an output current of $I_{\text{0max}}/2$ and as a function of the load current with $V_a = 220\,\text{V}$.

The operation of the secondary-side overcurrent limiting can be seen very clearly in Figure 6.13. Such a steep reduction of voltage with load currents in excess of the maximum cannot be achieved with only primary-side pulse limiting (see Figure 5.68).

Although the minimum load current was stated to be 1.5 A, the circuit is in fact open-circuit-proof, because with severely reduced load the driving pulses are suppressed. True

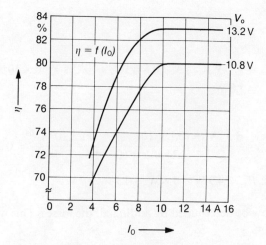

Fig. 6.11 Variation of efficiency η with load current I_0 with output voltages $V_0 = 13.2$ V and $V_0 = 10.8$ V (both versions)

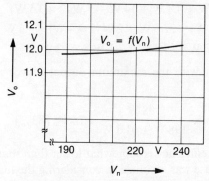

Fig. 6.12 Output voltage $V_0 = f(V_i)$ at $I_0 = 7.5$ A

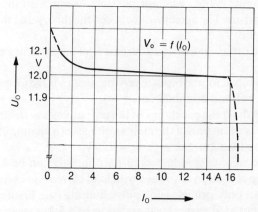

Fig. 6.13 Output voltage $V_0 = f(I_0)$ at $V_{mains} = 220$ V

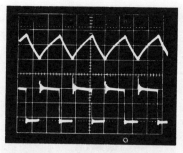

Fig. 6.14 Current ΔI_L in the inductor (upper trace) and voltage V_L across the inductor (lower trace) with $f = 40\,\text{kHz}$ (bipolar version), $V_0 = 13.2\,\text{V}$ and $I_0 = 15\,\text{A}$: Horizontal scale: $5\,\mu s$/division. Vertical scale (upper) 2 A/division. Vertical scale (lower) 10 V/division

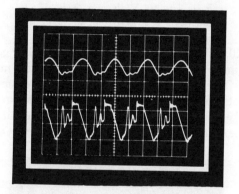

Fig. 6.15 Current I_L in the inductor (upper trace) and voltage V_L across the inductor (lower trace) with $f = 40\,\text{kHz}$ (bipolar version) and $I_0 < I_{0\text{min}}$ (discontinuous-current operation: $V_0 = 13.2\,\text{V}$ and $I_0 = 0.33\,\text{A}$). Horizontal scale: $5\,\mu s$/division. Vertical scale (upper): 0.5 A/division. Vertical scale (lower): 10 V/division

no-load operation never arises, because of the unavoidable losses, and is certainly not desirable. In Figure 6.14 the upper trace represents the alternating current component in the inductor (at $I_{0\text{max}}$) and the lower trace the inductor voltage. The same quantities are shown in Figure 6.15 at $I_0 \approx 0.4\,\text{A}$, which is below the limit of continuous current. The inductor voltage, particularly, departs considerably from the desired waveform, for which reason this mode of operation should be avoided as far as possible.

Figure 6.14 and 6.15 relate to the bipolar version; similar indications were obtained with the MOSFET version.

The upper trace in Figure 6.16 shows the collector current of a bipolar switching transistor (measured with a current probe in the collector lead) at $I_{0\text{max}} = 15\,\text{A}$ and $V_a = 220\,\text{V}$. ($I_{p\text{max}} = 3.25\,\text{A}$ was calculated for 15 per cent low mains voltage; here 3 A is

192

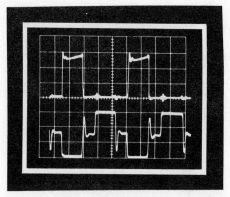

Fig. 6.16 Collector current I_C of Tr_1 (measured with a current probe (upper trace) and collector-emitter voltage V_{CE} of Tr_1 (lower trace) with $V_0 = 13.2$ V, $I_0 = 15$ A and $V_{mains} = 220$ V. Horizontal scale: 5 s/division. Vertical scale (upper): 1 A/division. Vertical scale (lower): 100 V/division. Switching frequency $f = 40$ kHz (bipolar version)

indicated with $V_a = 220$ V.) The small overshoots at the leading edges are caused by the (very short) recovery time of the rectifier diodes. The increasing magnetizing current can also be seen clearly. The lower trace shows the collector-emitter voltage of Tr_1. Considerable overshoots occur at the switching-off of the collector current, although they do not reach V_{imax}. The duty cycle $\delta_T = 0.33$ can also be measured from Figure 6.16.

Figure 6.17 shows the current and voltage waveforms in the MOSFET version, with the

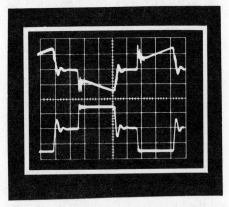

Fig. 6.17 Drain current I_D of Tr_1 and Tr_2 (measured with a current probe in the midpoint connection of the two transistors) (upper trace) and drain-source voltage of Tr_1 (lower trace) with $V_0 = 13.2$ V, $I_0 = 15$ A and $V_{mains} = 220$ V. Horizontal scale: 2 μs/division. Vertical scale (upper): 2.5 A/division. Vertical scale (lower): 100 V/division. Switching frequency $f = 60$ kHz (MOSFET version)

time scale extended by a factor of 2.5. In this case, in order to show both current half-cycles (in contrast to Figure 6.16), the current measurement was made in the lead to the mid-voltage point. The voltage was measured as before across Tr_1.

6.5 DRIVE CIRCUITS

Bipolar version

Figure 6.18 shows the drive circuit for the bipolar version; for clarity, only the circuit for Tr_1 is shown complete, that for Tr_2 being merely indicated.

The same drive circuit has already been shown in Figure 5.40 in connection with the single-ended forward converter and thoroughly discussed. Here it is only necessary to refer to some aspects of dimensioning. The difference between Figures 6.18 and 5.40 lies in the different connections of the control IC. For the drive circuit of Figure 5.40, a control IC (SG3524) with accessible collector and emitter was available, whereas in the present case only the collector connection is brought out, for which reason an additional *p-n-p* switching stage (Tr_5) has been inserted. It must be ensured that when the output transistor in the control IC turns on the switching transistor in the power stage also turns on.

To permit the dimensioning of the individual stages of the drive circuit, the calculations for the drive transformer T_3 should first be performed. 15 V is applied to the primary side; on the secondary side two identical windings are required. The negative auxiliary supply voltage on C_{32} should not exceed -7 V (maximum permissible reverse base-emitter voltage), but if possible should be as high as -5 V. A pulse voltage of 6.5 V at the transformer produces approximately -6 V on C_{32}. This determines the transformation ratio n:

$$n = N_s/N_p = V_s/V_p = 6.5\,\text{V}/15\,\text{V} = 0.4333$$

The maximum current in the switching transistors on the primary side has already been determined as 3.25 A. With a minimum current gain of at least ten, the maximum base current I_{Bmax} required for Tr_1 is 0.325 A. Allowing a small addition for the mean charging current of C_{32}, the maximum current (relative to secondary winding) $I_{smax}(T_3)$ is about 0.35 A. With $n = 0.4333$ the primary current is then $I_{pmax}(T_3) = 0.35\,\text{A} \times 0.4333 = 0.15\,\text{A}$.

The choice of core is not determined here by the low power to be transmitted (about 2 to 3 W) but by the required winding space. Two secondary windings have to be accommodated in addition to the primary winding. The EF25 core was therefore selected. If the power rating were the selection criterion, a much smaller core would certainly be adequate. In addition, good mains isolation must be maintained, since the secondary windings are at mains potential, while the primary is at the potential of the output voltage V_0 through the control IC. The essential data for the EF25 core are:

$$A_{min} = A_e = 0.525\,\text{cm}^2; \quad A_L = 600\,\text{nH with } s = 0.08\,\text{mm}$$

(As previously mentioned, a small air gap—e.g. 2×0.04 mm insulating material in both limbs = 0.08 mm total gap—results in a higher magnetizing current and hence a more rapid magnetization of the core, which leads to a shorter switching time.) Equation (2.51) gives the number of primary turns N_p with $\hat{B}_{max}$ restricted to 0.1 T (to preserve good pulse characteristics):

$$N_p \geqslant \frac{15\,\text{V} \times 0.4 \times 10^4}{0.525\,\text{cm}^2 \times 40 \times 10^3\,\text{s}^{-1} \times 0.1\,\text{T}} \geqslant 29$$

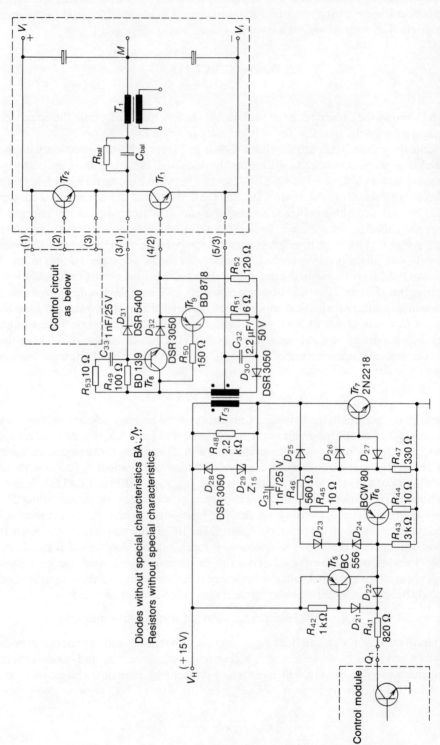

Fig. 6.18 Transformer-coupled drive circuit (with anti-saturation circuit for the switching transistor) for the bipolar version

To ensure that the primary inductance is not too low, and that the magnetizing current not too high, the primary winding is provided with 40 turns. The number of turns on each secondary winding is then obtained from the transformation ratio:

$$N_s = N_p n = 40 \times 0.4333 = 17.3, \quad \text{rounded up to } 18.$$

Equation (2.17):

$$L_p = N_p^2 A_L = 40^2 \times 600 \times 10^{-9}\,\text{H} = 1\,\text{mH}$$

Equation (2.54):

$$I_M = \frac{15\,\text{V} \times 0.4}{1 \times 10^{-3}\,\text{H} \times 40 \times 10^3\,\text{s}^{-1}} = 0.15\,\text{A}$$

This brings the total primary current in T_3 and Tr_7 to 0.3 A. Based on a current density of about 4 A/mm², the primary cross-sectional area is

$$A_p = \frac{0.3\,\text{A} \times \sqrt{0.4}}{4\,\text{A/mm}^2} = 0.048\,\text{mm}^2$$

This corresponds to a wire of 0.25 mm diameter; the standard diameter of 0.315 mm is selected (maximum outside diameter, from Figure 2.10, 0.352 mm). The usual procedure is reversed in this case, in order to use the smallest possible wire diameter and hence provide as much insulation as possible. Of the maximum winding width of 15.1 mm, a space of at least 3 mm should be left on either side, so that 15.1 mm − 6 mm remains for the winding. Thus one layer contains up to 9.1 mm/0.352 mm = 25 turns, and two primary layers are required to accommodate the 40 turns. The 20 turns per layer occupy only 20 × 0.352 mm = 7.04 mm, so that in fact there is 2 × 4 mm left for insulation.

Since the maximum secondary current, at 0.35 A, is almost as high as the primary current including the magnetizing current, the same wire diameter of 0.315 mm can be used on the secondary side. The two secondary windings are wound on with 18 turns per layer, only a thin interlayer insulation being required. The maximum depth of the four layers is then 4 × 0.352 mm = 1.41 mm. Out of the total winding depth of 3.7 mm, 3.7 mm − 1.41 mm is therefore available for insulation. With this amount of insulation, using puncture-proof film, a high standard of mains isolation can be achieved.

For the Zener diode D_{29} a Zener voltage of 15 V is chosen, to ensure resetting of the transformer during the interval between the pulses. From Equation (2.55), the maximum loss is

$$P_Z = 0.5 L_p I_M^2 f$$
$$= 0.5 \times 1 \times 10^{-3}\,\text{H} \times 0.15^2\,\text{A}^2 \times 40 \times 10^3\,\text{s}^{-1} = 0.45\,\text{W}$$

A low-power Zener diode may be adequate if it is ensured that the connecting leads can be attached to copper-clad sections of about 1 cm² in area on the circuit board; otherwise a higher-power type such as the ZX15 is preferable.

Diodes D_{28}, D_{30} and D_{32} should be fast diodes with a maximum forward current of about 1 A, e.g. Type DSR3050. The reverse voltage rating of 50 V is quite adequate. D_{31} must be a high-voltage fast diode, e.g. DSR5400 ($V_{Rmx} = 400$ V). The two transistors Tr_7 and Tr_8 should have a high transition frequency, to ensure fast switching. The maximum current loading is around 0.5 A; the voltage stress is insignificant. To turn on Tr_1, Tr_8 is turned on rapidly through C_{33} and R_{53}. The permanent base current to Tr_8,

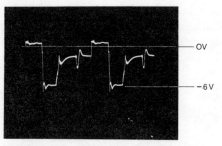

Fig. 6.19 Waveform of base voltage of the switching transistor Tr_1 in the bipolar version

corresponding to the base current required by Tr_1 (self-regulating anti-saturation circuit) is applied through R_{49}. So long as base current flows in Tr_1, C_{32} is charged negatively through D_{30} to about -6 V. When the output transistor in the control IC turns off, Tr_7 also turns off, and the voltage generated in the secondary winding reverses in polarity. Tr_9 then turns on rapidly and drives the base of Tr_1 to -6 V. The rate of rise of the negative base current is somewhat restricted by R_{51}, so that recombination can take place in the base/collector region before charge carriers are swept out of the base. The base of Tr_1 remains negatively biased until that transistor is turned on again. The waveform of the base voltage of Tr_1 is shown in Figure 6.19.

The extremely rapid turn-off of the base, with quite a brief overshoot (originating from stray inductances), can be seen clearly, as can also the somewhat higher base voltage (corresponding to a relatively large initial base current). The transients in the middle of each period, which have no effect on Tr_1 because of the negative bias voltage, are attributable to coupling between the two sides of the push–pull stage through the floating mid-point. Similar transients can be seen in Figures 6.9 or 6.16.

Tr_7 is turned on through the collector-emitter path of Tr_5, D_{24}, R_{45}, R_{46} and C_{31} in parallel and D_{26}. Turn-off is effected initially through D_{27}, C_{31}, R_{45}, Tr_6 and R_{44}, and then continuously through R_{47}. The voltage across R_{46} is about 12 V, as can be simply checked allowing about 0.7 V for the diode forward voltage drops. For a maximum collctor current of 0.3 A in Tr_7 and a minimum current gain of 20, R_{46} must be less than $800\,\Omega$; $560\,\Omega$ was chosen. R_{41} need only be above a certain minimum value so that the maximum permissible output current of the control IC is not exceeded (70 mA maximum). Sufficient base current is provided for Tr_5 with a value of $820\,\Omega$. Diodes D_{21} and D_{27} should be switching diodes of the fastest possible type, with low current and voltage ratings. Dimensioning is not critical for anything else in the circuit.

MOSFET version

The MOSFET version requires a different drive circuit from that used in the bipolar version, as shown in Figure 6.20.

This drive circuit corresponds, apart from small differences in the dimensioning of components, to the circuit of Figure 5.47, which has already been discussed in Chapter 5. The EF20 core was chosen for the pulse transformer T_7 in this case, since it is only required to deliver short charging and discharging pulses, but no continuous power. Thinner wire can thus be used for the windings. As before, the number of primary turns is given by

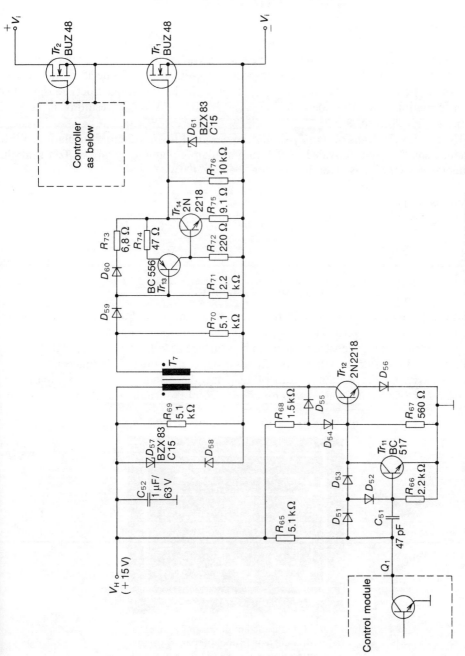

Fig. 6.20 Transformer-coupled drive circuit for the MOSFET version

Equation (2.51) (in this case $f = 60\,\text{kHz}$; $A_e = 0.335\,\text{cm}^2$):

$$N_p \geqslant \frac{15\,\text{V} \times 0.4 \times 10^4}{0.335\,\text{cm}^2 \times 60 \times 10^3\,\text{s}^{-1} \times 0.1\,\text{T}} \geqslant 29.9,$$

rounded up to 30 turns.

The BUZ48 MOSFET power transistor requires a typical gate voltage $V_{\text{GSmin}} = 5\,\text{V}$; for the sake of certainty 8 V is allowed. In view of the voltage drop of about 1.5 V across the two diodes D_{59} and D_{60}, an open-circuit secondary voltage of 10 V is decided on. Since the maximum permissible gate voltage V_{GSmax} is 20 V there is an ample safety margin. The transformation ratio thus becomes $n = V_s/V_p = 10\,\text{V}/15\,\text{V} = 2/3$, resulting in $N_s = 30$ turns $\times\ 2/3 = 20$ turns.

Here again the transformer T_7 should be designed for satisfactory insulation. With, again, a small air gap provided by 0.04 mm insulating material, giving $s = 0.08\,\text{mm}$, the inductance factor A_L is given in the data sheet as 400 nH.

Equation (2.17):

$$L_p = N_p^2 A = 30^2 \times 400 \times 10^{-9}\,\text{H} = 0.36\,\text{mH}$$

Equation (2.54):

$$I_M = \frac{15\,\text{V} \times 0.4}{0.36 \times 10^{-3}\,\text{H} \times 60 \times 10^3\,\text{s}^{-1}} = 0.28\,\text{A}$$

This is certainly a relatively large current, but it flows only for a short period, and the r.m.s. current is practically zero. Therefore, nothing needs to be changed.

A practical winding construction could be achieved as follows: wire diameter 0.25 mm, with an outside diameter of 0.284 mm. In this instance the primary winding can be wound in one layer. Of the winding width of 12.1 mm, the primary winding occupies $30 \times 0.284\,\text{mm} = 8.5\,\text{mm}$, leaving 1.8 mm on either side of the winding. If the two secondary

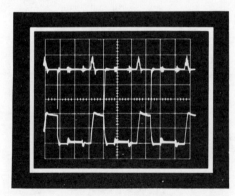

Fig. 6.21 Gate current pulses in Tr_1 (measured with a current probe) (upper trace) and gate-source voltage of Tr_1 (lower trace) with $V_0 = 13.2\,\text{V}$, $I_0 = 15\,\text{A}$ and $V_{\text{mains}} = 220\,\text{V}$. Horizontal scale: 5 µs/division. Vertical scale (upper): 0.2 A/division. Vertical scale (lower): 5 V/division

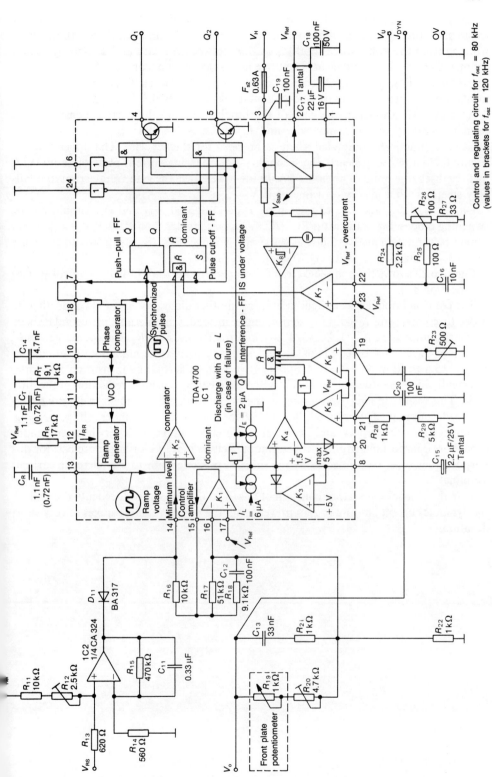

Fig. 6.22 TDA4700 control circuit with complete connections for the half-bridge push-pull converter

windings, each requiring $20 \times 0.284\,\text{mm} = 5.68\,\text{mm}$, are laid exactly in the middle, there is left an insulation space of 3.2 mm on either side. With the 1.8 mm separation of the primary winding from the bobbin, the total separation between primary and secondary is thus 4 mm, which should be sufficient. The three layers occupy a maximum of 0.85 mm out of the total winding depth of 2.8 mm. Almost 2 mm of film can thus be wound on as insulation between the primary and secondary windings, which is ample.

As previously shown in the case of the bipolar version, the current density in the primary winding with 0.25 mm wire and $I_p \approx 0.3\,\text{A}$ is $4\,\text{A/mm}^2$. Since the secondary winding is practically unloaded, there is no need for thicker wire. The upper trace in Figure 6.21 shows very clearly the secondary-side charging current pulse of about 0.2 A and the discharge pulse of about 0.45 A. In the lower trace the gate voltage rises and falls very steeply with a maximum value of 10 V.

Diodes D_{56}, D_{58} and D_{60} can again be Type DSR3050. D_{51} and D_{55} should be fast switching diodes. Since the ratio $R_{66}/(R_{66} + R_{65})$ is greater than $R_{67}/(R_{67} + R_{68})$, Tr_{12} cannot be turned on inadvertently as the supply voltage rises. D_{56} is introduced to ensure that Tr_{12} is turned off positively when Tr_{11} turns on.

Figure 6.22 shows the complete circuit arrangement for the TDA4700 control IC (see also Section 8.2).

The control amplifier is configured here, as in other diagrams, as a PID controller, in order to compensate rapidly for step changes in load. The method of secondary-side current limiting using IC_2 (1/4 CA324) is of interest. By means of R_{12} the voltage across R_s (R_{61}) (see Figures 6.4 or 6.5) is offset sufficiently to permit the maximum load current $I_{0\text{max}} = 15\,\text{A}$ to flow. At $I_0 = 16\,\text{A}$ the negative voltage across the measuring resistor is sufficiently high that the output of IC_2 is driven to earth, causing a reduction of the duty cycle. This results in effective limiting of the output current, as shown in Figure 6.13.

The decoupling of Terminals 20 and 19 for the over- and undervoltage sensing, Terminal 3 (positive supply) and Terminal 2 (reference voltage) with very low-inductance capacitors is important. Multilayer ceramic capacitors of $0.1\,\mu\text{F}$ or larger are best for this purpose; they must be connected by the shortest possible leads to the reference earth at Terminal 1.

As in the other circuits a time constant of $1\,\mu\text{s}$ (R_{25}/C_{16}) was introduced into the lead for the dynamic current limiting I_{dyn}, to prevent tripping by unavoidable spikes of very short duration.

7 Flyback converters for mains operation

Flyback converters are more suitable than forward converters for relatively low powers, because of their lower circuit complexity resulting from the elimination of the output inductor and the free-wheel diode. It is also possible at little cost to provide several electrically isolated outputs and to stabilize all the output voltages with a single control circuit. The power transformer is problematical, especially when high currents and low output voltages are involved, because the desirable low leakage inductances present difficulties (see Figure 1.1).

In the flyback converter energy is supplied to the transformer from the rectified mains supply during the conducting period of the switching transistor and stored. During the non-conducting period of the transistor the stored energy is transferred to the load. Figure 7.1 shows a flyback converter with a single output, omitting the mains rectification and the control, with its driver stage, but including the necessary protective networks.

In comparison with the circuit of a single-ended forward converter (see Figure 5.1), the most noticeable feature here is the reversed polarity of the secondary winding relative to the primary winding.

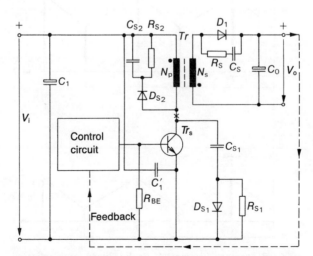

Fig. 7.1 Flyback converter with one output, including protective networks but omitting mains rectification and the control IC

201

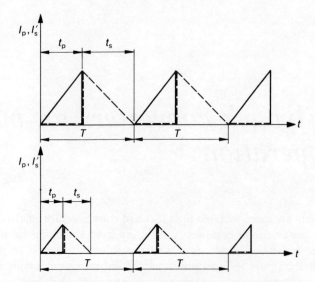

Fig. 7.2 Primary current I_p (solid lines) and reflected secondary current I_s (dashed lines) in <u>discontinuous-mode</u> operation, with maximum load (upper diagram) and with no load (lower diagram)

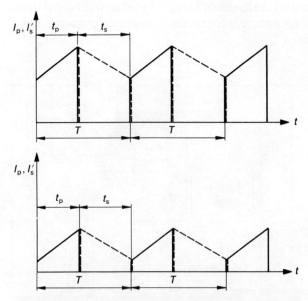

Fig. 7.3 Primary current I_p (solid lines) and reflected secondary current I_s (dashed lines) in <u>continuous-mode</u> operation, with maximum load (upper diagram) and with no load (lower diagram)

As a first step the mode of operation must be decided. There are two possible modes: with a triangular or with a trapezoidal current waveform.

Advantages of the trapezoidal current waveform:

- Simple operation at fixed frequency with small variation of duty cycle with changes of load.
- Lower current ratings of the power transistor and the rectifier diode, because for a given load current the peak current is lower than with a triangular waveform.

Disadvantages of the trapezoidal current waveform:

- A larger transformer is required than with a triangular current waveform.
- The switching losses are higher (with a considerable turn-on loss).
- The rectifier diode is subject to more stringent requirements in regard to recovery time.
- The primary inductance and hence the leakage inductance are higher, making the transformer more difficult to design.

With most direct off-line flyback converters a recovery or protection winding is included on the transformer. This recovery winding is placed physically on the mains side of any isolation barrier and usually has the same number of turns as the primary winding. It is tightly coupled to the primary winding and is connected, via a series diode, to the power supply input reservoir capacitor.

The recovery winding handles very little energy serving only to limit peak voltage levels at instants of switching and to provide additional protection in the event of open-circuit failure of the load.

Common applications:

- continuous current mode for small load variations;
- discontinuous current mode for large load variations.

In calculating the maximum current I_{pmax} on the primary side, the input power must first be determined. An efficiency of 80 per cent may be assumed as a basis for a first approximation excluding the loss in the rectifier diode, the latter being allowed for separately. The maximum input power is then given by: [4]

$$P_{imax} = 1.25 P_{0max} \left(1 + \frac{V_F}{V_0} \right) \quad (W) \tag{7.1}$$

where P_{0max} is the maximum output power (W), V_0 is the output voltage (V) and V_F is the forward voltage drop of rectifier diode (V) at I_{0max}.

With a continuous current mode it is a requirement that at P_{imin} and V_{imin} the current waveform should just become discontinuous:

$$p = \frac{P_{imax}}{P_{imin}} = \frac{\hat{I}_p + A\hat{I}_p}{\hat{I}_p - A\hat{I}_p} = \frac{1 + A}{1 - A} \tag{7.2}$$

where P_{imax} is the maximum input power (W), P_{imin} is the minimum input power (W) and A is the multiplication factor for $\hat{I}_p$ (initial value of current pulse).

Equation (7.2) can be rearranged to give A:

$$A = \frac{p - 1}{p + 1} \tag{7.3}$$

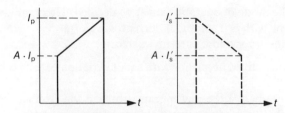

Fig. 7.4 Primary current pulse I_p (solid lines, left) and secondary current pulse I_s (dashed lines, right) with multiplication factor A

In contrast to the forward converter, in which the current I_p is of approximately rectilinear waveform, the flyback converter is so designed that A lies between 0.3 and 0.7, so that $p \approx 2$ to 5 (this applies only to the trapezoidal waveform). The dimensioning of a flyback converter is most satisfactory, according to [21] and [4], when $\delta_T = 0.5$ with minimum input voltage.

$$\delta_{Tmax} = 0.5 \text{ at } V_{imin} \tag{7.4}$$

(This only applies so long as the control IC is not of the push–pull type. As mentioned in Chapter 2, all push–pull controllers exhibit a dead time at $\delta_T = 0.5$; they cannot, therefore, be operated in this region. In this case δ_{Tmax} must be modified to about 0.42 to 0.45.) Once the maximum duty cycle δ_{Tmax} is decided, the minimum duty cycle δ_{Tmin} can be calculated. [7]

$$\delta_{Tmin} = \cfrac{1}{1 + \cfrac{V_{imax}}{\cfrac{\delta_{Tmax}}{1 - \delta_{Tmax}} V_{imin}}} \tag{7.5}$$

or, with $\delta_{Tmax} = 0.5$,

$$\delta_{Tmin} = \cfrac{1}{1 + \cfrac{V_{imax}}{V_{imin}}} \tag{7.6}$$

Equations (7.5) and (7.6) apply only to the case of I_{Omax}.

The duty cycle according to Equations (7.5) or (7.6) takes into account only the varying input voltage, not the loading conditions. In the forward converter, and in the flyback converter with continuous operation, the duty cycle varies slightly in practice with varying current, as a result of the finite internal impedance of the circuit, but is almost independent of the load current.

In the flyback converter with discontinuous mode operation, Equations (7.11), (7.13) and (7.1) lead to the relationship

$$\delta_T = \frac{\sqrt{2P_{imax}fL_p}}{V_i} = \frac{\sqrt{2.5fL_p(V_0 + V_F)I_0}}{V_i} \tag{7.7}$$

Since L_p, V_0 and, approximately, V_F are constant, if the drive circuit operates at a constant frequency and the input voltage V_i is constant the duty cycle must decrease as the output current I_0 falls. The situation where $\delta_T = 0$ at $I_0 = 0$ does not arise, because there are

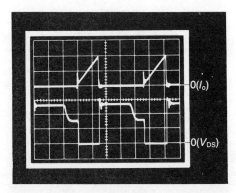

Fig. 7.5 Current waveform $I_D = f(t)$ (upper trace) and voltage waveform $V_{DS} = f(t)$ (lower trace) with $V_{imax} = 242$ V r.m.s. and $I_{0max} = 2$ A for the BUZ50A MOSFET. Horizontal scale: 5 μs/division. Vertical scale (upper): 0.5 A/division. Vertical scale (lower): 200 V/division

always losses, but δ_T does become very small under this condition. With a drive circuit that operates with variable frequency, e.g. Type TDA4600, the frequency increases as the current decreases, so that from Equation (7.7) the duty cycle remains approximately the same. The reduction in δ_T with I_0 has to be taken into account in designing the RCD suppression circuit.

Figure 7.5 shows the discontinuous mode of the primary current I_D and the voltage V_{DS} (MOSFET version) at I_{0max} and V_{imax}, with δ_{Tmin} in accordance with Equation (7.6). The same quantities are shown in Figure 7.6, but at $I_0 = 0$; the 'on' time in this case can be seen to be only about 1 μs.

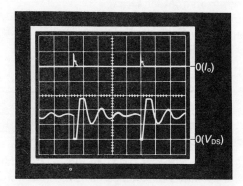

Fig. 7.6 Current waveform $I_D = f(t)$ (uppter trace) and voltage waveform $V_{DS} = f(t)$ (lower trace) with $V_{imax} = 242$ V r.m.s. and $I_0 = 0$ for the BUZ50A MOSFET. Horizontal scale: 5 μs division. Vertical scale (upper): 0.3 A/division. Vertical scale (lower): 200 V division

7.1 MAINS RECTIFICATION

To facilitate a quantitative appreciation, figures will be worked out for an example, which has been developed in both MOSFET and bipolar versions.

Example

A flyback converter is to be designed in accordance with Figure 7.7, with controlled output voltage, to meet the following requirements:

Mains input voltage: $V_a = 220\,\text{V}_\sim + 10$ per cent, -15 per cent $= 187$ to $242\,\text{V}$ r.m.s.
Output voltage: $V_0 = 24\,\text{V}$
Output current: $I_0 = 0$ to $2\,\text{A}$, hence discontinuous current mode
Output ripple: $V_0 \leqslant 0.25 V_{pp}$
Maximum ambient temperature: $\theta_{amax} = 50\,^\circ\text{C}$
Insulation test voltage, primary/secondary: $\geqslant 2.5\,\text{kV}$
Stabilization against mains voltage: < 0.5 per cent
Stabilization against load variation: < 0.5 per cent

The first step, again, is to dimension the reservoir capacitor C_1, for which purpose the input power P_{imax} must be determined.

Equation (7.1):

$$P_{imax} = 1.25 \times 24\,\text{V} \times 2\,\text{A} \times \left(1 + \frac{1\,\text{V}}{24\,\text{V}}\right) = 62.5\,\text{W} \approx 63\,\text{W}$$

Alternatively, as was done with the forward converter, it is possible to calculate the maximum input power from an estimated efficiency of 75 per cent (which should certainly be achieved):

$$P_{imax} = 24\,\text{V} \times 2\,\text{A}/0.75 = 64\,\text{W}\text{—practically the same.}$$

At 1.5 to $2\,\mu\text{F/W}$, C_1 becomes (1.5 to 2) $\mu\text{F/W} \times 64\,\text{W} = 96$ to $128\,\mu\text{F}$. From the series in Figure 5.10, $95\,\mu\text{F}/350\,\text{V}$ is barely sufficient, while $190\,\mu\text{F}$ is generous. A capacitor of $150\,\mu\text{F}/385\,\text{V}$ was chosen from another series.

The data for this capacitor are: ESR $(100\,\text{Hz}, 20\,^\circ\text{C}) = 1.2\,\Omega$; $I_{C\sim}$ $(100\,\text{Hz}, 85\,^\circ\text{C}) = 0.62\,\text{A}$. This gives a final figure for C_1/P_i of $150\,\mu\text{F}/64\,\text{W} = 2.34\,\mu\text{F/W}$. From Figure 5.5, with $a = 1$ (loss of one half-cycle),

$$V_{imin} + 2V_F = 233\,\text{V} \quad \text{or} \quad V_{imin} = 230\,\text{V} \ (V_F = 1.5\,\text{V})$$

and

$$V_{ir} = 32 V_{pp}$$

As will be shown, in a converter of this low power it is not necessary to short-circuit the protective resistor R_1 after starting-up. The voltage waveform can thus be observed simply across R_1 with an oscilloscope, as shown in Figure 7.8. (Figure 7.8 corresponds to Figure 5.7, except that in the latter case, since no protective resistor was available, the current in the capacitor was detected with a current probe.)

Mean converter input current I_{iAV}, from Equation (5.5):

$$I_{iAV} = \frac{64\,\text{W}}{187\,\text{V} \times \sqrt{2} - 0.5 \times 32\,\text{V}} = 0.257\,\text{A}$$

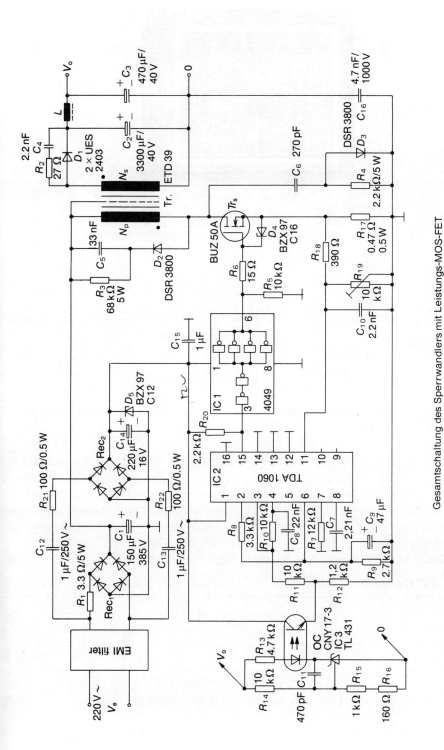

Fig. 7.7 Complete circuit of the MOSFET version of the flyback converter for $V_0 = 24$ V and $I_0 = 0$ to 2 A, with TDA1060 control circuit

Gesamtschaltung des Sperrwandlers mit Leistungs-MOS-FET

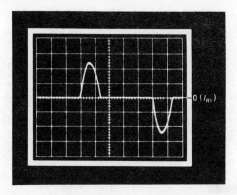

Fig. 7.8 Current waveform $I_{R1} = f(t)$ with $V_a = 187$ V r.m.s. and $I_{0max} = 2$ A. Horizontal scale: 2 ms/division. Vertical scale: 0.5 A/division

Equation (5.3):

$$\delta_{Ti} = \frac{2.6 \text{ ms}}{10 \text{ ms}} = 0.26 \ (\alpha_i \text{ and } D \text{ taken from Figure 7.8})$$

Equation (5.2):

$$\bar{i}_i^2 = \left(\frac{9.87}{8 \times 0.26} - 1 \right) \times 0.257^2 \text{ A}^2 = 0.25 \text{ A}^2$$

Equation (5.4) applies only to forward converters, and cannot be used here. The similar Equation (7.8) is applicable:

$$\bar{i}_0^2 = \left(\frac{4}{3\delta_{Tmin}} - 1 \right) I_{iAV}^2 \quad (\text{A}^2) \tag{7.8}$$

The maximum voltage applied to C_1 with mains voltage 10 per cent high is $V_{imax} = 220$ V $\times \sqrt{2} \times 1.1 - 2V_F = 340$ V. This voltage is reached only on no-load, and is significnat from the point of view of voltage rating. With maximum load, $a = 0$ and mains voltage 10 per cent high, from Figure 5.5, $V_{imax} = 250$ V $\times 1.1/0.85 = 320$ V.

This figure of 320 V is used in Equation (7.6) for the calculation of δ_{Tmin}:

Equations (7.4) and (7.6):

$$\delta_{Tmin} = \frac{1}{1 + \dfrac{320 \text{ V}}{230 \text{ V}}} = 0.42$$

Equation (7.7):

$$\bar{i}_i^2 = \left(\frac{4}{3 \times 0.42} - 1 \right) \times 0.257^2 \text{ A}^2 = 0.143 \text{ A}^2$$

Equation (5.1):

$$I_{C1rms} = \sqrt{0.25 \text{ A}^2 + 0.143 \text{ A}^2} = 0.63 \text{ A} \ [0.56 \text{ A}]$$

The type chosen, according to the data sheet, has a permissible alternating current $I_{C\sim}$ of 0.62 A at 85 °C. It would thus meet the requirements at this high temperature. At $\theta_{amax} = 50\,°C$, applying the correction factor from Figure 5.11, twice the magnitude—1.24 A—would be permissible. The r.m.s. capacitor current may now be calculated in accordance with Equations (5.9) and (5.7).

With $V_{ir} = 32V_{pp}$ and $\hat{V}_a = 265\,V$, the form factor $k_f = 4.1$.

Equation (5.7):

$$I_{C1rms} = 0.257\,A \times \sqrt{0.5 \times 4.1^2 + 1.5^2 - 2} = 0.75\,A$$

This figure is about 20 per cent high, as was to be expected from the neglect of the total supply impedance, including R_1. Since, however, the permissible current is much higher, the measurement can be dispensed with; the second method always gives a result on the safe side.

Equation (5.9):

$$I_{R1rms} = \frac{4.1}{1.41} \times 0.275\,A = 0.75\,A$$

The two bridge rectifiers Rec_1 and Rec_2 must be rated for an r.m.s. input voltage of 250 V. The smallest type, e.g. B250C700 (not in the series shown in Figure 5.12), is adequate. The I^2t value is

$$I^2\,dt = 6.3\,A^2\,s$$

The calculation of the protective resistor R_1 can be attempted first using Equation (5.12). $C_{1max} = 150\,\mu F \times 1.5 = 225\,\mu F$.

$$R_1 \geqslant \frac{0.5 \times 225 \times 10^{-6}\,F \times (220 \times 1.41\,V)^2}{6.3\,A^2\,s} \geqslant 1.7\,\Omega$$

Equation (5.16):

$$R_1 \geqslant \frac{220\,V \times 1.41 \times 1.1}{100\,A} \geqslant 3.4\,\Omega;\ \text{selected},\ 3.3\,\Omega/5\,W$$

$$P_{R1} = 0.75^2\,A^2\,3.3\,\Omega = 1.85\,W$$

The maximum loss in R_1 is less than 4 per cent which is acceptable. It is thus not worth providing, in this case, a complicated shorting circuit as in a larger converter. If a forward voltage drop of 0.8 V is postulated for a diode in the bridge rectifier, the total loss in the rectifier is

$$P_{Rec} = 2V_F I_{iAV} = 1.6\,V \times 0.275\,A = 0.4\,W$$

The auxiliary power supply is dimensioned in accordance with Figure 5.55; this is not affected by the different values of the protective resistors R_{21} and R_{22}. As shown in Section 5.8, the current flowing in D_5 is between 25 and 36 mA, allowing for the whole range of input voltages. With a minimum load current of about 10 mA (TDA1060 + R_{12}), the Zener diode is not overloaded, and with the maximum load current of about 20 mA there is still sufficient current left in the diode. The residual current in IC_1 is less than 1 mA, which is negligible; the pulse current is covered by C_{14} and C_{15}. The maximum loss in C_1 is $P_{C1} = ESR I_{C1rms}^2 = 1.2\,\Omega \times 0.63^2\,A^2 = 0.48\,W$. The total loss in the mains power supplies is thus $P_{R1} + P_{Rec1} + P_{D5} + P_{C1} = 1.85\,\text{W} + 0.4\,\text{W} + 0.45\,\text{W} + 0.48\,\text{W} = 3.18\,\text{W} \approx 3.2\,\text{W}$.

7.2 CALCULATIONS FOR THE MAIN TRANSFORMER

Since a transformer without leakage is not a practical possibility, and the lowest possible leakage is very important in a flyback converter, it is useful to consider the equivalent circuit of a transformer in Figure 7.9, including the secondary-side suppression.

It is apparent from this equivalent circuit that when the switch S is opened only the energy stored in the inductance $L_p (1 - \sigma)$ can be transferred to the secondary side. The energy stored in the leakage inductance $L_p \sigma$ must be absorbed by the capacitor C_6 if an excessive overvoltage is not to be generated.

During the conducting period of the switching transistor (S) $t_1 = \delta_{Tmax}T$, an amount of energy W_1 is stored in the primary inductance L_p.

$$W_1 = L_p I_{pmax}^2/2 \quad \text{(W s)}$$

where I_{pmax} is the current in L_p immediately before interruption.

From Faraday's law, $V_{1/2} = L_p dI_p/dt = L_p I_{max}/t_1$. From this relationship the stored energy is derived as

$$W_1 = \frac{0.5\,V_{1/2}^2 t_1^2}{L_p}$$

In the same way, on the secondary side, during the non-conducting period of the switching transistor $t_2 = (1 - \delta_{Tmax})T$,

$$W_2 = \frac{0.5\,V_{3/4}^2 t_2^2}{L_s}$$

where L_s is the secondary inductance.

From the energy conservation law $W_1 = W_2$,

$$\frac{V_{3/4}}{V_{1/2}} = \frac{t_1\sqrt{L_s}}{t_2\sqrt{L_p}} = \frac{t_1 N_2}{t_2 N_1} \quad \text{or} \quad \frac{N_2}{N_1} = \frac{V_{3/4} t_2}{V_{1/2} t_1} = n$$

With $V_{3/4} \approx V_0 = \text{const.}$ and $V_{1/2} \approx V_i = \text{const.}$, $t_2 = t_1$.

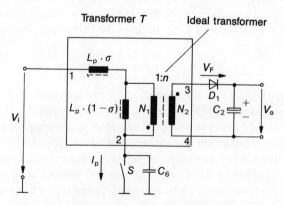

Fig. 7.9 Equivalent circuit of transformer T including
secondary connections

With the tolerances, n is then given by

$$n = \frac{(1 - \delta_{Tmax})(V_0 + V_F + V_{Ls})}{\delta_{Tmax}(V_{imin} - V_{CE} - V_{Lp})} \tag{7.9}$$

or, putting $\delta_{Tmax} = 0.5$,

$$n = \frac{V_0 + V_F + V_{Ls}}{V_{imin} - V_{CE} - V_{Lp}} \tag{7.10}$$

where V_0 is the output voltage (of one winding) (V), V_F is the forward voltage drop of a rectifier diode (V) (0.5 to 1 V), V_{Ls} is the voltage drop in the secondary winding (0.3 to 0.5 V), V_{imin} is the minimum input voltage (V), V_{CE} is the collector-emitter voltage (V) in the conducting state (about 0.5 V in saturated operation; 1.5 to 2 V in the quasi-saturated mode; 1 to 2 V with a MOSFET), V_{Lp} is the voltage drop in the primary winding (1 to 3 V) and δ_{Tmax} is the maximum duty cycle (about 0.5).

If Equation (7.10) is compared with the corresponding calculation of the transformation ratio for the single-ended forward converter, n is about half, or $1/n$ about double (see Equation (5.19)). This indicates a lower secondary voltage in comparison with the forward converter. This relationship is also apparent from the consideration that in the flyback converter no voltage has to be built up across an inductor. The necessary primary inductance is obtained for the discontinuous current mode from the energy stored in the inductance:

$$W_{imax} = 0.5 L_p I_{pmax}^2 f,$$

or

$$L_p = \frac{2 P_{imax}}{I_{pmax}^2 f} \quad \text{(H)} \tag{7.11}$$

where P_{imax} is the maximum input power (W) from Equation (7.1), I_{pmax} is the maximum primary current (A) from Equations (7.13) or (7.14) and f is the switching frequency (Hz).

For a continuous current mode, according to [4],

$$L_p = \frac{W_{imax}(p + 1)^2}{2 I_{pmax}^2 f p} \quad \text{(H)} \tag{7.12}$$

where P is the ratio of maximum to minimum input power.

(The factor A in Equation (7.3) only becomes equal to zero when $p = 1$; i.e. for discontinuous-mode operation p is always equal to 1 as calculated on the basis of continuous-mode operation, even though the ratio of maximum to minimum power is in fact large with a discontinuous-mode. If $p = 1$ is put into Equation (7.12), the result is Equation (7.11). This means, however, that the operating mode is not continuous but discontinuous.) The maximum primary current in discontinuous mode operation can be obtained by considering that the mean value of a discontinuous area ($P = VI$) is equal to the base multiplied by half the height, so that $W_{imax} = 0.5 I_{pmax} \delta_{Tmax} V_{imin}$. Solved for I_{pmax}, this gives

$$I_{pmax} = \frac{2 W_{imax}}{\delta_{Tmax} V_{imin}} \quad \text{(A)} \tag{7.13}$$

or, putting $\delta_{Tmax} = 0.5$,

$$I_{pmax} = \frac{4 W_{imax}}{V_{imin}} \quad \text{(A)} \tag{7.14}$$

With the continuous current mode the current peaks are somewhat lower. [4]

$$I_{pmax} = 2\frac{P_{imax} + P_{imin}}{V_{imin}} \quad \text{(A)} \tag{7.15}$$

The number of primary turns is governed by a relationship similar to that for the single ended forward converter (Equation (5.20)), but in this case with the usual correspondence between δ_{Tmax} and V_{imin}. In the event of a step change in load, the switch is always closed on to a discharged transformer.

$$N_p \geqslant \frac{\delta_{Tmax} V_{imin} \times 10^4}{\hat{B}_{max} A_{min} f} = \frac{L_p I_{pmax} \times 10^4}{\hat{B}_{max} I_{min}} \tag{7.16}$$

where $\hat{B}_{max}$ is the maximum permissible flux density (T) (about 0.3 T with a discontinuous current mode) and A_{min} is the minimum cross-sectional area (cm^2) (data sheet; possibly also A_e).

The number of secondary turns follows from the already determined transformation ratio $n = N_s/N_p$ and the value calculated from Equation (7.16):

$$N_s = nN_p \tag{7.17}$$

If N_s is not a round number, it must be rounded up or down and N_p is adjusted accordingly.

Since in Equation (7.16) the quantity A_{min} (or A_e) is determined by the core to be used this must now be picked from the available sizes. A starting point is given by Figure 2.38 for $f = 20$ kHz. The low leakage required in the flyback converter means that a larger core must be selected, in order to keep the numbers of turns low (A_{min} larger). In addition as high a value of $\hat{B}$ as possible must be assumed. Figure 2.38 gives the maximum flux density $\hat{B}_{max}$ as 0.2 T. This flux density was also the basis of the examples calculated previously for the forward converter. This was the more justified in thoe cases because it was no necessary to provide an air gap. Figure 2.12 shows that the core losses increase with increasing flux density. If an air gap is provided, however, the losses are reduced further in the ratio μ_e/μ_i.

$$P'_{core} = P_{core}\frac{\mu_e}{\mu_i} \quad \text{(W)} \tag{7.18}$$

where μ_e is the effective permeability with air gap and μ_i is the initial permeability without air gap (measured on a ring core) = 2000 for N27 material

$$\mu_e = \frac{A_L \Sigma l_e/A_e}{\mu_0} \tag{7.19}$$

where A_L is the inductance factor (H) (from data sheet), l_e is the effective length of magnetic path (cm) (data sheet), A_e is the effective magnetic cross-sectional area (cm^2), A_{min} is the minimum cross-sectional area of core (cm^2) and μ_0 is the magnetic constnat = 12.57×10^{-9} H/cm.

A further guide for the selection of a suitable core can be obtained from the following relationship [4] (for $f = 20$ kHz and $\hat{B} = 0.3$ T).

Discontinuous current mode:

$$V_{cd}(\text{cm}^3) \approx (0.1 \text{ to } 0.5)P_{imax} \quad \text{(W)} \tag{7.20}$$

where V_{cd} is the effective magnetic volume (cm³) (data sheet) + winding volume V_L (cm³) calculated from the bobbin dimensions);

$$V_L \approx A_N L_N.$$

Continuous current mode:

$$V_{ctrap} = V_{ctri}\left[\frac{(p+1)^2}{4p}\right]^2 \text{(cm}^3\text{)} \tag{7.21}$$

where V_{ctri} is the volume according to Equation (7.20) and p is the ratio of maximum to minimum input power (Equation (7.2)).

The required air gap can be determined approximately as

$$s \approx \mu_0 N_p^2 \frac{A_{min}}{L_p} = \mu_0 \frac{N_p I_{pmax} \times 10^4}{\hat{B}_{max}} \text{(cm)} \tag{7.22}$$

where $\mu_0 = 12.57 \times 10^{-9}$ H/cm.

The air gap thus determined must now be compared with the available lengths (in the data manual for the chosen core). If necessary N_p must then be corrected. Basically, from Equation (2.17),

$$N_p = \sqrt{\frac{L_p}{A_L}}$$

where L_p is the primary inductance (H) (calculated from Equations (7.11) or (7.12) and A_L is the inductance factor (H) (obtained from the curve $A_L = f(s)$ in the data book for the chosen core).

The bobbin is now wound with the primary and secondary windings in such a way that out of the available winding width a space of about 4 mm is left on either side in order to provide good mains isolation (the primary winding is at mains potential). If w is the total winding width, the usable width is thus $w' = w - 8$ mm.

The possible wire diameter (normal diameter) is given by

$$d = 0.8 \frac{w'}{N_L} \text{ (mm)} \tag{7.23}$$

where w' is the winding width w (mm) $- 8$ mm and N_L is the assumed number of turns per layer.

The factor 0.8 ensures that the calculated number of turns can be applied, even if the winding is not completely regular. Other important aspects of the choice of wire diameter are the current density, which should not exceed 3 to 5 A/mm² (see also Figure 2.38), and the skin effect. The increase in resistance owing to the skin effect should if possible be not more than 1 per cent and certainly not more than 5 per cent. The chosen wire diameter should therefore be checked against Figure 2.11 for the switching frequency f.

If the number of secondary turns is not too low, the leakage inductance can be estimated approximately from [4]

$$L_\sigma = \mu_0 N_p^2 \frac{l_N}{w}\left(x + \frac{y}{3}\right) \text{ (H)} \tag{7.24}$$

where μ_0 is the magnetic constant = 12.57×10^{-9} H/cm, N_p is the number of primary turns, l_N is the mean length of turn (cm) (see Figure 2.6), w is the maximum winding width (cm) (see Figures 2.6 and 7.10), x is the separation between primary and secondary winding (cm), (insulation) (see Figure 7.10) and y is the total depth of all windings (cm) (see Figure 7.10).

To obtain reasonably accurate results, the leakage inductance L_σ should be measured. This is done by measuring the inductance between the primary terminals with the secondary winding short-circuited. The primary inductance L_p is measured with the secondary terminals open. The ratio L_σ/L_p gives the leakage coefficient k_σ.

$$k_\sigma = \frac{L_\sigma}{L_p}$$ (7.25)

For discontinuous-mode operation it should be ensured, according to [4], that

$$k_\sigma \leqslant 0.5\%$$ (7.26)

For continuous-mode operation, similarly,

$$k_\sigma \leqslant \frac{p}{(p+1)^2} \times 2\%$$ (7.27)

If the windings are interleaved (see Figure 7.10), the leakage inductance is reduced. The relationship is [4]

$$L'_\sigma \approx \frac{L_\sigma}{4}$$ (7.28)

(In practice it has been found that the result according to Equation (7.28) cannot be achieved. To what extent it is necessary to employ a special winding technique, possibly based on multiple interleaving, is not clear at the present time.) The calculations for the transformer in the present example may now be carried out.

Equation (7.9):

$$V_{imin} = 230 \text{ V}$$

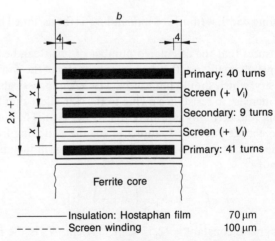

Fig. 7.10 Windng arrangement of transformer T

estimated values:

$$V_F = 0.8 \text{ V (epitaxial diode)}; \quad V_{Ls} = 0.3 \text{ V}; \quad V_{CE} = 2 \text{ V};$$
$$V_{Lp} = 2 \text{ V}$$

$$n = \frac{24 \text{ V} + 0.8 \text{ V} + 0.3 \text{ V}}{230 \text{ V} - 2 \text{ V} - 2 \text{ V}} = 0.111 \quad \text{or} \quad \frac{1}{n} = 9.04, \text{ rounded to 9};$$

$$n = 1/9$$

Equation (7.14):

$$I_{pmax} = \frac{4 \times 64 \text{ W}}{230 \text{ V}} = 1.1 \text{ A}$$

(Figure 7.5 indicates approximately 1 W).

As in the other chapters, calculations will be worked out here for a MOSFET and a bipolar version, the latter with two different control ICs. In view of the short switching times the MOSFET version is operated at $f = 40$ kHz; for the bipolar version 20 kHz was chosen.

Equation (7.11):

MOSFET version

$$L_p = \frac{2 \times 64 \text{ W}}{1.1^2 \text{ A}^2 \times 40 \times 10^3 \text{ s}^{-1}} = 2.64 \text{ mH}$$

Bipolar version

$$L_p = \frac{2 \times 64 \text{ W}}{1.1^2 \text{ A}^2 \times 20 \times 10^3 \text{ s}^{-1}} = 5.28 \text{ mH}$$

The choice of core must be based on a number of considerations. According to [4] the leakage is lower by about 20 per cent if the centre limb is round rather than square, the cores being similar in other respects. For this reason the selection is made from the ETD34 to ETD49 series. In the first place the core must be large enough to handle the maximum power (in this case about 65 W). According to Figure 2.38 the ETD34 core would be too small at $f = 20$ kHz; at $f = 40$ kHz it would (theoretically) be just adequate.

Equation (7.20):

$$V_{ctri} = (0.1 \text{ to } 0.5) \times 64 \text{ (cm}^3) = 6.4 \text{ to } 32 \text{ cm}^3, \text{ with an average of } 16 \text{ cm}^3$$

From the information given in the data sheet, the following volumes can be calculated for the various cores: ETD34, 15 cm^3; ETD39, 23 cm^3; ETD44, 34 cm^3. The ETD34 core is again too small; the other two are possibilities. Further considerations: in the flyback converter the leakage should be as small as possible. From Equations (7.25), (2.17) and (7.24) may be derived the relationship: $k_\sigma \propto 1/A_L$. From Equations (7.22) and (7.16), $s \propto N_p \propto 1/A_{min}$ and $\propto 1/f$, or in other terms $A_L \propto 1/s \propto A_{min} \propto f$. ($A_L \propto 1/s$ from the data sheet, similar to Figure 2.6.) That is, finally,

$$k_\sigma \propto \frac{1}{\text{core size}} \propto \frac{1}{f}$$

The leakage thus becomes smaller with increasing core size and also with increasing frequency (because N_p decreases). For these reasons, in the present example, the smaller ETD39 core is chosen for 40 kHz operation and the larger ETD44 core for 20 kHz.

Data:

ETD 39 core	ETD 44 core
$A_{min} = 1.23\,cm^2\,(A_e = 1.25\,cm^2)$	$A_{min} = 1.23\,cm^2\,(A_e = 1.25\,cm^2)$
$l_N = 69\,mm$	$l_N = 77.7\,mm$
$w_{max} = 25.7\,mm$	$w_{max} = 29.5\,mm$
$l_e = 9.22\,cm$	$l_e = 10.3\,cm$

(The ratio l_N/w in Equation (7.24) is somewhat smaller in larger cores, but the consequences are insignificant.)

MOSFET version with $f = 40\,kHz$ and ETD39 core

Equation (7.16):

$$N_p \geqslant \frac{0.5 \times 230\,V \times 10^4}{0.3\,T \times 1.23\,cm^2 \times 40 \times 10^3\,s^{-1}} \geqslant 78$$

Equation (7.22):

$$s = 12.57 \times 10^{-9}\,H/cm \times 78^2 \times \frac{1.23\,cm^2}{2.64 \times 10^{-3}\,H} = 0.036\,cm$$

$$= 0.36\,mm$$

From the data sheet, only air gaps of 0.2 mm or 0.5 mm are available. The combination of two half-cores, each with 0.2 mm gaps gives $s = 0.4$ mm, which is suitable. The corresponding inductance factor with 0.4 mm total air gap can be read off as 350 nH (data sheet, similar to Figure 2.6). A check on N_p using Equation (2.17) gives

$$N_p = \sqrt{\frac{2.64 \times 10^{-3}\,H}{350 \times 10^{-9}\,H}} = 87$$

selected, $N_p = N_1 = 81$ turns; this gives $L_p = 2.3\,mH$ [2.44 mH]; $\hat{B}_{max} = 0.29\,T$.

Equation (7.17): $N_s = N_2 = N_p n = 81/9 = 9$ turns.
　The r.m.s. current is given, according to [21], by

$$I_{pr.m.s.} = I_{pmax}\sqrt{\tfrac{1}{6}(1 + A + A^2)}$$
$$= 0.4082 I_{pmax}\sqrt{1 + A + A^2} \quad (A) \tag{7.29}$$

where A is the current factor in acordance with Equation (7.3) and Figure 7.4.
　Equation (7.29) applies equally to the secondary current, I_{smax} merely being substituted for I_{pmax}.
　For discontinuous-mode operation with $A = 0$, from Equation (7.29),

$$I_{pr.m.s.}(I_{sr.m.s.}) = 0.4082 I_{pmax}(I_{smax}) \quad (A) \tag{7.30}$$

The maximum secondary current is given, according to [21], by

$$I_{smax} = 2I_{0max}\frac{1}{\delta_{Tmax}(1+A)} = \frac{4I_{0max}}{1+A} \quad \text{(A)} \tag{7.31}$$

$$\delta_{Tmax} = 0.5$$

Hence, for discontinuous-mode operation,

$$I_{smax} = 4I_{0max} \quad \text{(A)} \tag{7.32}$$

Equation (7.29), in conjunction with Equation (7.31), can also be related to I_{0max}:

$$I_{sr.m.s.} = \frac{1.633I_{0max}}{1+A}\sqrt{1+A+A^2} \quad \text{(A)} \tag{7.33}$$

This leads, further, for discontinuous-mode operation $(A = 0)$ to

$$I_{sr.m.s.} = 1.633I_{0max} \quad \text{(A)} \tag{7.34}$$

(The maximum secondary current cannot be calculated on the basis of the transformation ratio n, because of the losses which are included in I_{pmax} but not in I_{smax}.)

Equation (7.23):

$$w' = w - 8\,\text{mm} = 25.7\,\text{mm} - 8\,\text{mm} = 17.7\,\text{mm}; \quad N_L = N_p/2 = 41$$

$$d_p = 0.8\frac{17.7\,\text{mm}}{41} = 0.345\,\text{mm}$$

Using the nearest nominal value from the wire table of Figure 2.10, $d_p = 0.355\,\text{mm}$. From Figure 2.11, the skin effect at $f = 40\,\text{kHz}$ is insignificant.

Primary cross-sectional area:

$$A_p = \frac{0.355^2\,\text{mm}^2 \times \pi}{4} = 0.099\,\text{mm}^2 \approx 0.1\,\text{mm}^2$$

Equation (7.30):

$$I_{pr.m.s.} = 0.4082 \times 1.1\,\text{A} = 0.45\,\text{A}$$
$$J_p = 0.45\,\text{A}/0.1\,\text{mm}^2 = 4.5\,\text{A/mm}^2, \text{ which is acceptable.}$$

If the secondary winding is to be applied in one layer, from Equation (7.23),

$$d_s = 0.8 \times \frac{17.7\,\text{mm}}{9} = 1.57\,\text{mm}$$

A litz wire of 1.5 mm outside diameter is used for the secondary winding.

Cross-sectional area:

$$A_s = \frac{1.5^2\,\text{mm}^2 \times \pi \times 0.42}{4} = 0.74\,\text{mm}^2$$

(Copper space factor $k_{Cu} = 0.42$, from Figure 2.42).

$$I_{sr.m.s.} = 1.633 \times 2\,A = 3.27\,A$$

$$J_s = 3.27\,A/0.74\,mm^2 = 4.4\,A/mm^2, \text{ which is acceptable.}$$

The total depth of all windings is $2 \times 0.395\,mm + 1.5\,mm = 2.3\,mm$. This is exactly one-third of the total winding depth of 6.9 mm, and sufficient room is left for insulation and screen. For the winding construction shown in Figure 2.10, the total winding depth including the screens and insulation becomes

$$h = 2x + y = 8 \times 0.07\,mm + 2 \times 0.395\,mm + 2 \times 0.1\,mm$$
$$+ 1.5\,mm = 3.05\,mm$$

Equation (7.24):

$$L_\sigma = 12.57 \times 10^{-9}\,H/cm \times 81^2 \times \frac{6.9\,cm}{2.57\,cm} \times (0.038\,cm + 0.229\,cm/3)$$

$$= 25.3\,\mu H$$

The measured value of $L_\sigma = 23.5\,\mu H$ is close to the calculated value; the leakage inductance according to Equation (7.28) should in fact be considerably less.

Equation (7.25):

$$k_\sigma = \frac{23.5\,\mu H}{2.44\,\mu H} = 0.96\%, \text{ which is relatively good}$$

Since in this case there is only one secondary output controlled via an optocoupler, the effect of leakage inductance is not very noticeable. The situation is different when there are several outputs, and particularly when, for the sake of economy, the regulated quantity is taken from a special additional winding. In such a case the leakage must be made much smaller still by use of careful winding techniques.

Transformer loss:

$$P_{Rp} = 0.45^2\,A^2 \times 81 \times 0.069\,m \times 0.1742\,\Omega/m = 0.19\,W$$

$$P_{Rs} = 3.27^2\,A^2 \times \frac{9 \times 0.069\,m}{57\,\Omega\,m/mm^2 \times 0.74\,mm^2} = 0.16\,W$$

Core loss without air gap, from Figure 7.11, at $\hat{B}_{max} = 0.29\,T$: $P_{core} = 8\,W$. From Figure 2.12, with $g = 60\,g$, $P_{core} = 160\,mW/g \times 60\,g = 9.8\,W$. Average for the purposes of calculation $= 9\,W$.

Equation (7.19):

$$\mu_e = \frac{350 \times 10^{-9}\,H \times \dfrac{9.22\,cm}{1.25\,cm}}{12.57 \times 10^{-9}\,H/cm} = 205$$

$$P'_{core} = 9\,W \times \frac{205}{2000} = 0.92\,W$$

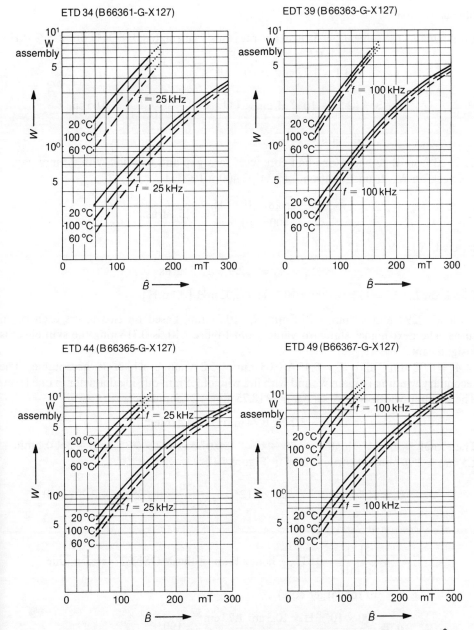

Fig. 7.11 Variation of loss in ETD cores (W/assembly) with alternating flux density $\hat{B}$ at various frequencies and temperatures (measured without air gaps) (Siemens)

Total transformer loss:

$$P_T = 0.19\,\text{W} + 0.16\,\text{W} + 0.92\,\text{W} = 1.27\,\text{W} \approx 1.3\,\text{W}$$

Bipolar version

$n = 9$ remains the same; $L_p = 5.28\,\text{mH}$ as previously calculated; ETD44 core; $f = 20\,\text{kHz}$.

$$N_p \geqslant \frac{0.5 \times 230\,\text{V} \times 10^4}{0.3\,\text{T} \times 1.72\,\text{cm}^2 \times 20 \times 10^3\,\text{s}^{-1}} \geqslant 111.4$$

$$s = 12.57 \times 10^{-9}\,\text{H/cm} \times 111^2 \times \frac{1.72\,\text{cm}^2}{5.28 \times 10^{-3}\,\text{H}} = 0.05\,\text{cm}$$

$$= 0.5\,\text{mm}$$

Selected from data sheet: one half-core with zero gap and one half with 0.5 mm gap.
Check on N_p, from Equation 2.17: $A_L = 400 \times 10^{-9}\,\text{H}$

$$N_p = \sqrt{\frac{5.28 \times 10^{-9}\,\text{H}}{400 \times 10^{-9}\,\text{H}}} = 115\ \text{turns};$$

selected, $N_p = N_1 = 112$ turns

$$N_s = N_2 = 112/9 = 12.5\ \text{turns}$$

Check on L_p: $L_p = 112^2 \times 400 \times 10^{-9}\,\text{H} = 5.02\,\text{mH}\ [5.1\,\text{mH}]$

With $w' = 29.5\,\text{mm} - 8\,\text{mm} = 21.5\,\text{mm}$, $d_p = 0.31\,\text{mm}$, based on two layers each of 56 turns. The next higher standard value, from Figure 2.10, is 0.315 mm; the skin effect is insignificant.

$A_p = 0.08\,\text{mm}^2$ and $J_p = 0.45\,\text{A}/0.08\,\text{mm}^2 = 5.8\,\text{A/mm}^2$, which is acceptable. The secondary winding is again wound with litz wire of 1.5 mm outside diameter in one layer. The winding width is $12.5 \times 1.5\,\text{mm} = 18.75\,\text{mm} < 21.5\,\text{mm}$.

$$J_s = 3.27\,\text{A}/0.74\,\text{mm}^2 = 4.4\,\text{A/mm}^2$$

The total winding depth is $2x + y = 8 \times 0.07\,\text{mm} + 2 \times 0.352\,\text{mm} + 2 \times 0.1\,\text{mm} + 1.5\,\text{mm} = 2.96\,\text{mm}$ (construction as in Figure 7.10).

$$L_\sigma = 12.57 \times 10^{-9}\,\text{H/cm} \times 112^2 \times \frac{7.77\,\text{cm}}{2.95\,\text{cm}} \times (0.038\,\text{cm} + 0.22\,\text{cm}/3)$$

$$= 46.2\,\mu\text{H}\ [31\,\mu\text{H}]$$

$$= \frac{31\,\text{H}}{5.1\,\text{mH}} = 0.61\% \text{---better than in the MOSFET version (due}$$

to the larger core)

$$\mu_e = \frac{400 \times 10^{-9}\,\text{H} \times 10.3\,\text{cm}/1.73\,\text{cm}^2}{12.57 \times 10^{-9}\,\text{H/cm}} = 190$$

$$P'_\text{core} = 6\,\text{W} \times 190/2000 = 0.6\,\text{W}$$

$$P_{Rp} = 0.45^2\,\text{A}^2 \times 12 \times 0.077\,\text{m} \times 0.2212\,\Omega/\text{m} = 0.38\,\text{W}$$

$$P_{Rs} = 3.27^2\,\text{A}^2 \times 0.023\,\Omega = 0.25\,\text{W}$$

Total transformer loss:

$$P_T = 0.6\,\text{W} + 0.38\,\text{W} + 0.25\,\text{W} = 1.23\,\text{W} \approx 1.3\,\text{W}$$

The transformer losses in the two versions are thus practically the same.

7.3 CALCULATIONS FOR THE SECONDARY CIRCUIT

Since there is no inductor in the secondary circuit of the flyback converter, calculations are limited to the rectifier diode and the output capacitor. The maximum secondary current has already been determined from Equations (7.31) and (7.32). (Here again a comparison can be made with the forward converter. If A is put equal to 1 in Equation (7.31), Equation (5.25) is obtained, giving the maximum current in the forward converter.)

In the present example with discontinuous current mode (Equation (7.32)),

$$I_{\text{smax}} = 4 \times 2\,\text{A} = 8\,\text{A}\,[8\,\text{A}]$$

In dimensioning the rectifier diode in relation to the maximum permissible current I_{FAVmax} and the maximum conduction loss W_{Dc}, a substantial safety margin has to be allowed in the flyback converter for the overload condition. Since the flyback converter is only used in the relatively low-power region, the circuit should be as uncomplicated as possible. Static secondary current limiting is therefore dispensed with (see Figures 2.25 and 6.4) and only the primary current is monitored, on a pulse basis. How rapidly the secondary current decreases after the maximum load current I_{Omax} has been exceeded, and the magnitude attained by the overload current, is very dependent upon the control IC used. Figures 7.12 to 7.14 show the overload characteristics with control ICs Types TDA1060, TDA4600 and TEA1001 (see also Figure 6.13).

The most satisfactory characteristic $V_0 = f(I_0)$ is that obtained with the TDA4600. This is not surprising, since this circuit has an additional control facility in comparison with the others, namely a variation in frequency according to the control conditions.

If, therefore, an overload condition can arise, it should be taken into account in dimensioning the rectifier diode. It is indeed safe to assume that other control ICs behave

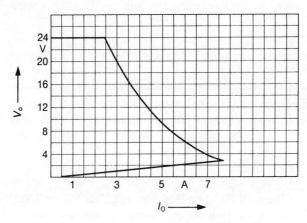

Fig. 7.12 Output voltage $V_0 = f(I_0)$ (complete characteristic) for the MOSFET version with the TDA1060 control circuit

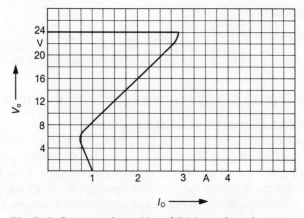

Fig. 7.13 Output voltage $V_0 = f(I_0)$ (complete characteristic) for the bipolar version with the TDA4600 control circuit

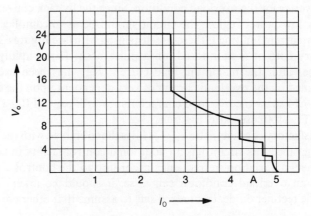

Fig. 7.14 Output voltage $V_0 = f(I_0)$ (complete characteristic) for the bipolar version with the TEA1001 control circuit

in a less satisfactory way than the TDA1060. It may thus be taken that

$$I_{FAVmax} = (2 \text{ to } 4)I_{0max} \quad (A) \tag{7.35}$$

With an unspecified control IC the higher multiplication factor should be assumed; for the TDA4600 a factor of two is sufficient. The conduction loss (without regard to the overload condition) is

$$P_{Dc} = V_F I_{0max}(1 - \delta_{Tmin})(2 - A) \quad (W) \tag{7.36}$$

where V_F is the forward voltage drop at $(2 - A)I_{0max}$ (V), I_{0max} is the maximum output current (A) and A is the multiplication factor for I_{smax} (see Figure 7.4). (If A is put equal to 1, as it must be for the forward converter, the corresponding Equation (2.42) is obtained.)

For discontinuous-mode operation, Equation (7.36) can be simplified to

$$P_{Dc} = V_F 2I_{0max}(1 - \delta_{Tmin}) \quad (W) \tag{7.37}$$

(If the overload behaviour is to be taken into account the higher value given by Equation (7.35) should be substituted for I_{Omax}, or at least the figure should be doubled.)

The maximum reverse voltage could be calculated from Equation (5.24), using the value of δ_{Tmin} obtained from Equation (7.6), but, since δ_{Tmin} becomes very low and ill-defined at minimum load or no-load (which condition is likely to arise in a flyback converter), the result given by Equation (5.24) is too inaccurate. It is better to use the following relationship, which was adduced in the case of the forward converter for the purpose of comparison:

$$V_{Rmax} = V_{imax}n + V_0 \quad (V) \tag{7.38}$$

It is necessary to calculate the switching loss in accordance with Equation (2.43), as for the forward converter, only in the case of the continuous current mode; in discontinuous-mode operation (as in the present example), the rectifier diode does not switch from full forward current to the reverse blocking condition. Since the forward current falls to zero before reverse voltage is applied, the reverse recovery charge is not a factor to be considered. This applies equally in the event of step changes of load, even if continuous-mode operation can occur briefly. Step-load changes are relatively infrequent, and do not, therefore, give rise to appreciable switching losses.

Thus, in the example,

Equation (7.35):

$$I_{FAVmax} = (2 \text{ to } 4) \times 2\,A = 4 \text{ to } 8\,A$$

Equation (7.38):

$$V_{Rmax} = 340\,V/9 + 24\,V = 62\,V\,[61\,V]$$

A diode must therefore be selected with a maximum forward current I_{FAV} of about 8 A and a reverse voltage rating of about 80 V. There is no question of using a Schottky diode here. Since V_0 is relatively high, epitaxial diodes are the most satisfactory. The Unitrode UES2402 was selected ($V_{Rmax} = 100\,V$; $I_{FAV} = 16\,A$; $t_{rr} = 35\,ns$).

Equation (7.37):

$$V_F = 1\,V \text{ at } I_F = 16\,A$$
$$P_{Dc} = 1\,V \times (4 \text{ to } 16\,A) \times (1 - 0.42) = 2.3 \text{ to } 9.3\,W$$

Equation (2.45):

$$R_{thca} \leqslant \frac{150\,°C - 50\,°C}{9.3\,W} - 1.75\,K/W \leqslant 9\,K/W$$

selected, $R_{thca} = 5\,K/W$.

An RC protective network is not necessary under conditions of constant or slowly varying load, but cannot be dispensed with for step load changes, being required for the suppression of possible overshoot voltages. Since no stored charge is quoted for the UES2402 in the data sheet, a figure may be taken from that for a similar type from another manufacturer (Thomson BYW51). The estimated leakage inductance on the secondary side is $L_{\sigma s} = 6\,\mu H$.

Equation (5.31):

$$\frac{dI_F}{dt} = \frac{62\,V}{6 \times 10^{-6}\,H} = 10\,A/\mu s;$$

from the data sheet, $Q_{rr} = 20\,nA\,s$; $t_{rr} = 80\,ns$.

Equation (5.28):

$$I_{RM} = \frac{2 \times 20 \times 10^{-9}\,A\,s}{80 \times 10^{-9}\,s} = 0.5\,A$$

Equation (5.27):

$$C_s \geqslant \frac{6 \times 10^{-6}\,H \times 0.5^2\,A^2}{62^2\,V^2} \geqslant 0.4\,nF$$

selected, $C_s = C_4 = 2.2\,nF/100\,V$.

Equation (5.26):

$$R_s \leqslant \frac{62\,V}{0.5\,A} \leqslant 124\,\Omega; \text{ selected, } R_s = R_2 = 27\,\Omega/0.5\,W$$

$$P_{Rs} = 0.5 \times 2.2 \times 10^{-9}\,F \times 62^2\,V^2 \times 40 \times 10^3\,s^{-1} = 0.17\,W$$

The dimensioning of the output capacitor C_0 must take into account both the current
rating and the permissible output ripple. According to [21],

$$I_{Crms} = \sqrt{I_{srms}^2 - I_0^2} = I_0 \sqrt{\frac{8}{3} \times \frac{1 + A + A^2}{(1 + A)^2} - 1}\quad (A) \tag{7.39}$$

Hence, for the discontinuous mode, with $A = 0$,

$$I_{Crms} = 1.3 I_0\quad (A) \tag{7.41}$$

The output ripple voltage (neglecting the inductance of the output capacitor, which i
usually legitimate) is given by

$$\Delta V_0 = ESR\,I_{smax} = Z_{max} I_{smax}\quad (V_{pp}) \tag{7.41}$$

In the example, with $I_0 = 2\,A$ and $I_{smax} = 8\,A$, $I_{Crms} = 1.3 \times 2\,A = 2.6\,A$. In making
choice from the capacitor tables, it should be observed that in most tables e permissibl
alternating current is quoted for $f = 100\,Hz$ and $\theta_a = 85\,°C$. For $f \geqslant 10\,kHz$ th
correction factor is given by Figure 2.19 as 1.4; a further factor of 1.7 is given b
Figure 2.20 for $\theta_{amax} = 50\,°C$. The total correction factor is then $k = 1.4 \times 1.7 = 2.38$. Th
maximum current I_{Crms} flowing in the capacitor can thus be reduced by this factor, so tha
the capacitor needs to have a rating of only $I_{Crms}/k = 2.6\,A/2.38 = 1.1\,A$.

According to Figure 2.16 the 2200 $\mu F/25\,V$ unit could be used, but the voltage safet
margin seems barely adequate. A 3300 $\mu F/40\,V$ type was therefore chosen from anothe
series. The data for this capacitor is: Z_{max} (10 kHz) = 85 mΩ; $I_{C\sim} = 2\,A$ (100 Hz and 85 °C
B41455 heavy-duty series (Siemens). It would alternatively be possible to use severa
capacitors of a cheaper type connected in parallel—e.g. three 2200 $\mu F/40\,V$ units in th
B41070 series with $Z_{max} = 0.12\,\Omega$ at $f \geqslant 10\,kHz$ and $I_{C\sim} = 0.37\,A$ (100 Hz/85 °C). Thi

would similarly give $I_{C\sim} = 1.1$ A and a still lower value for $Z_{max} = 120\ m\Omega/3 = 40\ m\Omega$. A further possibility would be to use a special electrolytic capacitor from the list of Figure 5.32. The $6300\ \mu F/28$ V unit could be considered, and possibly also the $4500\ \mu F/35$ V; both have approximately similar characteristics. There is no need to apply the frequency correction factor from Figure 2.19 to these capacitors, since the quoted current rating applies with $f \geqslant 20$ kHz; the permissible current is so high, however, that no difficulty arises in this regard. Also the extremely low impedance of about $7\ m\Omega$ should satisfy most requirements.

Calculation may now proceed based on the $3300\ \mu F/40$ V capacitor selected from the B41455 series (Siemens), with $Z_{max} = 85\ m\Omega$ at $f \geqslant 10$ kHz. The output ripple is given by Equation (7.41):

$$\Delta V_0 = 85 \times 10^{-3}\ \Omega \times 8\ A = 680\ mV_{pp}$$

This output ripple is much too large in relation to the specification in the example of $250\ mV_{pp}$ (the normal level of 1 per cent of the direct output voltage V_0). If a special electrolytic of e.g. $4500\ \mu F/35$ V, were used ΔV_0 would be sufficiently low at $7 \times 10^{-3}\ \Omega \times 8\ A = 56\ mV_{pp}$. Another solution was found, however, in the addition of an output filter, consisting of a small inductor and a further small electrolytic capacitor. A ring core Type $R_{25/10}$ with a single turn was used; i.e. the connecting lead between C_2 and C_3 was passed through the core. N30 core material was selected, giving an inductance factor $A_L = 400$ nH. The inductance is thus $L = 1^2 \times A_L = 4.4\ \mu H$. The voltage division between L and C_3 is given in general terms by

$$\frac{V_{C3}}{V_{C2}} = \frac{Z_{max}}{Z_{max} + X_L} \tag{7.42}$$

where Z_{max} is the equivalent impedance of capacitor C_3 (Ω) and X_L is the inductive reactance of inductor L (Ω).

At the switching frequency of 40 kHz,

$$X_L = 6.28 \times 40 \times 10^3\ s^{-1} \times 4.4 \times 10^{-6}\ H = 1.1$$

From the data sheet for the $470\ \mu F$ capacitor, $Z_{max} = 0.13\ \Omega$. Based on these values of Z_{max} and X_L, the attenuation is

$$\frac{V_{C3}}{V_{C2}} = \frac{0.13\ \Omega}{0.13\ \Omega + 1.1\ \Omega} = 0.1$$

The output ripple then becomes

$$\Delta V_0' = 680\ mV_{pp} \times 0.1 = 68\ mV_{pp}\ [25\ mV_{pp}]$$

It remains to check the inductor L in respect of d.c. magnetization. With $l_e = 63 \times 10^{-3}$ m (data sheet), Equation (2.18) gives

$$H = \frac{2\ A \times 1}{63 \times 10^{-3}\ m} = 32\ A/m$$

Assuming a characteristic similar to those of Figure 2.9 (no corresponding curves were available for the N30 core material), at $H = 30$ A/m the beginning of a reduction in μ_{rev} can be discerned, but it is still very small. Figure 7.15 shows the effect of L and C_3 at $f = 20$ kHz (bipolar version). Some observations on the dimensioning of the output

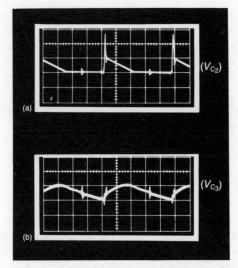

Fig. 7.15 Output ripple V_0 across capacitor C_2 (a) and voltage V_0 across capacitor C (b), after LC filter (bipolar version). Horizontal scale: $10\,\mu s$/division. Vertical scale (upper): $200\,mV$/division. Vertical scale (lower): $50\,mV$/division

capacitor in switched-mode power supplies may be presented here. In forward converters the alternating component of the output current ΔI_L is relatively small (around 10 per cent of I_0), and hence the capacitor current I_{Crms} is also small. The size of the output capacitor is thus determined by the permissible alternating voltage ΔV_0, which is particularly critical at low output voltages ($\Delta V_0 \approx 0.01\ V_0$). The alternating current loading is hardly a factor in the dimensioning of the capacitor. In flyback converters, on the other hand, the current that flows in the capacitor is much greater than I_0, so that both the output ripple and the heating current in the capacitor are considerably greater, and both must certainly be taken into account in the dimensioning. However, since the flyback converter is used only at relatively low power levels, it is the absolute quantities that are the ruling factors. The

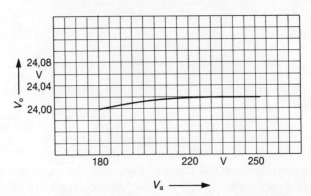

Fig. 7.16 Output voltage $V_0 = f(V_a)$ at $I_0 = 2\,A$ (MOSFET version)

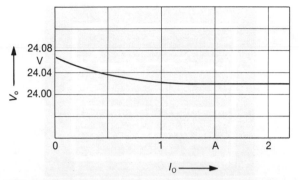

Fig. 7.17 Output voltage $V = f(I)$ (part characteristic) in the MOSFET version; $V_a = 220$ V r.m.s.

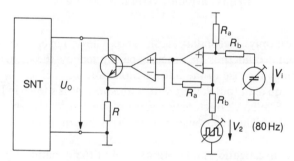

Fig. 7.18 Basic circuit of an electronic load for testing the dynamic response of a switched-mode power supply

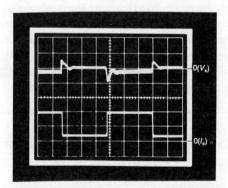

Fig. 7.19 Transient response of the output voltage V_0 to a step load change $I_0 = 1.8$ A with optimized control loop in the MOSFET version. Horizontal scale: 2 ms/division. Vertical scale (upper): 100 mv/division. Vertical scale (lower): 1 A/division

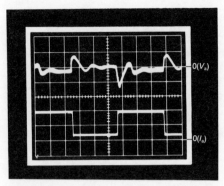

Fig. 7.20 Transient response of the output voltage V_0 to a step load change $I_0 = 1.8$ A with control loop not optimized in the MOSFET version. Horizontal scale: 2 ms/division. Vertical scale (upper): 100 mV/division. Vertical scale (lower): 1 A/division

actual capacitance is only of secondary significance. The design of a flyback converter for a high power output would thus in practice be limited by the critical stray (leakage) inductance as well as the size of the output capacitor.

The static variation of output voltage with mains voltage is shown in Figure 7.16 for the MOSFET version. Similar characteristics were obtained for the two bipolar versions with different control ICs. The relationship $V_0 = f(I_0)$ for the same circuit is shown in Figure 7.17.

To facilitate the investigation and optimization of the dynamic regulation characteristics, an electronic load was devised, as shown in Figure 7.18 [31].

The static load current drawn from the output of the power supply unit can be set by means of the voltage V_1. In addition, the load current can be varied step-wise under the control of V_2 at a repetition frequency of 80 Hz. Figure 7.19 shows the load response with the optimum choice of $R_{10} = R_{11} = 10$ kΩ (giving the control amplifier in the control IC a gain of unity) and $C_{11} = 470$ pF. For Figure 7.20, C_{11} was increased to 1.5 nF.

An increase in both the voltage overshoot and the recovery time is seen here. Although increasing the gain of the control amplifier produced a further improvement in the already good static regulation $V_0 = f(I_0)$ (Figure 7.16), the dynamic regulation was impaired. There is also an increase in the output voltage fluctuation in IC3 (TL431) due to the optocoupler.

7.4 CALCULATIONS FOR THE POWER TRANSISTOR

The power transistor in the flyback converter is subjected to the maximum input voltage plus the reflected secondary voltage plus the overshoot voltage. Since the transformation ratio is calculated according to Equation (7.10) on the basis of V_{imin}, the reflected secondary voltage can be taken for practical purposes as V_{imin}. Thus, allowing a safety factor of 0.8,

$$V_{imax} + V_{imin} + \Delta V \leqslant 0.8 V_{CBO} \quad \text{or} \quad \leqslant 0.8 V_{DSmax} \quad \text{(V)} \tag{7.43}$$

where V_{imax} is the maximum input voltage (V) (normally 340 V), V_{imin} is the minimum input voltage (V) (normally 220 to 230 V), ΔV is the overshoot voltage, from Equation (7.44), V_{CBO} is the collector-base off-state voltage with $I_E = 0$ (data sheet) $\approx V_{CEX}$ and V_{DSmax} maximum permissible drain-source voltage (data sheet).

The overshoot voltage is calculated as

$$\Delta V = \sqrt{\frac{L_\sigma}{C_{s1}}} I_{pmax} \quad \text{(V)} \tag{7.44}$$

where L_σ is the stray (leakage) inductance (H) (Equations (7.24) and (7.28)), C_{s1} is the capacitance in RCD transistor protection network (Equation (5.48)) and I_{pmax} is the maximum primary current (A) ((7.14) and (7.15)).

A low overshoot voltage ΔV, and consequently a not-excessive voltage stress on the transistor, can be achieved either through a low stray inductance (where possible) or a relatively large capacitor C_{s1}. The latter, however, implies a large loss in the associated resistor. Another possible way of limiting the overshoot is by means of a further protection network in parallel with the primary winding (R_{s2}, C_{s2} and D_{s2} in Figure 7.1). In the latter case the effect is reduced to that of the small stray inductances outside the transformer, resulting in an overshoot voltage of no more than 100 V. It would, of course, be possible to choose a transistor with a very high voltage rating, but that is certainly not the cheapest solution. Since the flyback converter, particularly in the discontinuous-waveform operating mode, may operate down to very low load currents, even on open circuit, the duty cycle δ_{Tmin} becomes very small and ill-defined. Equation (5.41) therefore needs to be expressed somewhat differently:

$$R_{s1} \leqslant \frac{T/10}{4C_{s1}} \quad (\Omega) \tag{7.45}$$

(If the 'on' time becomes extremely short with no load on the secondary side—see Figure 7.6—C_{s1} may not discharge within one cycle; i.e. the protective network may be only partially effective, or not effective at all. This is not very significant, however, because in that case the turn-off loss [see Equation (5.37)] also becomes very small.) To suppress undesirable oscillations between L_p and C_{s1}, R_{s1} should comply with the following relationship:

$$R'_{s1} \geqslant 2 \sqrt{\frac{L_p}{C_{s1}}} \quad (\Omega) \tag{7.46}$$

If Equation (7.46) is not satisfied, an additional resistor, which makes R_{s1} according to Equation (7.45) up to the value required by Equation (7.46), may be inserted at point x in Figure 7.1, i.e. in series with L_p and C_{s1}. This resistor must be shunted by a diode similar to

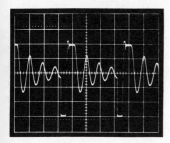

Fig. 7.21 Collector-emitter voltage in a flyback converter with too low a resistance in the RCD suppression network, showing pronounced oscillations in the off' periods of the switching transistor ($V_a = 220$ V r.m.s.; $I_0 = 7$ per cent I_{0max}). Horizontal scale: $10 \, \mu s$/division. Vertical scale: 100 V division

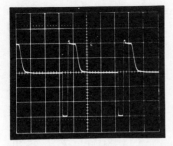

Fig. 7.22 Collector-emitter voltage in a flyback converter with optimum resistance in the RCD suppression net work, obtained by means of an additional RD network (similar to R_5 and D_4 in Figure 7.31) operative in the 'off' periods of the switching transistor (operation as in Figure 7.21). Horizontal scale: 10 μs/division. Vertical scale: 100 V/division

D_{s1} in the forward direction. The additional resistance is then effective only in the blocking direction of the transistor, and suppresses the undesired oscillations very effectively. In Figure 7.21, in the work reported in [32], this RD element was short-circuited; the normal operation is shown in Figure 7.22 (see also R_5 and D_4 in Figure 7.31).

MOSFET version

Assuming the protective network consisting of C_{s2}, R_{s2} and D_{s2} to be connected in parallel with L_p, ΔV in Equation (7.43) is estimated initially at 100 V.

Equation (7.43):

$$340 \text{ V} + 230 \text{ V} + 100 \text{ V} = 670 \text{ V};$$

$$I_p \text{ previously calculated as 1.1 A}$$

A MOSFET transistor with $V_{DSmax} = 700$ to 800 V and $I_{Dmax} = 1.5$ A is thus required. With some safety margin, the BUZ50A ($V_{DSmax} = 1000$ V, $I_{Dmax} = 2.5$ A) was selected. The BUZ80 ($V_{DSmax} = 800$ V, $I_{Dmax} = 2.6$ A) would also have met the requirements, but was not available.

Equation (5.40) (in the denominator, half the voltage V_{DSmax} is inserted, as previously in connection with MOSFET's: t_f (data sheet) $= 0.1$ μs):

$$C_{s1} \geqslant \frac{1.1 \text{ A} \times 0.1 \times 10^{-6} \text{ s}}{2 \times 500 \text{ V}} \geqslant 0.11 \text{ nF};$$

selected, $C_{s1} = C_6 = 270$ pF/1000 V.

Equation (7.45):

$$R_{s1} \leqslant \frac{0.1 \times 25 \times 10^{-6} \text{ s}}{4 \times 270 \times 10^{-12} \text{ F}} \leqslant 2.3 \text{ k}\Omega$$

Equation (5.42):

$$R_{s1} \geqslant \frac{340 \text{ V}}{1 \text{ A}} \geqslant 340 \text{ }\Omega$$

(Since in this case there is no current flowing, apart from the small charging current into the transformer winding capacitance, calculation could also be based on the maximum permissible drain current: it is sufficient to take a figure of 1 A.) It is convenient to determine $R_{s1} = R_4$ as the next lower standard value to 2.3 kΩ, i.e. 2.2 kΩ/3 W.

Equation (5.43):

$$P_{Rs1} = 0.5 \times 270 \times 10^{-12}\,\text{F} \times 570^2\,\text{V}^2 \times 40 \times 10^3\,\text{s}^{-1} = 1.75\,\text{W}$$

Equation (7.46):

$$R'_{s1} \geqslant 2\sqrt{\frac{2.44 \times 10^{-3}\,\text{H}}{270 \times 10^{-12}\,\text{F}}} = 6\,\text{k}\Omega$$

It would thus be necessary to add to the chosen R_{s1} (R_4 in Figure 7.7) a further resistor of about 3.9 kΩ (shunted by a diode), in order to suppress the oscillations visible in Figure 7.6, but since operation was not affected the additional circuit was dispensed with.

Equation (7.44):

$$\Delta V = \sqrt{\frac{23.5 \times 10^{-6}\,\text{H}}{270 \times 10^{-12}\,\text{F}}} \times 1.1\,\text{A} = 325\,\text{V}$$

This overshoot voltage added to the maximum voltage $V_{imax} + V_{imin} = 570$ V would give about 900 V, which is close to the absolute maximum permissible value of 1000 V. As can be seen from Figure 7.5, the maximum voltage across the transistor (at V_{imax}) is about 620 V; the overshoot is thus about 50 V.

Equation (5.47):

$$R_{s2}C_{s2} \geqslant 10\,\text{T} \geqslant 250\,\mu\text{s}$$

If $C_{s2} = C_5$ is determined as 33 nF/400 V, then

$$R_{s2} \geqslant \frac{250 \times 10^{-6}\,\text{s}}{33 \times 10^{-9}\,\text{F}} \geqslant 7.5\,\text{k}\Omega$$

From considerations of power loss, $R_{s2} = R_3$ was selected as 68 kΩ/3 W.

$$P_{Rs2} = P_{R3} = \frac{340^2\,V^2}{68 \times 10^3} = 1.7\,\text{W}$$

For the diodes $D_{s1} = D_2$ and $D_{s2} = D_3$ Type DSR3800 was chosen ($V_R = 800$ V, $I_F = 1$ A, $t_{rr} = 50$ ns).

Transistor losses:

The loss is mainly the conduction loss W_{Trc}. Equation (2.56) cannot be applied in this case, however, because in the example under consideration the current waveform is not rectilinear but triangular. (It would be possible to write Equation (2.56) as

$$P_{Trc} = R_{DSon}I_{prms}^2$$
$$P_{Trc} = 7.5\,\Omega \times 0.45^4\,\text{A}^2 = 1.52\,\text{W}$$
$$(R_{DSonmax} = 7.5\,\Omega\ \text{at}\ \theta_j \approx 100\,^{\circ}\text{C})$$

Equation (5.44) (from the data sheet, $t_f = 0.1\,\mu$s):

$$P_{Trsoff} = \frac{1.1^2\,\text{A}^2 \times (0.1 \times 10^{-6}\,\text{s})^2}{24 \times 270 \times 10^{-12}\,\text{F}} \times 40 \times 10^3\,\text{s}^{-1} = 75\,\text{mW}$$

which is insignificantly small. W_{Trson} can be neglected, since it is similarly small.

232

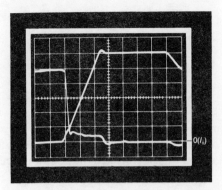

Fig. 7.23 Turn-off waveforms of the BUZ50A MOSFET with $V_{imax} = 242$ V r.m.s. and $I_{0max} = 2$ A, showing 'current tail'. Horizontal scale: 100 ns/division. Vertical scale for I_D: 0.2 A/division. Vertical scale for V_{DS}: 100 V/division

After the 'inner MOSFET' has turned off, a current flows briefly in the output capacitance C_{oss} (according to the data sheet about 50 pF at high drain-source voltages). With a rate of rise of voltage, from Figure 7.23, $dV_{DS}/dt = 600\,V/250\,ns = 2.4\,kV/\mu s$, this 'current tail' can be calculated as

$$I'_D = C_{oss}\,dV_{DS}/dt = 50 \times 10^{-12}\,F \times 2.4 \times 10^3\,V/s \times 10^{-6}$$
$$= 0.12\,A$$

Approximately the same figure can be read off from Figure 7.23.

Here again an efficiency calculation will be carried out for the MOSFET version and compared with the measured figure.

Mains rectification:

$$P_{C1} = 0.48\,W; \quad P_A = 0.45\,W; \quad P_{Rec1} = 0.4\,W$$
$$P_{filter} \approx 0.5\,W; \quad P_{R1} = 1.85\,W; \quad P_{tot} = 3.7\,W$$

Rectifier diode:

$$P_{Dc} = 2.4\,W \text{ (without overload)}; \quad P_{C2} = 0.57\,W$$
$$P_{tot} = 2.97\,W$$

Transistor:

$$P_{Tr} = 1.6\,W; \quad \text{transformer:} \quad P_T = 1.3\,W$$

Protective networks:

$$P_{Rs1} = 1.75\,W; \quad P_{Rs2} = 1.7\,W; \quad P_{Rs} = 0.17\,W; \quad P_{tot} = 3.62\,W$$

Total loss:

$$P_{tot} = 3.7\,W + 2.97\,W + 1.6\,W + 1.3\,W + 3.62\,W = 13.2\,W$$

Efficiency $\eta = \dfrac{48\,W}{48\,W + 13.2\,W} = 79\%\ [83\%]$

Because the losses are always calculated on principle for the worst-case conditions, the calculated efficiency is here, as for the other converters, somewhat worse than that measured.

Bipolar version

The bipolar technique was applied, in the work discussed in [31], in two different variants, in one case with the switching transistor operated in the quasi-saturated mode with the TEA1001 control IC and in the other with the transistor saturated, using the TDA4600 control IC.

Quasi-saturated switching transistor

Figure 7.24 shows the switching transistor together with (part of) the output stage of the control IC and the anti-saturation diode D_7. The complete circuit of this version will be discussed in Section 7.6. The BUV46 transistor chosen for the bipolar version has the following characteristics:

$$V_{CE0} = 400 \text{ V}$$

$$V_{CEX} \text{ (with } I_C = 0 \text{ and } R_{BE} \leqslant 25\,\Omega) = 800 \text{ V}$$

$$V_{BE} \text{ (with } I_C = 1 \text{ A)} = 0.9 \text{ V}$$

$$t_f \text{ (saturated operation)} \leqslant 0.4 \,\mu\text{s}$$

To be able to select the power transistor correctly it is essential to refer to the safe operating area (SOAR) diagrams in regard to the turn-on and turn-off characteristics. These diagrams are shown in Figures 7.25 and 7.26; they apply in principle to all switching transistors. The turn-on diagram yields an equation for the determination of the resistor R_{s1} in the protective network.

$$R_{s1} \geqslant \frac{V_{CEmax} - V_{CE0}}{I_x} \quad (\Omega) \tag{7.47}$$

Where V_{CEmax} is the maximum applied voltage (V), V_{CE0} is the collector-emitter voltage with $I_B = 0$ (data sheet) and I_x is the transition current (2.3 A in Figure 7.25).

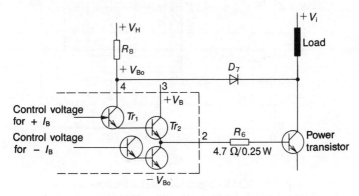

Fig. 7.24 Driver stage of the TEA1001 control circuit with a bipolar switching transistor and an anti-saturation diode D_7 (part-circuit)

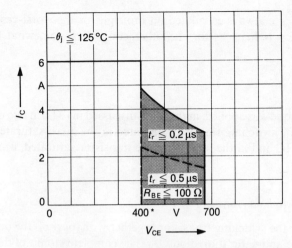

Fig. 7.25 SOAR diagram for the turn-on process using
the BUV46 bipolar switching transistor (Thomson CSF)

Equation (7.43): ΔV estimated at 60 V maximum having regard to the protective network R_{s2}, C_{s2}, D_{s2}. 340 V + 230 V + 60 V = 630 V < 640 V. The transistor is thus suitable. The same applies to the maximum current $I_{pmax} = 1.1$ A ($I_{Cmax} = 6$ A). Since, as can be seen from Figure 7.5, both the overshoot voltage and the reflected voltage V_{imin} have died away long before the instant of turn-on, the maximum voltage at turn-on is $V_{imax} = 340$ V. Thus the turn-on process, according to Figure 7.25, is uncritical.

At turn-off, a voltage of over 600 V occurs initially; at this point, from Figure 7.26, the current must have fallen below 0.3 A. Without a snubber network the turn-off process would not be possible with this transistor. Ameliorating the power loading of the transistor is thus not the only purpose of the RCD protective network.

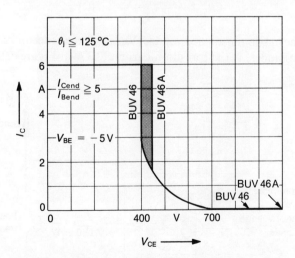

Fig. 7.26 SOAR diagram for the turn-off process using
the BUV46 bipolar switching transistor (Thomson
CSF)

Equation (7.47):

$$R_{s1} \geqslant \frac{570\,\text{V} - 400\,\text{V}}{2.3\,\text{A}} \geqslant 75\,\Omega$$

Equation (5.42) ($\Delta I_C = 3\,\text{A}$):

$$R_{s1} \geqslant \frac{340\,\text{V}}{3\,\text{A}} = 113\,\Omega$$

Equation (5.40):

$$C_{s1} \geqslant \frac{1.1\,\text{A} \times 0.4 \times 10^{-6}\,\text{s}}{2 \times 400\,\text{V}} \geqslant 0.55\,\text{nF};$$

Selected, $C_{s1} = C_6 = 0.68\,\text{nF}/1000\,\text{V}$
Equation (7.45):

$$R_{s1} \leqslant \frac{0.1 \times 50 \times 10^{-6}\,\text{s}}{4 \times 0.68 \times 10^{-9}\,\text{F}} = 1.84\,\text{k}\Omega;$$

selected, $R_{s1} = R_4 = 1.8\,\text{k}\Omega/5\,\text{W}$

$$P_{Rs1} = 0.5 \times 680 \times 10^{-12}\,\text{F} \times 570^2\,\text{V}^2 \times 20 \times 10^3\,\text{s}^{-1} = 2.2\,\text{W}$$

Equation (7.46):

$$R'_{s1} = 2 \times \sqrt{\frac{5.28 \times 10^{-3}\,\text{H}}{680 \times 10^{-12}\,\text{F}}} = 5.5\,\text{k}\Omega$$

$$R'_{s1} - R_{s1} = 5.5\,\text{k}\Omega - 1.8\,\text{k}\Omega = 3.7\,\text{k}\Omega$$

A resistor of about $3.7\,\text{k}\Omega$ would thus have to be incorporated to suppress the collector voltage oscillations completely. In practice a value of $2.7\,\text{k}\Omega$ (R_5) has been found to be sufficient. The turn-off waveforms so achieved are shown in Figure 7.27.

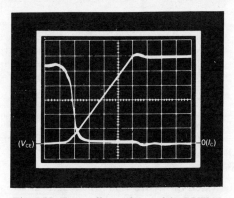

Fig. 7.27 Turn-off waveform of the BUV46 switching transistor with $V_{a\,max} = 242\,\text{V}$ r.m.s. and $I_{0\,max} = 2\,\text{A}$ in quasi-saturated operation. Horizontal scale: 100 ns/division. Vertical scale for I_C: 0.2 A/division. Vertical scale for V_{CE}: 100 V/division

236

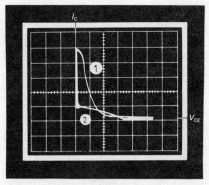

Fig. 7.28 SOAR diagram $I_C = f(V_{CE})$ with $V_{amax} = 242$ V r.m.s. and $I_{0max} = 2$ A for the BUV46 switching transistor in saturated operation. (1) Turn-off locus; (2) turn-on locus. Horizontal scale for V_{CE}: 100 V/division. Vertical scale for I_C: 0.2 A/division

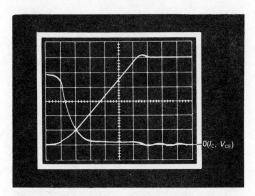

Fig. 7.29 Turn-off waveforms of the BUV46 switching transistor with $V_{amax} = 242$ V r.m.s. and $I_{0max} = 2$ A in saturated operation. Horizontal scale: 100 ns/division. Vertical scale for I_C: 0.2 A/division. Vertical scale for V_{CE}: 100 V/division

Due to the anti-saturation circuit, the current falls not, as assumed, in 0.4 μs, but in about 0.15 μs. At $V_{CE} = 100$ V the current is only 0.2 A. The turn-off behaviour is thus much better than required by Figure 7.26, as is confirmed by the recorded current-voltage diagram of Figure 7.27.

The base drive is illustrated in Figure 7.30 (quasi-saturated operation). At turn-on the output stage of the drive IC delivers a high-amplitude current pulse of about 0.6 A, to enable the transistor to turn on very rapidly and with low loss. The base current is interrupted for about 2 μs to dispel this over-saturation and then rises to about 0.16 A. Because the current in Figure 7.30 was measured in the lead from Terminal 2 of IC1, it includes the component through R_{10}. At $V_{BE} = 0.9$ V $I_{R10} = 0.9$ V/24 Ω = 0.04 A. There remains for the base about 0.12 A, implying a current gain of 1.1 A/0.12 A = 9.2, or approximately 10. The voltage across R_6 is $0.16\,\mathrm{A} \times 4.7\,\Omega = 0.75\,\mathrm{V}$. From Figure 7.24, $V_{CE} = 0.9\,\mathrm{V} + 0.75\,\mathrm{V} + 0.7\,\mathrm{V} + 0.3\,\mathrm{V} - 0.7\,\mathrm{V} = 2.05\,\mathrm{V}$, or approximately $2\,\mathrm{V}$. Thus the transistor is certainly operating in the quasi-saturated mode. It can also be seen from Figure 7.30 that between the switching-off of the positive base current and the initiation of the negative sweep-out current there is a gap of about $2\,\mu s$, so that recombination can take place in the collector-base region before it occurs in the base-emitter region. The high sweep-out current results in the very short measured fall time of the collector current.

The transistor loss in this variant also can be taken to be principally the conduction loss P_{Trc}. Calculation of the small turn-on and turn-off losses may be dispensed with.

The conduction loss is given by the relationship

$$P_{\mathrm{Trc}} = 0.5 V_{CE} I_{\mathrm{pmax}} \delta_{\mathrm{Tmax}} \quad (\mathrm{W}) \tag{7.48}$$

(The factor of 0.5 in comparison with Equation (5.35) arises from the discontinuous mode.)

With $V_{CE} = 2\,\mathrm{V}$ and $I_{\mathrm{pmax}} = 1.1\,\mathrm{A}$, the conduction loss in the quasi-saturated mode is $P_{\mathrm{Trc}} = 0.5 \times 2\,\mathrm{V} \times 1.1\,\mathrm{A} \times 0.5 = 0.55\,\mathrm{W}$, which may be rounded up to $1\,\mathrm{W}$ to represent the total loss.

In saturated operation, the conduction loss may be expected to be considerably lower (about $0.2\,\mathrm{W}$), but the switching losses are somewhat higher because of the less satisfactory switching conditions, and the very small total loss will again amount to around $1\,\mathrm{W}$. An exact calculation will be omitted here. It should be mentioned finally that the turn-off characteristics shown in Figure 7.26 apply only to turn-off with negative base-emitter voltage. The negative base voltage ($-1.5\,\mathrm{V}$ minimum) can either be provided by a negative auxiliary supply (see Figures 7.31, 6.18 and 5.40) or obtained from a positively charged capacitor switched to earth—e.g. C_7 in Figure 7.32.

7.5 DRIVE CIRCUITS

In the circuit of Figure 7.7 it is necessary to provide a MOSFET driver stage as shown in Figure 5.24, so that the MOSFET power transistor can be driven from a source of sufficiently low impedance to ensure short switching times. In the bipolar versions, drive

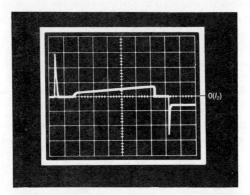

Fig. 7.30 Waveform of drive current in quasi-saturated operation (current from Terminal 2 of the TEA1001 control IC). $V_i = 220\,\mathrm{V}$ r.m.s. Horizontal scale: $2\,\mu s$/division. Vertical scale: $0.2\,\mathrm{A}$/division

ICs with integrated driver stages were employed in order to make the circuit as simple as possible. These drive circuits incorporate a small cooling tab to dissipate the small drive losses. It is particularly important in flyback converters in the low-power range to keep complexity to a minimum, since it has to be cheaper to provide a low output power than a higher.

7.6 CIRCUIT EXAMPLES

Flyback converter for $V_0 = 24\,V$ at $I_0 = 0$ to 2 A with bipolar switching transistor operating in the saturated mode

This circuit, shown in Figure 7.31, has the same characteristics as the circuit of Figure 7.7, already discussed in detail, with the BUZ50A MOSFET transistor. The mains supply, the transformer T (primary and secondary winding), the secondary-side rectification and the selection and protection of the power transistor have been described previously, and only some special features are considered here.

The optocoupler, not so far discussed, is dimensioned as follows. The TL431 integrated circuit represents approximately a programmable Zener diode with high internal gain (about 56 dB), such that at the reference terminal there is a very accurately controlled reference voltage $V_{ref} = 2.495$ V. Thus if the desired output voltage V_0 (in this case 24 V) is divided down to V_{ref} by means of an accurate voltage divider $R_{21} + R_{22}/R_{20} + R_{21} + R_{22}$, a certain cathode voltage appears on IC2. This can be between 2 V and 36 V; the possible range of current is 1 to 100 mA. For the purpose of dimensioning the circuit a voltage $V_K = 11$ V will be assumed, with $I_K = 2.5$ mA. With $V_F \approx 1.5$ V, $R_{19} = (24$ V $- 1.5$ V $- 11$ V$)/2.5$ mA $= 4.6$ kΩ; the standard value of 4.7 kΩ is selected. The current gain of the optocoupler is approximately 1, so the current in the transistor is also 2.5 mA. From the point of view of control and response characteristics, it is more satisfactory to provide the control amplification in advance of the light-emitting diode of the optocoupler. The same principle was applied in the MOSFET circuit of Figure 7.7 using the TDA1060 control IC; in that case, however, the circuit arrangement was simpler, because in the TDA1060 the 'plus input' of the control amplifier is connected to V_{ref}, whereas in the TEA1001 the 'minus input' is connected through an internal 1 kΩ resistor. Since the feedback resistor has always to be taken to the 'minus input' (and to obtain a gain of unity must similarly be of 1 kΩ), the voltage at that point is affected by the output voltage, owing to the low feedback resistance. From the data sheet (see also Section 8.3), the output voltage V_{11} of the error amplifier (Terminal 11) with an average supply voltage of 10 V is approximately 5 V. With $V_{12} = V_{13}$,

$$V_{13} = V_{ref} + (V_{11} - V_{ref})\frac{1\,k\Omega}{1\,k\Omega + R_{12}}$$

$$= 2.5\,V + (5\,V - 2.5\,V) \times \frac{1\,k\Omega}{2\,k\Omega} = 3.75\,V$$

R_{11} can then be calculated as

$$R_{11} = 3.75\,V/2.5\,mA = 1.5\,k\Omega$$

Due to the closed control loop, the duty cycle at any given time is just that required to keep the output voltage constant at $V_0 = 24$ V, which through the voltage divider

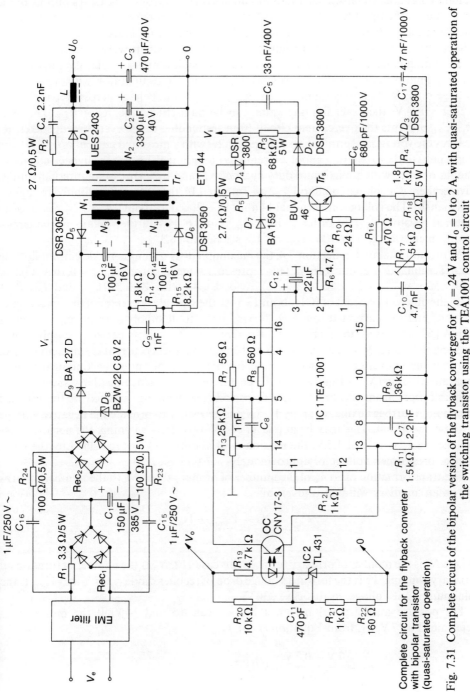

Fig. 7.31 Complete circuit of the bipolar version of the flyback converger for $V_0 = 24$ V and $I_0 = 0$ to 2 A, with quasi-saturated operation of the switching transistor using the TEA1001 control circuit

$R_{21}+R_{22}/R_{21}+R_{22}+R_{20}$ produces exactly the reference voltage of IC2. If V_0 increases (e.g. due to an increase in input voltage), the voltage applied to V_{ref} in IC2 rises, and the cathode voltage consequently falls. This results in a higher current through the light-emitting diode and hence in the optotransistor. The voltage across R_{11} therefore increases and the duty cycle is reduced. By this means the postulated increase in V_0 is reduced to a small amount. The control of the duty cycle at Terminal 13 is the same as if there were a direct connection to the output voltage. The latter, however, is not possible for reasons of isolation. Primary current limiting is effected at Terminal 15, through a comparator with a threshold voltage of 0.2 V. At a maximum current $I_{pmax} = 1.1$ A, the voltage across R_{18} is 1.1 A $\times$ $0.22\,\Omega = 0.24$ V; the exact limiting value can be adjusted by means of R_{17}. The filter R_{16}/C_{10} prevents the overcurrent circuit from responding to the ever-present spikes. R_7 reduces the voltage at Terminal 3 (see Figure 7.24) with a measured current of 0.16 A from Terminal 2 to the extent that about 2 V remains on Terminal 3. Since the flyback converter should operate with a maximum duty cycle of 0.5, and by virtue of an internal voltage divider there is a voltage equal to $V_A^+/2$ on Terminal 14, the duty cycle is limited to slightly less than 0.5. The voltage on Terminal 14 must be reduced to about 0.45 V_A^+ by means of an external resistor (R_{13}), in order to limit δ_{Tmax} to 0.6. For further details of the circuit arrangement for the TEA1001 control IC, see Section 8.3.

As previously stated in Section 7.4, the maximum discontinuous-mode current flowing from Terminal 2 is about 0.16 A. This represents a mean current $I_{2AV} = 0.16$ A $\times$ 0.5 $\times$ $0.44 \times 50 \times 10^{-6}$ s $\times 20 \times 10^3$ s^{-1} = 35 mA ($\delta_T \approx 0.44$ with $V_a = 220 V_{rms}$ and $I_0 = 2$ A). With the quiescent current of IC1 (about 18 mA), the secondary current of the optocoupler (2.5 mA) and the current through D_7 (6.5 mA), the total supply current requirement is 62 mA [52 mA]. Since the auxiliary power supply via $C_{16} + C_{15}$ can only supply 20 mA with mains voltage low (see Section 5.8 and Figure 5.55), this current is sufficient only to start up the converter. The actual auxiliary power supply must be obtained from a separate winding (N_3 in Figure 7.31). To ensure that the positive auxiliary supply V_A^+ is available right from the first switching period, this winding is poled in the forward direction. A further auxiliary supply V_A^- (N_4) is necessary to provide the negative voltage for base sweep-out. This may be at a level of 3 V. (For fast switching, the negative base voltage must be at least 1.5 V; the maximum of about -6 V is prescribed by the base-emitter breakdown voltage of approximately -7 V.)

The transformation ratio n_A of the number of primary turns N_p to the number of turns N_A on an auxiliary winding is given by

$$n_A = \frac{N_A}{N_p} = \frac{V_{Amin} + V_F}{V_{imin}} \qquad (7.49)$$

Where N_p is the number of primary turns, in this case 112, N_A is the number of turns on an auxiliary winding, V_F is the forward voltage drop of rectifier diode ≈ 0.7 V and V_{imin} is the minimum input voltage, in this case 230 V.

The positive auxiliary voltage must be between 6.5 and 14 V. If the minimum is determined as 7 V, n_A, from Equation (7.49), is

$$n_A^+ = \frac{7\,V + 0.7\,V}{230\,V} = 0.0335$$

$$n_A^+ = 112 \times 0.0335 = 3.75; \quad \text{selected,} \quad N_A^+ = N_3 = 4$$

Check on V_A^+:

$$V_{A\min}^+ = \frac{4}{112} \times 230\,\text{V} - 0.7\,\text{V} = 7.5\,\text{V}$$

$$V_{A\max}^+ = \frac{4}{112} \times 340\,\text{V} - 0.7\,\text{V} = 11.4\,\text{V}\,[11.2\,\text{V}]$$

Similarly, for the negative auxiliary supply,

$$n_A^- = \frac{-3\,\text{V} - 0.7\,\text{V}}{230\,\text{V}} = 0.0161$$

$$N_A^- = 112 \times 0.0161 = 1.8; \quad \text{selected,} \quad N_A^- = N_4 = 2$$

Check on V_A^-:

$$V_{A\min}^- = -\frac{2}{112} \times 230\,\text{V} + 0.7\,\text{V} = -3.4\,\text{V}$$

$$V_{A\max}^- = -\frac{2}{112} \times 340\,\text{V} + 0.7\,\text{V} = -5.35\,\text{V}\,[-4.7\,\text{V}]$$

The rectifier diodes must be fast-switching types, with a fairly high surge current rating, because of the large reservoir capacitors. Peak rectification takes place here, owing to the small load current. The type used was the DSR3050 ($V_R = 50\,\text{V}$, $I_F = 1\,\text{A}$, $I_{\text{surge}} = 20\,\text{A}$ for 1 ms).

The mean negative current is essentially determined by the resistors $R_6 + R_{10} + R_{18} = 29\,\Omega$. With $V_A^- = 4.65\,\text{V}$ at $V_s = 220\,V_{\text{rms}}$.

$$I_{2\text{AV}}^- = \frac{-4.65\,\text{V} + 0.3\,\text{V}}{29\,\Omega}$$

$$= -84\,\text{mA}\,[-89\,\text{mA}]$$

During the non-conducting period winding N_4 produces a positive voltage which (reduced to about a fifth by R_{14} and R_{15}) inhibits the turning-on the power transistor until it has fallen below 0.1 V. This means that the transistor can be turned on only when the magnetic flux in the transformer has decayed. An overshoot of the collector voltage as in Figures 7.6 or 7.21 would interfere with the correct functioning of this monitoring circuit, and R_5 and D_4 were therefore inserted.

Flyback converter for $V_0 = 24\,\text{V}$ at $I_0 = 0$ to 2 A with bipolar transistor operating in the saturated mode

This circuit, shown in Figure 7.32, conforms again to the same data as the two previous circuits (Figures 7.7 and 7.31). The mains supply, the transformer T and the auxiliary power supply (N_3) have already been discussed, and only some special features will be considered here.

Since the supply voltage must reach 12 V for IC1 to be fully energized, the Zener voltage of the Zener diode D_5 was increased to 15 V. Similarly an increase in the number of turns on N_3 to five would be recommended. The calculated auxiliary supply voltage of 7.5 V in operation is hardly sufficient with low mains voltage, but in fact the minimum voltage

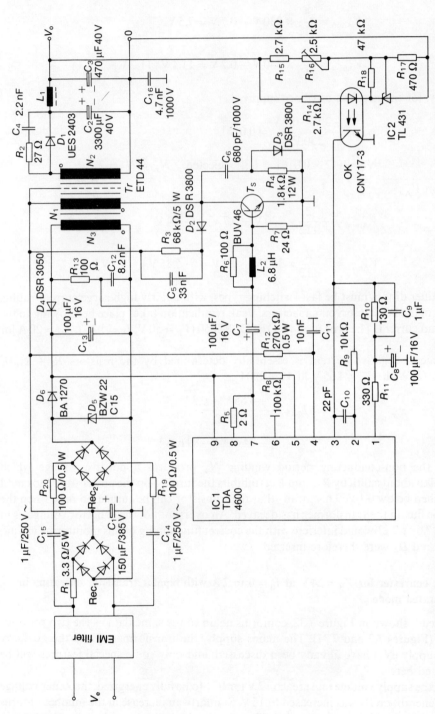

Fig. 7.32 Complete circuit of the bipolar version of the flyback converter for $V_0 = 24\,V$ and $I_0 = 0$ to $2\,A$, with saturated operation of the switching transistor using the TDA4600 control circuit

V_{imin} was determined for the case of one missing mains voltage half-cycle, which occurs very infrequently. In the normal situation, i.e. $a = 0$, the minimum voltage V_{imin} is not 230 V, but, with the chosen capacitor C_1 in Figure 5.5, substantially higher, namely 250 V. This increases V_{Amin}^+ to 8.2 V, which is sufficient. The Zener voltage selected for the Zener diode D_5 in Figure 7.32, as in Figure 7.31, is of secondary importance; it fulfils mainly a protective function. In some applications this diode is in fact superfluous, because after the start-up period (current $< 5 \, mA$) the voltage on Terminal 9 of IC1 is determined by the voltage of the permanently-connected auxiliary winding N_3, due to the much higher current consumption in operation (60 to 160 mA). The control voltage on Pin 3 should be about 2 V and is obtained from the 4 V reference voltage on Pin 1. The maximum load current at this point should not exceed 5 mA. With the chosen dimensioning, a collector current of 4.35 mA flows in the optotransistor, and the same in the light-emitting diode. As there is no facility in the TDA4600 control IC for varying the gain of the control amplifier, the control characteristic was optimized by adjusting the gain of IC2 through R_{18} and C_9. Consequently the exact output voltage must be adjusted by means of R_{16}. In this case the optotransistor is connected differently in comparison with Figure 7.31. An increase in V_0 produces an increase in the collector current of the optotransistor and hence a reduction in the voltage at Pin 3 of IC1, which effects a reduction in the duty cycle. The voltage produced by winding N_3 on the one hand provides the auxiliary power supply, while on the other it provides for monitoring of the resetting of the transformer at Terminal 2. The low-pass filters R_{13}/C_{12} (time constant about $1 \, \mu s$) and R_9/C_{10} prevent a response to superimposed voltage spikes. The voltage is reduced to the required level by R_9 in combination with an internal resistance. The rate of rise of the voltage on C_{11} (Terminal 4) determines the tilt of the base-drive current pulse as shown in Figure 7.33.

$$\frac{\Delta V_4}{\Delta T} = \frac{V_{imin}}{\tau_s} \quad (V/s) \tag{7.50}$$

where ΔV_4 is the voltage difference between initial and final values on Terminal 4 = 2 V, Δt is the maximum 'on' time of switching transistor at V_{imin} (with $\delta_{Tmax} = 0.5$ and $f = 20 \, kHz$ $\Delta t = 0.5 \times 50 \, \mu s = 25 \, \mu s$), V_{imin} is the minimum direct input voltage, in this case 230 V and τ_s is the charging time constant $= R_{12}C_{11}$.

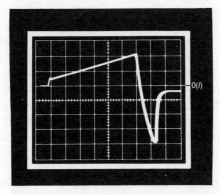

Fig. 7.33 Waveform of base current in the circuit of Figure 7.32. Horizontal scale: $2 \, \mu s$/division. Vertical scale: 0.1 A/division

244

Equation (7.50):

$$\tau_s = \frac{25 \times 10^{-6}\,\text{s} \times 230\,\text{V}}{2\,\text{V}} = 2.87\,\text{ms}$$

If C_{11} is chosen as 10 nF, R_{12} becomes 287 kΩ (standard value 270 kΩ). If V_4 reaches a limit of 4 V, the turn-on process is inhibited and the primary current thereby limited. To ensure adequate transistor saturation, an overdrive factor of about two is provided. Compared with the previous version, with the same output, this implies a current gain β_{sat} of 5, as opposed to 10. During the conducting period capacitor C_7 is charged from an initial voltage of 1.5 V to a maximum of 3.5 V, representing a charging voltage of 2 V. The positive base current delivered from Terminal 8 is limited by resistor R_5; this may be between 0.33 Ω and 2.2 Ω, and is given approximately by

$$R_5 \approx \frac{0.17\beta_{sat}L_p}{\tau_s} \quad (\Omega) \tag{7.51}$$

With the calculated values $\beta_{sat} = 5$, $L_p = 5.28$ mH and $\tau_s = 2.87$ ms, R_5 becomes

$$R_5 = \frac{0.17 \times 5 \times 5.28 \times 10^{-3}\,\text{H}}{2.87 \times 10^{-3}\,\text{s}} = 1.6\,\Omega$$

The fall time could be reduced somewhat by increasing R_5 slightly to 2 Ω. The inductor L_2 was optimized at 6.8 μH to provide the shortest fall time (dI_B/dt matching).

The instant of turn-off is determined by the voltage on Terminal 3. This was adjusted, by means of R_{16}, so that the required output voltage of 24 V was just attained. Turn-on can then occur only when the transformer is reset—i.e. the voltage on Pin 2 is practically zero.

Since the TDA4600 IC used controls the duty cycle mainly in response to mains voltage variations, but also, in response to load variations, controls the frequency (at $I_0 = 2$ A the frequency was measured as 23 kHz, but at $I_0 = 0$ as 100 kHz), very good regulation is to be expected. The stabilization against both mains variations (about 10 mV over the whole input voltage range) and load variations (about 10 mV with $I_0 = 2$ A to zero), and also the dynamic response, were optimal in this version. The overload characteristic shown in Figure 7.13 can also be regarded as excellent.

The negative sweep-out current, flowing in Terminal 7, has no limiting resistor, and will therefore be higher (see Figure 7.33). Inductor L_2 reduces the rate of fall, so that the collector-base region can be cleared of charge carriers before the onset of the negative base current. This enables the minimum fall time to be obtained in saturated operation, signifying a low turn-off loss. It is worth noting that resistor R_4 becomes hot under no-load conditions, i.e. at maximum frequency. In many applications, therefore, only the capacitor C_6 is provided, which means that at turn-on the transistor is loaded by the discharge current of the capacitor. The efficiency, however, is improved.

As is clear from the explanatory notes, the TDA4600 control circuit is suitable only for flyback converters operating with a discontinuous current mode.

Flyback converter with multiple outputs using a bipolar switching transistor in the saturated mode; tailored for a television receiver

A circuit of this type is shown in Figure 7.34; it differs little from that of Figure 7.32 in its operation. The Zener diode is omitted from the start-up circuit. The output is stabilized

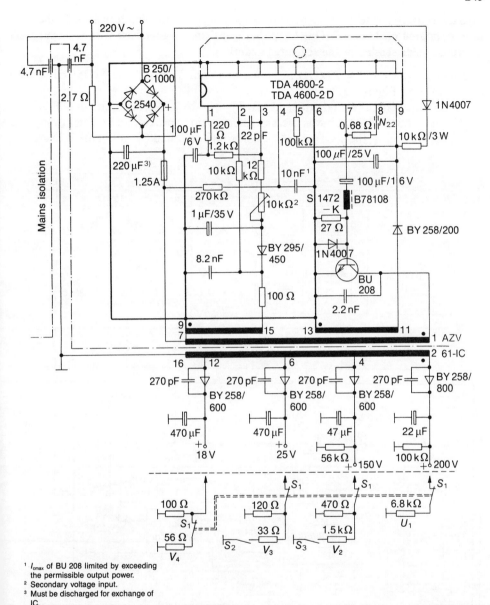

[1] I_{cmax} of BU 208 limited by exceeding the permissible output power.
[2] Secondary voltage input.
[3] Must be discharged for exchange of IC.

Fig. 7.34 Flyback converter with discontinuous current mode in the switching transistor in saturated operation, dimensioned with several outputs for a television receiver, usng the TDA4600-2 control circuit (Siemens)

not through an optocoupler, but through the rectified voltage of an auxiliary winding on the transformer, regulation being effected by virtue of the negative polarity of this voltage, as in the optocoupler circuit of Figure 7.32. The overall regulation in this case cannot, of course, be so good as that afforded by the version of Figure 7.32; this makes it particularly important that the transformer (differently dimensioned) should have an extremely low leakage reactance.

Flyback converter with multiple outputs using a bipolar switching transistor in the quasi-saturated mode or a MOSFET power transistor (in each case with secondary-side or primary-side sensing of the regulated quantity)

The bipolar version ($f = 25\,kHz$) is shown in Figure 7.35 and the MOSFET version ($f = 50\,kHz$) in Figure 7.36).

The outputs in both versions are $\pm 15\,V$ at $I_0 = \pm 0$ to 3 A. As in the circuit of Figure 7.7 (and Figures 7.31 and 7.32), it was necessary here again to provide an LC filter to suppress spikes. As previously, the (a.c.) gain of the control IC (R_{20} and C_{14} in Figure 7.35 and R_{20}

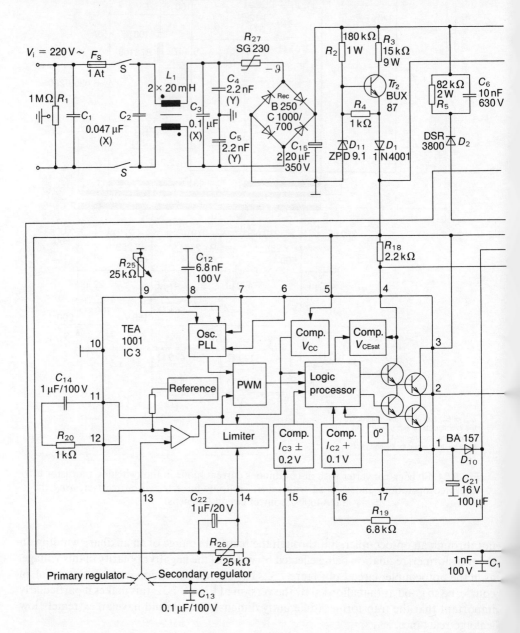

and C_{16} in Figure 7.36) was kept low; the d.c. gain was left at its maximum value. In these examples the regulated quantity can be taken alternatively from the secondary side through an optocoupler (as in the circuit of Figure 7.7), but of course from only one of the two outputs. The second output (in this case the negative supply) does not then have the same constancy of output voltage as the regulated positive supply. The same applies to primary-side sensing from a separate winding. The latter alternative makes for a simpler circuit, but gives inferior results compared with the optocoupler because of the leakage inductances. In the bipolar version shown in Figure 7.35 it was possible to make the feedback quantities for the error amplifier in the control IC (R_{20} and C_{14}) the same for

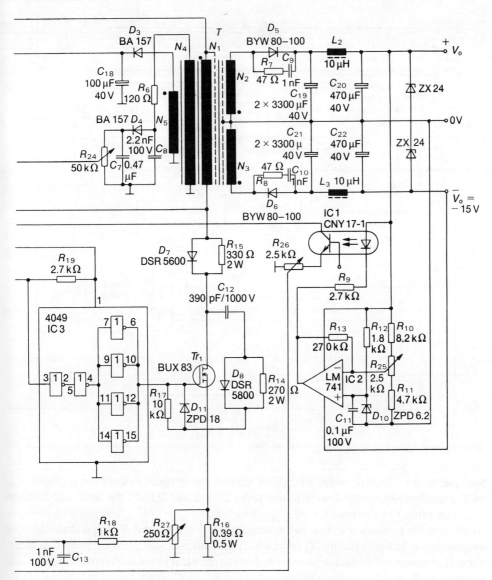

Fig. 7.35 Flyback converter with discontinuous current mode in the bipolar version described in [32] for $V_o = \pm 15\,\text{V}$ and $I_0 = \pm 0$ to 3 A, with quasi-saturated operation of the switching transistor using the TEA1001 control IC

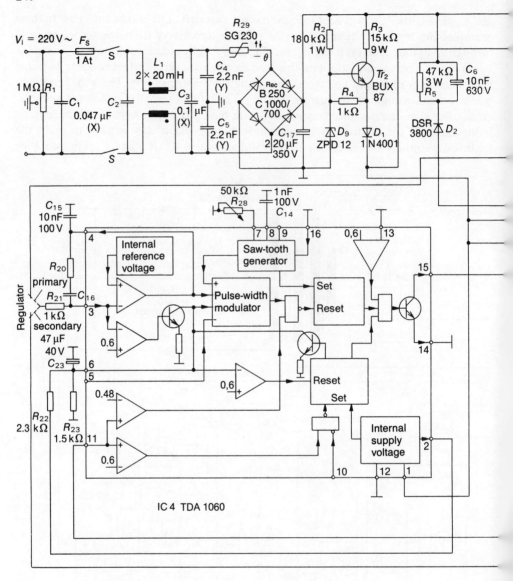

both methods of control; in the MOSFET version the optimum values for R_{20} and C_{16} with secondary-side control were found to be $2.2\,k\Omega$ and $0.1\,\mu F$; the most satisfactory values for primary-side control were $R_{20} = 63\,k\Omega$ and $C_{16} = 1\,nF$. The transformer has, in addition to the primary winding, two secondary windings N_2 and N_3 to produce the two outputs and a further winding N_4 from which the feedback voltage is derived. The latter winding is connected flyback-converter fashion to enable it to sense overshoot. Spikes are removed by R_6 and C_8, so that the controlled voltage on C_7 is not raised to the level of the spikes, which would result in poorer regulation characteristics. The time constant $C_7 R_{21}$ in Figure 7.35 ($C_7 R_{24}$ in Figure 7.36) must be relatively long so that the feedback voltage

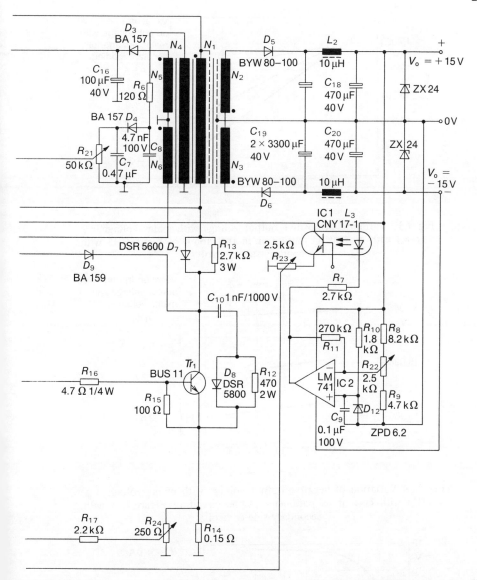

Fig. 7.36 Flyback converter with discontinuous current mode in the MOSFET version described in [32] for $V_0 = \pm 15\,\text{V}$ and $I_0 = \pm 0$ to 3 A, using the TDA 1060 control IC

does not exhibit a sawtooth waveform. The problem arises with primary-side control that a complete open circuit on the outputs is not permissible. Since the minimum 'on' time of the switching transistor is in the region of a few microseconds, with the outputs completely unloaded the voltage at the outputs and from N_4 rises. The duty cycle cannot be reduced indefinitely, and the transistor consequently receives no drive pulses for a certain period. The voltage on C_7 now falls more rapidly than the output voltages, so that drive pulses are again applied to the transistor. The result is that the output voltage increases slowly to an unacceptably high level. This increase in output voltage can be limited by means of the Zener diode shown, e.g. to 24 V, with a 24 V diode, in the case of a 15 V output, the diode

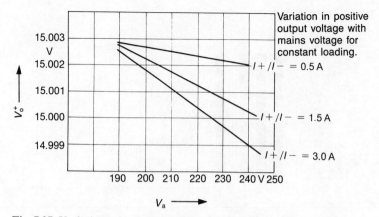

Fig. 7.37 Variation of positive output voltage with mains voltage V_0^+ = $f(V_a)$ with current as parameter in the circuit of Figure 7.35 with secondary-side control

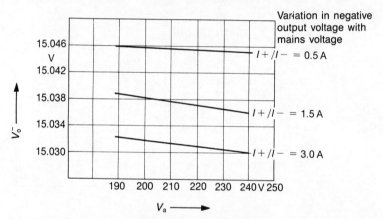

Fig. 7.38 Variation of negative output voltage with mains voltage V_0^- = $f(V_a)$ with current as parameter in the circuit of Figure 7.35 with secondary-side control

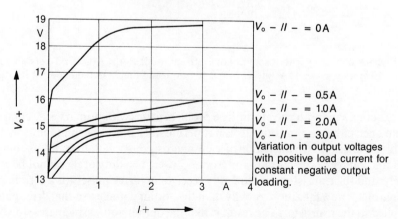

Fig. 7.39 Variation of output voltages V_0^+ and V_0^- with positive load current I_0^+ with negative load current I_0^- as parameter in the circuit of Figure 7.35 with primary-side control; $V_a = 220\,\text{V}$ r.m.s.

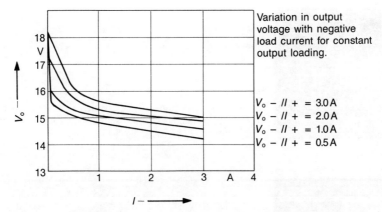

Fig. 7.40 Variation of negative output voltage V_0^- with negative load current I_0^- with positive output current I_0^+ as parameter in the circuit of Figure 7.35 with primary-side control; $V_a = 220$ V r.m.s.

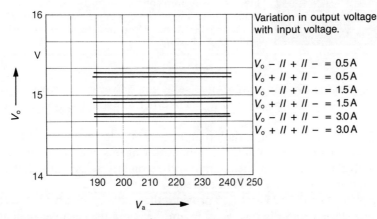

Fig. 7.41 Output voltages V_0^+ and $V_0^- = f(V_{\text{mains}})$ with current as parameter in the circuit of Figure 7.35 with secondary-side control; $V_a = 220$ V r.m.s.

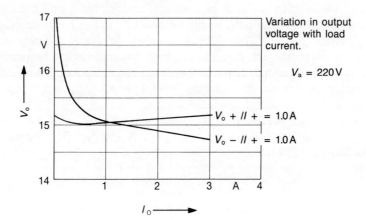

Fig. 7.42 Output voltages V_0^+ and V_0^- as functions of negative load current I_0^- with $I_0^+ = 1$ A in the circuit of Figure 7.35 with secondary-side control; $V_a = 220$ V r.m.s.

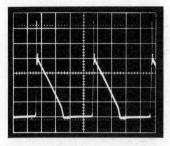

Fig. 7.43 Waveform of current in the secondary-side rectifier diode (D_5 in Figure 7.35; positive output) with $I_0^- = 0$, $V_a = 187\,\text{V}$ r.m.s. Horizontal scale: $10\,\mu s$/division Vertical scale: $5\,\text{A}$/division ($I_0^+ = 3\,\text{A}$)

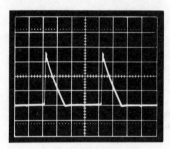

Fig. 7.44 Waveform of current in the secondary-side rectifier diode (D_5 in Figure 7.35; positive output) with $I_0^+ = I_0 = I_0^-$ $= 3\text{A}$, $V = 187\,\text{V}$ r.m.s. Horizontal scale: $10\,\mu s$/division. Vertical scale: $2.5\,\text{A}$/division

normally carrying hardly any current. It is sufficient, however, to provide a very small residual load ($I_0 = 0.1$ to 1 per cent of $I_{0\text{max}}$) in order to avoid the voltage rise (see also Figure 7.34).

Figures 7.37 to 7.42 show the variation in the output voltages V_0^+ and V_0^- with mains voltage and with load current. While the stabilization against mains voltage variations is very good, only the positive regulated output with secondary-side control shows the usual degree of constancy with varying load current.

Winding N_5 in Figures 7.35 and 7.36 provides the auxiliary supply for the control IC and the driver stage. From winding N_6 in Figure 7.35 both the negative auxiliary supply for turning off the switching transistor and a signal to indicate the resetting of the transformer are obtained. This supply is not necessary in the circuit of Figure 7.36.

A flyback converter with two or more outputs presents a further problem in regard to the maximum current flowing in the rectifier diodes (D_5 and D_6). From Equation (7.32), $I_{s\text{max}} = 4I_{0\text{max}}$ with $\delta_{T\text{max}} = 0.5$. This is valid for a number of windings if they are *all* loaded with their maximum permissible currents. A critical condition arises, however, when one winding is fully loaded and the other(s) is (are) practically unloaded. From Equation (7.7), with their maximum permissible currents. With $I_0^+ = 3\,\text{A}$, $I_{s\text{max}}^+ = 5.71 \times 3\,\text{A} = 17.1\,\text{A}$; with $I_0^+ = 3\,\text{A}$ and $I_0^- = 3\,\text{A}$, $I_{s\text{max}}^+ = 4 \times 3\,\text{A} = 12\,\text{A}$.

Figure 7.43 illustrates the first case (V_0^- unloaded) with $I_{s\text{max}}^+ = 17.5\,\text{A}$ and $\delta_T = 0.35$; Figure 7.44 shows the full-load case with $I_{s\text{max}}^+ = 11\,\text{A}$ and $\delta_T = 0.5$. The measurements thus substantiate the calculated results closely.

8 Integrated control circuits

A large number of integrated control circuits are available on the market, which are to a large extent similar in their modes of operation but which in some aspects incorporate peculiarities that determine their ultimate usefulness. They can be divided broadly into three categories: single-ended and push–pull controllers without integrated driver stages and single-ended controllers with integrated driver stages.

The *single-ended controllers without intergrated driver stages* can be used in all inductor-coupled converters and in single-ended forward and flyback converters; the necessary driver stage has to be added, but can be designed optimally for the particular application. This type of controller has a maximum duty cycle of almost 100 per cent. To limit the power available under a possible fault condition, the maximum duty cycle should be limited to a value a little higher than the maximum required in operation.

The *push–pull controllers without integrated driver stages* are provided specifically for all push–pull and bridge converters, but also find application in single-ended converters. If the two outputs are connected in parallel, the maximum possible duty cycle is almost 100 per cent, as in the single-ended controller, but cannot be limited. If the maximum duty cycle in operation is only about 45 per cent, as in the single-ended forward converter and sometimes in the flyback converter, the push–pull controller can be used with only one output. Limitation of the maximum duty cycle is then unnecessary.

The *single-ended controllers with integrated driver stages* are particularly specified for the requirements of the flyback converter. The inclusion of the integrated driver stage avoids the need for additional circuits, and so saves cost. Since flyback converters are restricted to the lower power range, there is a limit to the complexity that is acceptable. There are no push–pull flyback converters, because flyback converters are not practicable for high powers and the complication of a push–pull design can only be justified at very high powers.

Obviously the types discussed in the following pages cannot represent a complete enumeration and evaluation of all the available integrated control circuits; the intention is only to show a selection of the various units. Only the information necessary for dimensioning and for an understanding of the circuits will be presented; everything else can be obtained from the manufacturers' data manuals.

8.1 SINGLE-ENDED CONTROLLERS WITHOUT INTEGRATED DRIVER STAGES

Integrated control circuit Type TDA 1060 (Valvo) [23]

This control circuit has a single output with free access to the collector and emitter of the output transistor, and can be used in all single-ended circuits (forward, flyback and

253

inductor-coupled converters, particularly with high duty cycles). The functional blocks of the IC are shown in Figure 8.1. Figure 8.2 shows a schematic plan with an example of external connections.

The following is a discussion of the individual functional blocks. Note should in all cases be taken of the characteristic values and the operating values.

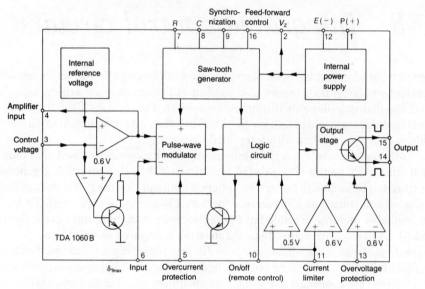

Fig. 8.1 Block diagram of integrated control circuit TDA1060 (B) (Valvo)

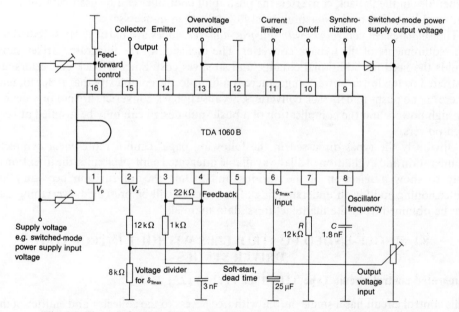

Fig. 8.2 Plan view of the TDA1060 (B) IC with typical connections (Valvo)

For $V_{P/E1/2} = 12$ V and $\theta_j = 25\,°C$, all voltages correspond to terminal 12 (E) unless otherwise stated

(a) *Internal voltage supply, protection against low voltage supply*
Supply voltage for current
$I_{P(1)} = 15$ mA: $\qquad\qquad\qquad\qquad V_{P/E(1/2)} = 23\,(18\dots27.5)$ V
$I_{P(1)} = 30$ mA: $\qquad\qquad\qquad\qquad V_{P/E(1/12)} = 24\,(19\dots29.5)$ V
Current absorption for feed-forward voltage,
for $R_7 = 25$ kΩ and $\delta_T = 0.5$; $-I_{z(2)} = 0$: $I_{P(1)} = (2.5\dots10) \leqslant 15$ mA
Stablizing voltage: $\qquad\qquad\qquad\qquad V_{z(2/12)} = 8.4\,(7.8\dots9.0)$ V
max drawing of current for external circuits: $-I_{z(2)} = 5$ mA

(b) *Internal reference voltage*
Reference voltage: $\qquad\qquad\qquad\qquad V_{Ref(4/12)} = 3.72\,(3.42\dots4.03)$ V

(c) *Saw-tooth generator, sychronization, forward regulation*
Frequency range
(adjustable by means of R_7 and C_8): $\qquad f = 50$ Hz$\dots100$ kHz
Range of external resistance: $\qquad\qquad R_7 = 5\dots40$ kΩ
Saw-tooth level HIGH: $\qquad\qquad\qquad V_H = 5.7$ V
Saw-tooth level LOW: $\qquad\qquad\qquad V_L = 1.3$ V
Sychronization level,
$\quad$ LOW, oscillator stopped $\qquad\qquad V_{9/12L} \leqslant 0.6$ V
$\quad$ HIGH, oscllator run $\qquad\qquad\qquad V_{9/12H} \geqslant 2.0$ V
Input current for $V_{9/12} = 0$: $\qquad\qquad -I_9 = 20\dots120\,\mu A$
Input current for the forward regulation $I_{16} \leqslant 5\,\mu A$

(d) *Control amplifier, protection against faults in control loop*
No-load gain of control amplifier $\qquad\qquad = \Delta V_4/\Delta V_3 = 60$ dB
Input voltage range: $\qquad\qquad\qquad\qquad V_{3/12} = 2.0\dots(V_2 - 1\text{ V})$
Input current for $V_{3/12} = 2$ V: $\qquad\qquad -I_3 = 12\,(1.5\dots35)\,\mu A$
Minimum value
Loop feedback resistor: $\qquad\qquad\qquad R_{4/3} = 10$ kΩ
Control amplifier threshold voltage $\qquad V_{93/12} = 460\dots720$ mA

(e) *V_{Tmax}– input, soft-start, on-off switch ($f = 10\dots100$ kHz)*
r_T input range: $\qquad\qquad\qquad\qquad\qquad r_T = 0\dots(T-1)/T$
Input voltage for $r_{Tmax} = 50\%$ $\qquad\qquad V_{6/12} = 0.4\,V_z$
Input current for $V_{6/12} = 2$ V: $\qquad\qquad -I_6 \leqslant 6\,\mu A$
Switching voltage for 'ON': $\qquad\qquad\quad V_{10/12ON} \geqslant 2.0$ V
Switching voltage for 'OFF': $\qquad\qquad\; V_{10/12OFF} \leqslant 0.5$ V
Input current for $V_{10/12} = 0$: $\qquad\qquad -I_{10} = 20\dots120\,\mu A$

(f) *Output stage*
Collector-emitter residual voltage
for $I_{15} = 40$ mA $\qquad\qquad\qquad\qquad\quad V_{15/14\,sat} \leqslant 0.4$ V

(g) *Current limiting*
Threshold voltage*
$\quad$ for one-off pulse suppression $\qquad\quad V_{s11/12} = 370\dots575$ mV
$\quad$ for switching during 'dead-time' $\qquad V_{s11/12} = 460\dots720$ mV
Input current for $V_{11/12} = 250$ mV: $\qquad -I_{11} \leqslant 10\,\mu A$
Switching delay for I15 = 30 mA: $\qquad\quad t_{doff} \leqslant 0.8\,\mu S$

(h) *Pulse wave modulator with overcurrent protection*
Input current
$\quad$ for $V_{5/12} = 2$ V, $V_{4/12}$ and $V_{6/12} > 2$ V: $-I_s \leqslant 6\,\mu A$

(i) *Overvoltage protection by interrupting the pulse*
Threshold voltage: $\qquad\qquad\qquad\qquad V_{s(13/12)} = 460\dots720$ mV
Input current for $V_{13/12} = 250$ mV: $-I_{13} \leqslant 7\,\mu A$

---

* The ratio of the threshold voltages is 4:5.

Fig. 8.3 Characteristic and operating values for the TDA1060(B) IC (Valvo)

256

Absolute limiting values: (Voltages corresponding to terminal 12)

Free-forward voltage (for voltage supply): $V_{P/E(1/12)} = \min 0.5\,\text{V, max}\,18\,\text{V}$

Current absorption (for current supply): $I_{P(1)} = \max 30\,\text{mA}$

Output current:* $I_{15}, -I_{14} = \max.\,40\,\text{mA}$

Voltage at the output terminals,

Emitter: $V_{14/12} = \max.\,5\,\text{V}$

Collector: $V_{15/12} = \max.\,V_P$

Voltage at the input terminals

Terminal 5: $V_{5/12} = \max.\,40\,\text{mA}$

Terminal 16: $V_{16/12} = \max.\,V_P^{\dagger}$

Terminals 3, 6, 9, 10, 11, 13: $V_{x/12} = \max.\,V_{z(2)}$

Total power loss[†]

for $T_a \leqslant 50\,°\text{C}$: $W_{tot} = \max.\,V_{2(2)}$

Junction temperature:

$$\theta_j = \max.\,150\,°\text{C}\,(125\,°\text{C})$$

Ambient temperature

$$T_a = \min.\,-55\,°\text{C}\,(-25\,°\text{C})$$
$$T_a = \max.\,125\,°\text{C}\,(85\,°\text{C})$$

---

* Peak current during 1 μs for $\delta_T < 0.1$ max. 200 mA.
† Therefore max. 24 V.

Fig. 8.4 Limiting values for the TDA1060(B) IC (Valvo)

The limiting values apply to the Type TDA1060B with an extended temperature range; the values in brackets apply to the TDA1060. The power loss is given by

$$W = V_s I_s + V_z I_z + V_{CEsat} I_C \delta_{Tmax} \quad \text{(W)} \tag{8.1}$$

where V_s is the supply voltage (V); a range from 11 to 18 V, I_s is the quiescent current consumption at V_s (A), V_z is the Zener voltage (V) (here 8.4 V), I_z is the Zener current (A) from Terminal 2, V_{CEsat} is the saturation output voltage (15) (V); here $\leqslant 0.4$ V, I_C is the output current (15) (A) and δ_{Tmax} maximum duty cycle.

Power supply

The circuit can be operated either with a voltage input (low-impedance source) within the range 11 to 18 V or with a current input (high-impedance source) of 15 to 30 mA; by virtue of the in-built Zener diode the latter arrangement results in a supply voltage of 23 to 24 V.

Control amplifier

A voltage proportional to the output voltage V_0 is fed back to the inverting input (3) of the control amplifier, while the non-inverting input is connected to the internal reference voltage source $V_{ref} = 3.72$ V $\pm$ 8.5 per cent. The difference voltage between Terminal 3 and V_{ref} is amplified and compared in the pulse-width modulator with a saw-tooth voltage.

The output (4) of the control amplifier is brought out, so that the gain required for a maximum permissible voltage change ΔV_0^* (V_0^* is the voltage of approximately 3.72 V produced by the output voltage divider) can be set by means of resistors R_4 and R_i. The open-loop gain is typically 60 dB = 1000 times. To improve the stability of the control loop a capacitor of between 2.2 and 20 nF should be connected between (4) and earth. A number of safeguards are incorporated against faults in the control loop. With an *open*

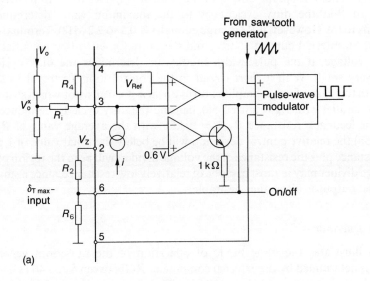

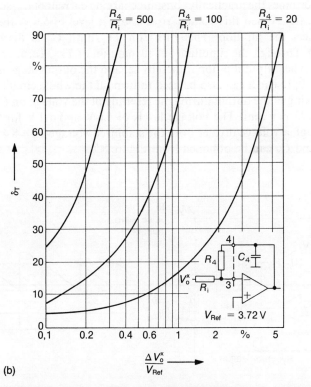

Fig. 8.5 (a) Control amplifier of the TDA1060 (B) (part diagram) and (b) variation of duty cycle δ_T with output voltage deviation V_0^+/V_{ref} (relative to the tap on the output voltage divider) (Valvo)

258

circuit in the control loop, the output of the control amplifier tends to be driven to high potential, so that the duty cycle goes to the maximum value determined by the connections to (6). However, if R_4 is large enough ($\geqslant 0.5$ to 2.2 MΩ), Terminal 3 is driven positive by an internal current source, and the output goes to low potential. Since the minimum voltage at the pulse-width modulator determines the duty cycle, δ_T then becomes very small. With a *short circuit* in the control loop Terminal 3 assumes 0 V potential. In this case a further amplifier turns on an *n-p-n* transistor which switches a 1 kΩ resistor in parallel with R_6 (Figure 8.5a), causing the duty cycle to be reduced.

Since the feedback resistance $R_{4\text{min}} = 10$ kΩ, with the average value of $R_4/R_i = 100$ (Figure 8.5b) the relative control deviation will be below the usual value of 1 per cent R_i (series resistance plus the resistance of the voltage divider) will be in the region of 1 to 3 kΩ. The voltage divider may in most cases be of relatively low resistance, since a small residual load on the output is altogether desirable.

Saw-tooth generator

Figure 8.6 illustrates the basic mode of operation of the saw-tooth generator. The frequency is determined by the external component R_7 (between 5 and 40 kΩ) and C_8; the range of adjustment is from 50 Hz to 100 kHz. The current flowing in R_7, with the current gain of Tr_4, determines the practically constant charging current of C_8, so that the voltage on C_8 rises linearly. When this voltage rises above a level of 5.6 V, the comparator H switches and sets the flip-flop. This turns on Tr_1 until C_8 is discharged to 1.1 V (comparator L). Through the simultaneous turning-on of Tr_2, Terminal 10 is driven to earth potential. The output flip-flop is thereby set and the output stage is turned off. The connection of (10) to earth can also be used to turn off the whole circuit (see also Figure 8.12). This blocking of the output during the resetting of the voltage on C_8 limits the duty cycle to about 95 per cent. The voltage levels of 5.6 V and 1.1 V for V_{C8} are derived internally through a voltage divider from the stabilized voltage $V_z = 8.4$ V. The required values for R_7 and C_8 can be obtained from Figure 8.7.

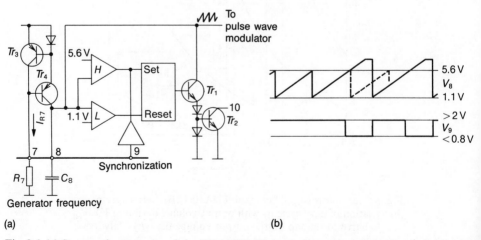

(a)   (b)

Fig. 8.6 (a) Saw-tooth generator of the TDA1060 (B) (part diagram) and (b) illustration of the synchronization of the saw-tooth voltage (Valvo)

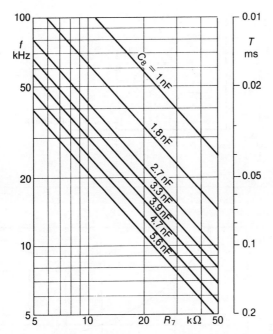

Fig. 8.7 Oscillator frequency $f_0 = f(R_7, C_8)$ in the TDA1060 (B) IC (Valvo)

Synchronization

The saw-tooth generator can be synchronized through the TTL-compatible input (9); $V_9 \geqslant 2$ V. With $V_9 \geqslant 0.8$ V the generator is stopped. The synchronizing frequency must be lower than the natural frequency (R_7/C_8). If no synchronization is required, (9) is not connected.

Reduction of duty cycle

According to Equation (2.4), for constant output voltage the duty cycle δ_T must vary in inverse proportion to the input voltage. This relationship can be fulfilled outside the control loop by connecting Terminal 16 to a voltage proportional to the input voltage. This so-called 'feed-forward' enables the ripple (100 Hz) at the output to be reduced. Since the input ripple voltage represents a voltage varying with the rectified mains voltage, and the stabilization of the output is a finite quantity, a 100 Hz component is present at the output. With no feed-forward (16) is connected to (12).

If because of a fault, for example an overload on the output, the voltage on (3) becomes too low (3.72 V $\geqslant V_3 \geqslant 0.6$ V), the duty cycle rises to the maximum value of about 0.95; more power is thus supplied to the output. To avoid this undesirable behaviour of the converter, it is useful to limit the maximum duty cycle δ_{Tmax}^* to a value around 10 per cent higher than the maximum normal operating value δ_{Tmax}. The maximum duty cycle δ_{Tmax} may be determined by the ratio of input to output voltage (in inductor-coupled converters); in transformer-coupled converters (single-ended and push–pull forward converters and flyback converters) δ_{Tmax} (at V_{imin}) is fixed at 0.45 to 06, according to the type of converter. Since the maximum possible duty cycle δ_{Tmax}^* is determined by

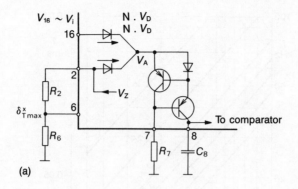

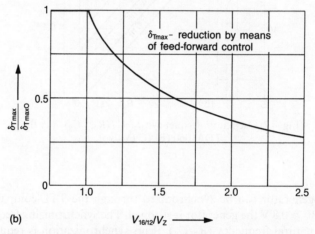

Fig. 8.8 (a) Part circuit for the reduction of the switching ratio
and (b) reduction in δ_{Tmax} by feed-forward control as a function
of $V_{16/12}/V_z$ in the TDA1060 (B) IC (Valvo)

the minimum voltage on Terminals 4, 5 or 6, the limiting value referred to above can
be set by means of a voltage divider on Terminal 6. If the voltage on Terminal 6 is
derived from the Zener voltage V_z, δ^*_{Tmax} is held substantially constant. The two voltage-
divider resistors can be selected with the aid of Figure 8.9.

The auxiliary quantity δ^*_{T0} may conveniently be set at 0.1 to 0.2.

Pulse-width modulator

The pulse-width modulator, by means of a saw-tooth voltage, converts the output voltage
of the control amplifier into a rectilinear-wave signal whose duty cycle depends upon the
lowest voltage on Terminals 4, 5 or 6.

Secondary current limiting (constant-current characteristic)

Terminal 5 makes available an unrestricted modulation input by means of which the duty
cycle can be reduced independently of the feedback loop (Terminal 3) and irrespective of

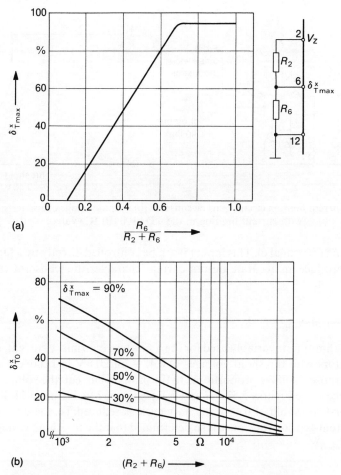

(a)

(b)

Fig. 8.9 (a) $\delta^*_{T\max}$ as a function of $R_6/(R_2 + R_6)$ and (b) r^*_{T0} as a function of $R_2 + R_6$ with $\delta^*_{T\max}$ as parameter in the TDA1060 (B) IC (Valvo-Signetics)

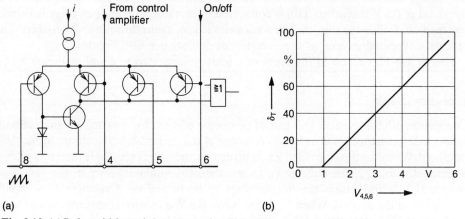

(a)

(b)

Fig. 8.10 (a) Pulse-width modulator (part circuit) and (b) variation of the duty cycle δ_T with the lowest of the voltages V_4, V_5 and V_6 in the TDA1060 (B) IC (Valvo)

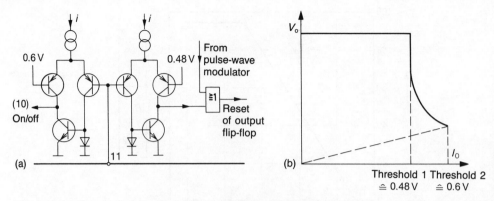

Fig. 8.11 (a) Current limiting circuit (part circuit) and (b) output voltage characteristic $V_0 = f(I_0)$ with current limiting in the TDA100 (B) IC (Valvo)

the δ^*_{Tmax} setting (Terminal 6). This input (5) can be controlled directly by a further control amplifier to produce an accurate output current characteristic (constant current or re-entrant).

Primary current limiting

Direct current limiting, invariably used for the limitation of primary current, is effected by two comparators with threshold voltages of 0.48 V and 0.6 V respectively.

For this purpose a voltage proportional to the primary current (the voltage drop across a low resistance—see Figures 5.52, 7.7 and 7.36) is applied to Terminal 11. If this voltage exceeds the first threshold of 0.48 V the duty cycle is reduced; i.e. the output flip-flop is reset. The current-sensing resistance is so determined that the threshold voltage is reached at about $1.2I_{pmax}$:

$$R_1 = \frac{0.48 \text{ V}}{1.2I_{pmax}} \quad (\Omega) \tag{8.2}$$

Since this method of current limiting is not adequate with very low duty cycles, because of the storage time of the transistor, the current continues to increase. When the second threshold of 0.6 V is reached, (10) is connected to earth and the output stage is thereby switched off. Following this, the circuit is switched on again through the soft-start. This process is repeated so long as the overcurrent persists (see also Figure 7.12).

This switch-off process in the event of a fault is illustrated in detail in Figure 8.12.

Start–stop circuit

Connection of (10) to earth (12) can of course be effected by a switch or an externally connected transistor as well as by the response of the current-limiting circuit at the 0.6 V threshold. When the flip-flop is set, Tr_1 is turned on and a 50 Ω resistor is thereby switched in parallel with R_6. This reduces V_6 to less than the minimum level of the saw-tooth voltage (1.3 V) and causes the output stage to be turned off. Capacitor C is rapidly discharged in the process. When V_6 drops below 0.6 V, a comparator turns on and resets the flip-flop. If V_6 rises above 1.3 V, the flip-flop is reset by the comparator, Tr_1 is turned off and the output stage is enabled.

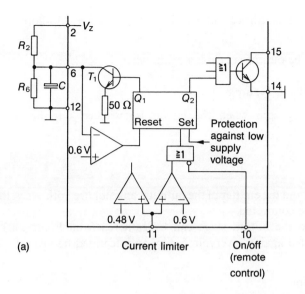

(a)

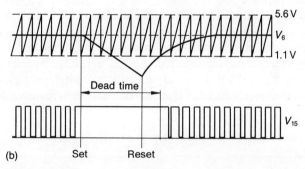

(b)

Fig. 8.12 (a) Start–stop circuit (remote control) ((part circuit) and (b) soft-start pulse diagram in the TDA1060 (B) IC (Valvo)

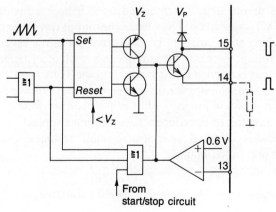

Fig. 8.13 Output stage (part circuit) of the TDA1060 (B) IC (Valvo)

The output stage is turned off by the setting of the flip-flop in the following circumstances:

- if $V_{10} \leqslant 0.8$ V; the output is enabled again when $V_{10} \geqslant 2$ V
- if $V_{11} \geqslant 0.6$ V
- if $V_1 \leqslant 10.8$ V (V_p in Figure 8.2)

Since V_6 is initially zero when the circuit is switched on, the soft start follows automatically.

Output stage

Since the collector and the emitter of the output transistor are both accessible, any kind of driver stage can be connected.

Data relating to the output stage can be gathered from Figures 8.3 and 8.4. The collector is protected against overvoltages by an integrated free-wheel diode.

Overvoltage shut-down

If a voltage greater than 0.6 V is applied to (13), a comparator connects the base of the output transistor to earth and the transistor is turned off. This circuit can also be dimensioned as a saturation protection circuit. For this purpose a winding poled in the flyback sense is provided on the main transformer and the voltage from it is rectified. So long as the transformer remains unsaturated, a (positive) voltage is present, and the output is inhibited because $V_{13} > 0.6$ V. It is only possible to switch on again when V_{13} has practically disappeared. According to Figure 8.4 the maximum permissible voltage on (13) is $V_z(2) = 8.4$ V.

The control loop

In this discussion the circuit of a forward converter in accordance with Figure 5.64 will be taken as an example.

The emitter (14) of the output stage is connected to earth. When the output transistor turns on, its collector (15) is driven towards earth potential. In this event Tr_1 (BD139) turns off. Since this driver stage is connected as a flyback converter, the switching transistor Tr_2 (BUX82) is turned on. Thus, when the output transistor of the control IC is turned on, the switching transistor Tr_2 is also turned on at the same time.

If now the output voltage increases, for example as a result of an increased input voltage, a higher positive voltage is applied to the input (3) of the TDA 1060. If the voltage at the inverting input of the control amplifier increases, its output voltage is reduced (see also Figure 8.5(a)). This, as shown in Figure 8.10(b), causes the duty cycle δ_T, and hence the 'on' time of the output transistor, to be reduced. Less power is supplied to the output and the output voltage increase is corrected (apart from a small residual error). The control loop thus operates in the prescribed manner.

Integrated control circuit Type MC35060/MC34060 (Motorola)

This type of integrated control circuit differs from the previously discussed TDA1060 mainly in the inclusion of two similar amplifiers, whose two inputs are available in each case. This affords greater freedom in the arrangement of the converter. The outputs of the amplifiers are connected to the positive input of the pulse-width modulator, so that the non-inverting (+) input must be connected to the (positive) output voltage. The MC34060 is the cheaper version in a plastic encapsulation with a restricted temperature range, while the MC35060 is in a ceramic housing and specified for a wider temperature range. The

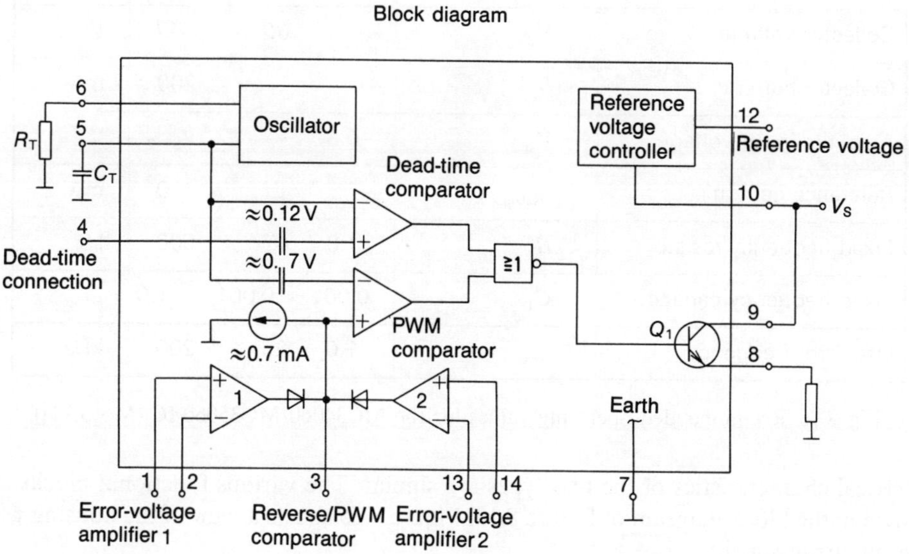

Block diagram

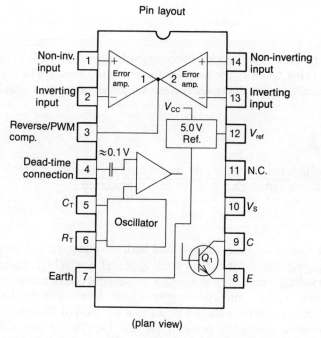

Fig. 8.14 Block diagram of the MC34060/MC35060 integrated control circuit (Motorola)

Pin layout

(plan view)

Fig. 8.15 Plan view of the MC34060/MC35060 IC (Motorola)

Recommended operating values:

	Symbol	MC35060/MC34060			Units
		Min.	Typ	Max.	
Power supply	V_S	7.0	15	40	V
Collector voltage	V_C	—	30	40	V
Collector current	I_C	—	—	200	mA
Amplifier input voltage	V_{in}	−0.3	—	V_{S-2}	V
Reference current	$I_{Ref max}$	—	—	10	mA
Fixed frequency resistor	R_T	1.8	47	500	kΩ
Fixed frequency capacitor	C_T	0.001	0.001	1.0	μF
Oscillator frequency	f_{osc}	1.0	25	200	kHz

Fig. 8.16 Recommended operating values for the MC34060/MC35060 IC (Motorola)

electrical characteristics of the two types are similar. The various functional blocks are shown in the block diagram of Figure 8.14; Figure 8.15 shows a view of the housing with the pin arrangement.

The recommended operating values are tabled in Figure 8.16; Figure 8.17 gives the principal characteristic values, and the limiting values are given in Figure 8.18.

Power supply

The circuit can be used over the wide operating voltage range of 7 V to 40 V, a nominal voltage of 15 V being taken as the basis for the operating values. The quiescent current consumption is only slightly dependent on the supply voltage, and varies between 5 and 6 mA.

Reference voltage

The reference voltage is 5 V ± 5 per cent, and is accessible on (12). It should not be loaded at more than 10 mA, although it is short-circuit protected.

Control amplifiers

The two similar control amplifiers can be used both as voltage-control amplifiers [e.g. (1) and (2)] and as comparators for current limiting [e.g. (13) and (14)]. Because of the large common-mode voltage range—up to $V_s - 2$ V—the inputs can if necessary be connected directly to V_{ref}, even with the minimum supply voltage. This is only worthwhile with $V_0 = 5$ V. The error-voltage amplifier should be set to a d.c. gain of 40 to 60 dB by means of external components connected between (2) and (3). The circuit includes internal compensation. It is useful to provide a PI characteristic by incorporating an RC network

Characteristic values $V_S = 15\,\text{V}$, $f_{osc} = 25\,\text{kHz}$, $\theta_a = 25\,^{\circ}\text{C}$

	Symbol	MC34060			Units
		Min.	Typ	Max.	
Reference voltage ($I_{ref} = 1\,\text{mA}$)	V_{ref}	4.75	5.0	5.25	V
Collector-emitter saturation voltage Common emitter ($V_E = 0\,\text{V}$, $I_C = 200\,\text{mA}$) Emitter follower $V_C = 15\,\text{V}$, $I_E = -200\,\text{mA}$	$V_{sat(C)}$ $V_{sat(C)}$ $V_{sat(E)}$	— —	1.1 1.5	1.3 2.5	V V
Output voltage rise time voltage Common emitter Emitter follower	t_r	— —	100 100	200 200	ns ns
Output voltage fall time voltage Commiter emitter Emitter follower	t_f	— —	25 40	100 100	ns ns
Error amplifier common mode range ($V_{S-2} = 7.0\,\text{V}\ldots 40\,\text{V}$)		−0.3	—	V_{S-2}	V
No load gain		70	95	—	dB
Threshold voltage for the PWM comparator for $\delta_T = 0$	V_{th}	—	3.5	4.5	V
Maximum switching ratio $V_4 = 0\,\text{V}$, $C_T = 0.1\,\mu\text{F}$, $R = 12\,\text{k}\Omega$ $V_4 = 0\,\text{V}$, $C_T = 0.001\,\mu\text{F}$, $R_T = 47\,\text{k}\Omega$	δ_{Tmax}	90 —	96 92	100 100	%
Threshold voltage for dead time $\delta_T = 0$ (4) δ_{Tmax}	V_{th}	— 0	2.8 —	3.3 —	V
Mean quiescent current $V_{[Pin4]} = 2.0\,\text{V}$, $C_T = 0.001$, $R_T = 47\,\text{k}\Omega$	I_s	—	7.0	—	mA

Fig. 8.17 Characteristic data for the MC34060/MC35060 IC with $V_S = 15\,\text{V}$, $\theta_a = 25\,^{\circ}\text{C}$ and $f_0 = 25\,\text{kHz}$ (Motorola)

in the feedback path (see also Figure 8.26). The external components are chosen differently for the comparator used for current limiting. With a capacitor connected between (3) and (13) this amplifier acquires an I characteristic. In conjunction with a series resistor this gives rise to a time constant of about 1 μs, which causes the very steep-fronted spikes at the current-sensing resistor to be suppressed. The bias voltage on (13) is obtained from the

Limiting values

	Symbol	MC35060	MC34060	Units
Power supply	V_S	42	42	V
Collector voltage	V_C	42	42	V
Collector current	I_C	250	250	mA
Amplifier input voltage	V_{in}	$V_S + 0.3$	$V_S + 0.3$	V
Power loss, $\theta_a \leqslant 45\,°C$	W	1000	1000	mW
Crystal temperature Plastic encapsulation Ceramic encapsulation	θ_j	 — 150	 125 150	°C
Crystal temperature	θ_a	-55 to 125	0 to 70	°C
		Ceramic encapsulation	Plastic encapsulation	
Thermal resistance at function	R_{thja}	100	80	K/W
Power loss- Reduction factor	$1/R_{thja}$	10	12.5	mW/K
Limiting temperature for powerless reduction	θ_a	50	45	°C

Fig. 8.18 Limiting data and thermal values for the MC34060/MC35060 IC (Motorola)

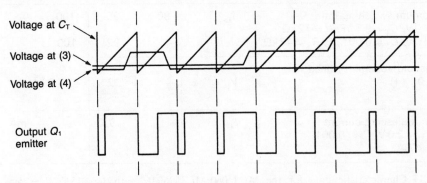

Fig. 8.19 Pulse diagram for the pulse-width modulator and the output stage of the MC34060/MC35060 IC with free emitter (Motorola)

reference voltage, and so chosen that a voltage of about 0.1 to 0.2 V is developed at $1.2I_{pmax}$.

In contrast to the principle adopted by most manufacturers, the duty cycle in this case is *reduced* as the voltage at the comparator is increased (see Figure 8.10(b)). The transistor Q_1 conducts only when the saw-tooth voltage exceeds the voltage on (3) or (4). The

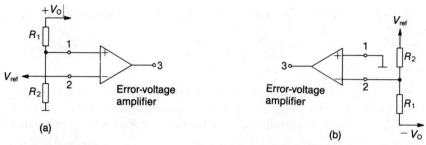

Fig. 8.20 Connections to the error-voltage amplifier (a) for positive output voltage and (b) for negative output voltage in the MC34060/35060 IC (Motorola)

external connections of the error-voltage amplifier must be determined according to whether the output voltage is positive or negative with respect to the reference point (earth). The alternative arrangements are shown in Figure 8.20.

For a positive output voltage,

$$V_0 = V_{ref}\left(1 + \frac{R_1}{R_2}\right) \quad (V) \tag{8.3}$$

For a negative output voltage,

$$-V_0 = V_{ref}\frac{R_1}{R_2} \quad (V) \tag{8.4}$$

If current limiting is not required, (14) is connected to earth and (3) is connected directly to

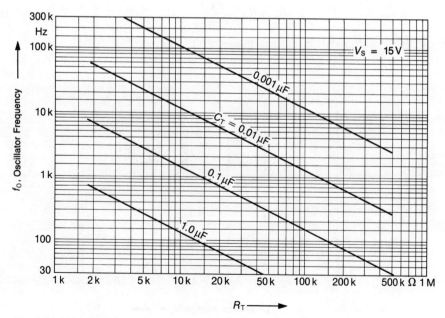

Fig. 8.21 Variation of oscillator frequency f_0 in the MC34060/35060 IC with timing resistance R_T, with capacitance C_T as parameter (Motorola)

270

V_{ref}. The output of this amplifier is then driven to a low potential and decoupled from the comparator.

Saw-tooth generator

The saw-tooth oscillation is obtained, in the usual manner, through the charging of the capacitor C_T by a current source which can be programmed by means of R_T. The frequency f_0 can either be determined by reference to Figure 8.21 or calculated from

$$f_0 \approx \frac{1.1}{C_T R_T} \quad \text{(Hz)} \tag{8.5}$$

where C_T is the frequency-determining capacitance (F) (plastic film) and R_T is the frequency-determining resistance (Ω).

Setting the maximum duty cycle δ^*_{Tmax}

Figure 8.22 gives the voltage V_4 to be provided for a maximum duty cycle δ^*_{Tmax} (conveniently about 10 per cent above the maximum value of δ_{Tmax} at V_{imin}). With $V_4 = 0$, δ^*_{Tmax} is in the region of 0.92; for $\delta_T = 0$ the threshold voltage $V_{th} = 2.8$ V. Since according to Figure 8.14 the dead-time comparator is biased at 0.12 V, while the PWM comparator is biased at about 0.7 V, the threshold voltage of the PWM comparator must be higher by about 0.6 V, i.e. in the region of 3.4 V (see Figure 8.17).

The voltage V_4 may conveniently be obtained from the 5 V reference supply by means of a potential divider.

For a given value of R_3 (e.g. 4.7 kΩ), R_4 is calculated from

$$R_4 = R_3 \frac{V_4}{5\,\text{V} - V_4} \quad (\Omega) \tag{8.6}$$

where V_4 is the voltage at δ^*_{Tmax} from Figure 8.22.

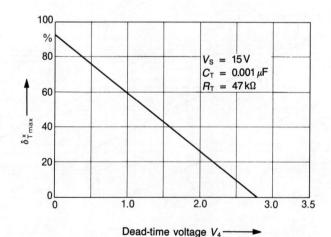

Fig. 8.22 Variation of duty cycle δ^*_{Tmax} (%) with dead-time voltage V_4 in the MC34060/35060 IC (Motorola)

271

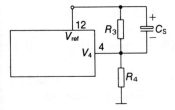

Fig. 8.23 Part circuit for the setting of δ^*_{Tmax} and the soft start in the MC34060/35060 IC (Motorola)

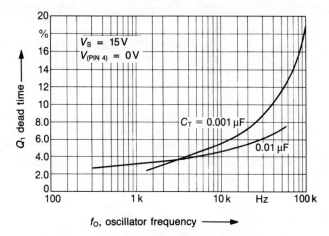

Fig. 8.24 Variation with frequency of dead time with $V_4 = 0$, with capacitance C_T as parameter in the MC34060/35060 IC (Motorola)

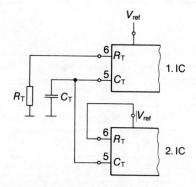

Fig. 8.25 Part circuit for the synchronization of a number of switched-mode power supplies to the same switching frequency with the MC34060/35060 IC (Motorola)

Alternatively, R_4 can be calculated without using Figure 8.22 from

$$R_4 = R_3 \frac{0.92 - \delta^*_{\text{Tmax}}}{1.6 + \delta^*_{\text{Tmax}} - 0.92} \quad (\Omega) \tag{8.7}$$

To obtain a soft start, V_4 must initially be in excess of 3 V. This can be brought about by a capacitor C_s of, say, 10 μF as shown in Figure 8.23. Since the maximum possible duty cycle

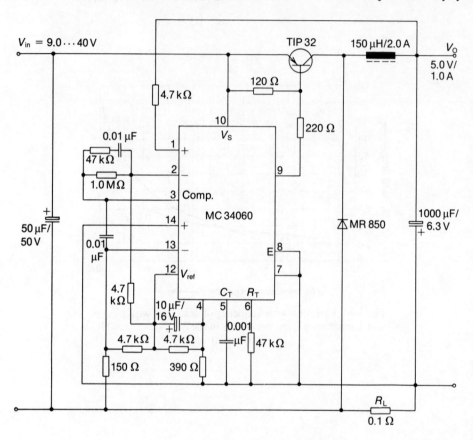

	Requirements	Result
Voltage deviation	V_{in} = 9.0 V ... 40 V, I_O = 1.0 A	25 mV 0.5%
Load deviation	V_{in} = 12 V, I_O = 1.0 mA ... 1.0 A	3.0 mV 0.06%
Output threshold	V_{in} = 12 V, I_O = 1.0 A	75 mV$_{\text{SS}}$
Short-circuit current	V_{in} = 12 V, R_L = 0.1 Ω	1.6 A
Efficiency	V_{in} = 12 V, I_O = 1.0 A	73%

Fig. 8.26 Inductor-coupled step-down converter with soft start and short-circuit protection as an example of the use of the MC34060/35060 IC (Motorola)

is related to the dead time of Q_1, the latter can be an important factor in dimensioning for a large duty cycle in, for example, an inductor-coupled converter. Figure 8.24 shows the variation of the dead time (= 'off' time of Q_1) with the value of C_T and the switching frequency.

The dimensioning of R_T and C_T in accordance with Figure 8.22 corresponds, according to Figure 8.21 or Equation (8.5), to a switching frequency of approximately 25 kHz. With $C_T = 0.001\ \mu F$ and $f = 25$ kHz, the dead time is given by Figure 8.24 as 8 per cent. From Figure 8.22, this corresponds to a maximum duty cycle of 0.92.

Synchronization

If two or more control ICs are required to operate at the same frequency, they can be interconnected as shown in Figure 8.25. The first IC, with its R_T–C_T network, determines the switching frequency, while the second, and any further, IC is fed with the same voltage from C_T. Since R_T is connected to the V_{ref} supply, no current is fed to C_T from the second IC.

Output stage

Since both the emitter and the collector connections of the output transistor are brought out, any kind of driver stage can be used without any difficulty.

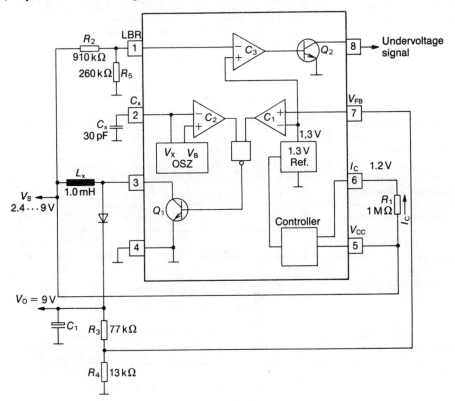

Fig. 8.27 Block diagram (including connections for a step-up converter) of the 4193 IC (Raytheon)

Application example

What has been said above concerning this circuit can be checked against the example shown in Figure 8.26. The short-circuit current of 1.5 A, which can be calculated retrospectively on the basis of the bias voltage on (13) and the given value of $R_L = 0.1\,\Omega$, seems somewhat high, but is quite acceptable. The MR850 free-wheel diode should, however, be replaced by a Schottky diode for the sake of better efficiency.

Integrated control circuit Type RC4193 (Raytheon)

This relatively simple control circuit is intended for battery operation, and is notable for its very low supply voltage (2.4 V minimum) and a very low quiescent current. Figure 8.27 shows the block diagram, including the pin connections, and illustrates the principal application to a step-up inductor-coupled converter.

With this circuit it is possible, for example, to use a 9 V battery down to a terminal voltage of 2.4 V before the loading on the battery has only noticeable effect. If the voltage

Characteristic data ($V_s = 6$ V, $I_C = 5\,\mu$A, $\theta_a = 25\,^{\circ}$C)

	Symbol	Requirements	Min.	Typ	Max.	Units
Supply voltage	V_s		2.4		24	V
Reference voltage	V_{ref}		1.24	1.31	1.38	V
Collector current (3)	I_{SW}	$V_3 = 400$ mV	75	100		mA
Quiescent current	I_s	gemessen an (5) $I_3 = 0$		135	200	μA
Efficiency				80		%
Voltage deviation		$0.5V_0 < V_s < V_0$		0.08	05	$\%V_0$
Load deviation		$V_s = 0.5V_0$ $W_L = 150$ mW		0.2	0.5	$\%V_0$
Oscillator frequency	f_0		0.1	25	150	kHz
Reference input current	I_C		1	5	50	μA
Collector saturation voltage	V_s	$I_3 = 100$ mA		0.4	1	V
Collector quiescent voltage	I_{CO}	$V_3 = 24$ V		0.01	5	μA
Input current (1)	I_1	$V_1 = 1.2$ V		0.7		μA
Input current (7)	I_{FB}	$V_7 = 1.3$ V		0.1		μA
Output current (8)	I_{LBD}	$V_8 = 0.4$, $V_1 = 1.1$ V	100	600		μA

Fig. 8.28 Absolute limits and thermal data for the 4193 IC (Raytheon)

Thermal values

	8-pole plastic encaps.	8-pole ceramic encaps.
Max. junction temperature	125 °C	175 °C
Max. $W: \theta_a < 50\,°C$	468 mW	833 mW
Therm. Res. R_{thjc}	—	45 °C/W
Therm. Res. R_{thja}	160 °C/W	150 °C/W
For $\theta_a > 50\,°C$: reduction of	6.25 mW per °C	8.33 mW per °C

Absolute limits
Power loss 500 mW
Supply voltage 24 V
Ambient temperature range 0 °C...+70 °C
Collector current (I_{SW}) 150 mA
Ref. input current (I_C)
V_1, V_2, V_7 1 mA
V_3, V_8 6 V
24 V

Fig. 8.29 Characteristic data for the 4193 IC with $V_s = 6\,V$ and $\theta_a = 25\,°C$ (Raytheon)

divider ratio for the battery undervoltage indicator is selected so that the output (8) is turned on at a substantial undervoltage (e.g. $V_{batt} \leqslant 6$ V), an indication can be given shortly before the battery fails completely. Figure 8.26 gives operating values for a nominal supply voltage $V_s = 6$ V; the absolute limiting values are given in Figure 8.29.

The oscillator frequency is determined by a capacitor C_x connected to (2). C_x is calculated from

$$C_x = \frac{2.14 \times 10^{-6}}{f_0} \quad \text{(F)} \tag{8.8}$$

where f_0 is the oscillator frequency (Hz).

8.2 PUSH–PULL CONTROLLERS WITHOUT INTEGRATED DRIVER STAGES

A push–pull controller is fundamentally necessary for any push–pull circuit—e.g. the simple push–pull circuit of Figure 1.9, the half-controlled bridge of Figure 1.10 or the fully-controlled bridge circuit of Figure 1.11; it can, however, be used also in single-ended circuits.

While in push–pull circuits it is of course necessary to use both outputs (here referred to as A and B), in single-ended circuits either one output (A or B) or both outputs in parallel can be used. The relationships are shown in Figure 8.30.

An oscillator frequency $f_0 = 40$ kHz is assumed, i.e. the cycle period T is 25 μs. The duty cycle δ_T is shown as 0.4 in all three diagrams.

In Figure 8.30(a) the two outputs are connected in parallel (the permissible collector current of the output transistor is not thereby doubled, since the two outputs do not switch simultaneously, but alternately) and the switching frequency is equal to the oscillator frequency. With $T = 25$ μs and $\delta_T = 0.4$ the conducting period of the switching transistor is 10 μs. The rectangles drawn in heavy lines symbolize the collector current of the switching transistor, and the dashed triangles represent the inductor current. If only one output, A or B, is used (Figure 8.30(b)) (the other being unconnected), the switching frequency is halved relative to the oscillator frequency, because of the inbuilt logic, i.e. the period is doubled. With $\delta_T = 0.4$ the conducting period is then 20 μs.

The maximum possible duty cycle with only one output is about 0.45, which is sufficient for a forward converter, and in some cases for a flyback converter. In inductor-coupled converters, duty cycles of 0.6 to 0.8 are often required. These higher ratios must be obtained through parallel connection of both outputs. In view of possible difficulties in the transition through $\delta_T = 0.5$ to higher values, however, it is better to use single-ended controllers. This is particularly so, since with push–pull controllers no limitation of the duty cycle is possible under fault conditions.

Thus when a push–pull controller is used in a single-ended circuit and a maximum duty cycle of about 0.45 is sufficient, only one output should be used. The oscillator frequency should then be set to twice the required switching frequency.

In a push–pull application (Figure 8.30(c)), the total period is similarly doubled, i.e. in this case 50 μs. The switching frequency at the power stage is thus half the oscillator frequency, but the frequency in the inductor is again equal to the oscillator frequency (double-way rectification).

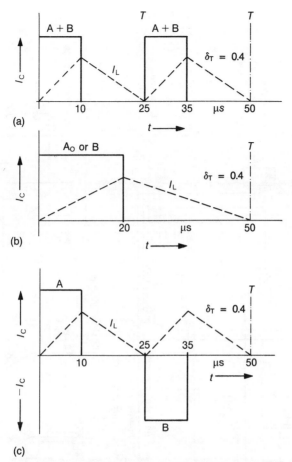

Fig. 8.30 (a) Push–pull control circuit for a forward converter used for single-ended operation with the two outputs connected in parallel: $\delta_T = 0.4$ (collector current in solid lines, inductor current in dashed lines). (b) Push–pull cntrol circuit for single-ended forward-converter operation with only one output; $\delta_T = 0.4$ (collector current in solid lines, inductor current in dashed lines) (c) Push–pull control circuit for push–pull forward-converter operation (collector current in solid lines, inductor current in dashed lines)

Integrated control circuit Type TDA4700(A) (Siemens) [24]

The TDA4700 circuit is a complex push–pull control circuit which can be used in many switching-regulator applications. This IC is the most comprehensive of a family of similar control circuits, which are discussed below. The various functional blocks are shown in Figure 8.31. A plan of the 24-pin housing with the connection arrangement is shown in Figure 8.32.

The range of functions, together with the permissible limits, is given in Figure 8.33, while characteristic data for the whole operating voltage range of 11 to 30 V and the

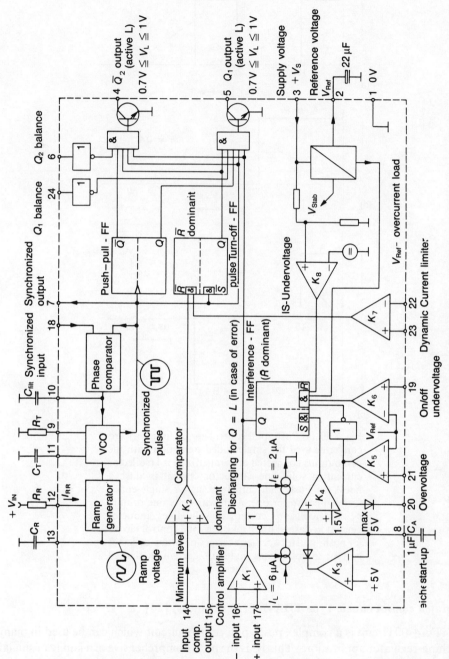

Fig. 8.31 Block diagram of the TDA4700(A) integrated control circuit (Siemens)

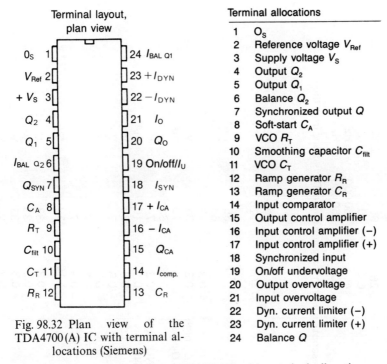

Terminal layout, plan view

O_S	1	24	$I_{BAL\ Q1}$
V_{Ref}	2	23	$+I_{DYN}$
$+V_S$	3	22	$-I_{DYN}$
Q_2	4	21	I_O
Q_1	5	20	Q_O
$I_{BAL\ Q2}$	6	19	On/off/I_U
Q_{SYN}	7	18	I_{SYN}
C_A	8	17	$+I_{CA}$
R_T	9	16	$-I_{CA}$
C_{filt}	10	15	Q_{CA}
C_T	11	14	$I_{comp.}$
R_R	12	13	C_R

Terminal allocations

1	O_S
2	Reference voltage V_{Ref}
3	Supply voltage V_S
4	Output Q_2
5	Output Q_1
6	Balance Q_2
7	Synchronized output Q
8	Soft-start C_A
9	VCO R_T
10	Smoothing capacitor C_{filt}
11	VCO C_T
12	Ramp generator R_R
13	Ramp generator C_R
14	Input comparator
15	Output control amplifier
16	Input control amplifier $(-)$
17	Input control amplifier $(+)$
18	Synchronized input
19	On/off undervoltage
20	Output overvoltage
21	Input overvoltage
22	Dyn. current limiter $(-)$
23	Dyn. current limiter $(+)$
24	Balance Q

Fig. 98.32 Plan view of the TDA4700(A) IC with terminal allocations (Siemens)

Fig. 8.32 Plan view of the TDA4700(A) IC with terminal allocations (Siemens)

permissible temperature range of -25 to $+85\,°C$, separated according to the functional blocks, are listed in Figure 8.34. The TDA4700A itself is specified only from 0 to $+70\,°C$.

Power supply

The power-supply voltage range extends from 11 to 30 V; depending on the tolerances the operating range may start at about 9 V. If the supply voltage drops below 10 to 10.5 V, the undervoltage monitor responds and turns off the output stage. The voltage should not, therefore, fall below the undervoltage limit of 10.5 V in the worst case with the lowest permissible mains voltage. The circuit switches on again with a soft start. The no-load current consumption is between 8 and 20 mA without the reference current.

Internal reference voltage source

The reference voltage is $2.5\,V \pm 6$ per cent (2.35 to 2.65 V); the maximum current loading should not exceed 5 mA. If the reference source is loaded to more than 10 mA the fault flip-flop switches off the outputs instantaneously.

Control amplifier K_1

The voltage proportional to the output voltage is fed back to the inverting input (16). Input (17) can be connected directly to the 2.5 V V_{ref} (2). The output (15) is connected to the input (14) of the comparator K_2. By means of the external connections to K_1 in the feedback path—preferably with PI characteristics—the d.c. gain should be reduced to about 40 to 50 dB (see also Figure 8.5b).

Range of functions		Observations	Lower limit B	Upper limit A	
Supply voltage	V_s		10.5	30	V
Ambient temperature in operation					
TDA 4700	θ_α		−25	85	°C
TDA 4700 A	θ_a		0	70	°C
max. VCO frequency	f		40	250 000	Hz
Ramp generator frequency	f_{RG}		40	250 000	Hz

The specified functions are complied with in the circuit description within the given range. Deviations from the characteristic data are possible.

Limiting data

			Lower limit B	Upper limit A	
Junction temperature	θ_j			125	°C
Substrate temperature	θ_s		−40	125	°C
Thermal resistances:					
system surroundings TDA 4700	R_{thSS}			65	K/W
TDA 4700 A	R_{thSS}			65	K/W
Operating voltage	V_S		−0.3	33	V
Voltage at Q_1, Q_2	V_Q	Q_1, Q_2 high	−0.3	33	V
Current at Q_1, Q_2	I_Q	Q_1, Q_2 low		70	mA
Balance 1, 2	V_{BAL}		−0.3	33	V
Synchronized output	V_{SYNQ}	SYNQ high	−0.3	7	V
		SYNQ low	0	10	mA
Synchronized input	V_{SYNI}		−0.3	33	V
Input C_{filt}	V_{ICF}		−0.3	7	V
Input R_T	V_{IRT}		−0.3	7	V
Input C_T	V_{ICT}		−0.3	7	V
Input R_R	V_{RR}		−0.3	7	V
Input C_R	I_{ICR}		10	10	mA
Input comparator					
K2, K5, K6, K7	I_{IK}		−0.3	33	mA
Output K5	V_{QK5}		−0.3	33	V
Input control amplifier	V_{IRV}		−0.3	33	V
Output control amplifier	V_{QRV}		−0.3	V_{s-1} max. 7	V
Reference voltage terminal	V_{ref}		−0.3	V_{ref}	V
Input C_a	V_{ICA}		−0.3	7	V

Characterstic data $V_s = 11$ to 30, $\theta_a = -25$ to $85°C$		Test conditions	Lower limit B	typ.	Upper limit A	
Current absorption	I_S	$C_T = 1\,nF$, $f_{VCO} = 100\,kHz$	8		20	mA
Reference						
Reference voltage	V_{Ref}	$0\,mA < I_{Ref} < 5\,mA$	2.35	2.5	2.65	V
I_{Ref}-overcurrent	I_{Ref}			10		ma
Oscillator (VCO)						
Frequency range	f_{VCO}	$C_T = 1\,nF$	40		200000	Hz
Saw-tooth period	t	$C_T = 1\,nF$		1		μs
	t	$C_T = 10\,nF$		10		μs
Switching current VCO	C_T		0.82		47	nF
	R_T		5		700	$k\Omega$
Ramp generator						
Frequency range	f		40		100000	Hz
Maximum voltage at C_R	V_H			5.5		V
Minimum voltage at C_R	V_L			1.8		V
Input current across R_R	I_{RR}		0		400	μA
Current transmission	I_{RR}/I_{CR}			4		
Synchronization						
Synchronized output	V_{QH}	$I_{QH} = -200\,\mu A$	4			V
	V_{QL}	$I_{QL} = 1.6\,mA$			0.4	V
Synchronized input	V_{IH}		2			V
	V_{IL}				0.8	V
Input-current	$-V_{IL}$				5	μA
Comparator K_2						
Input current	$-I_{1K2}$				2	μA

(Cont.)

Characteristic data
V_s = 11 to 30, θ_a = −25 to 85 °C

		Test conditions	Lower limit B	typ.	Upper limit A	
Turn-off delay	t_{OUT}				500	ns
Input voltage	V_{IK2}	for switching ratio $r = 0$		1.8		V
		$r = $ max		5		V
Common mode range	V_{IC}		0		5.5	V
Soft start-up K3, K4						
Charge current for C_A	I_{CCA}			6		µA
Discharge current for C_A	I_{DCA}			2		µA
Upper limiting voltage	V_{LIm}			5		V
Switching voltage	V_{K4}			1.5		V
Control amplifier						
No load gain	V_{IC}		60	80		dB
Commun mode range	V_{IC}		0		5	V
Output voltage	$V_{QH/L}$	$-3\,\text{mA} < I < 1.5\,\text{mA}$	1.5		5.5	V
Balance						
Input voltage	V_{IH}		2.0			V
	V_{IL}				0.8	V
Output stages Q_1 Q_2						
Output voltage	V_{QH}	$I_Q = 20\,\text{mA}$			30	V
	V_{QL}	$V_{QH} = 30\,\text{V}$			1.1	V
Output residual current	I_Q				2	µA

Figure 8.33 Range of functions and limit data for the TDA4700 (A) IC (Siemens)

Voltage-controlled oscillator (VCO)

A current source programmed by means of R_T supplies the charging current for C_T, the duration of the trailing edge being determined by the choice of C_T. The relationship of switching frequency to R_T and C_T is shown in Figure 8.35.

Since in principle the switching frequency should be above the audible range (i.e. higher than about 20 kHz), in Figure 8.35 with push–pull operation or with single-ended operation using both outputs in parallel operation should be restricted to the region above the chain-dotted line, while with single-ended operation using just one output only the region above the dashed line should be used. Attention must be paid to the dead time, for in push–pull operation the two switching transistors must not conduct simultaneously in any circumstances. The dead time, during which neither of the output transistors in the control IC conducts, must be longer than the sum of the storage and fall times of the two switching transistors. If the dead time according to Figure 8.36 is not sufficiently generous, the switching transistors must be operated in a anti-saturation circuit, so that the storage and fall times become very short (see also Chapter 6).

Even though the duty cycle should not exceed 0.45 in a transformer-coupled forward converter, its maximum value with the chosen value of R_T should be checked from Figure 8.37. This is essential for the dimensioning of the transformer (see Equations (5.19), (5.20) and (6.3)).

The capacitor C_{filt} should have a value of 4.7 nF with $f_0 \geqslant 10$ kHz. In free-running operation, with Terminals 18 and 7 connected together, C_{filt} can be omitted.

Ramp generator/feed-forward control

The ramp generator is triggered by a synchronizing pulse from the VCO. For the purpose of pulse-width control its leading edge is compared with a direct voltage in K_2. The steepness of the leading edge is determined by the current through R_R. This affords the possibility of an additional control of the duty cycle in dependence upon the input voltage V_i (feed-forward).

The capacitance C_R may have a value up to a maximum equal to that of C_T.

$$C_R \leqslant C_T \quad \text{(F)} \tag{8.9}$$

The resistance R_R is given by

$$R_R \approx \frac{3 C_T R_T (V_{RRmin} - 0.7 \, \text{V})}{C_R \Delta V_{CR}} - 4 \times 10^3 \quad (\Omega) \tag{8.10}$$

where V_{RRmin} is the minimum input voltage V_{imin} (V) and ΔV_{CR} is the peak/peak ramp voltage (V); selected between 2 and 2.9 V.

If feed-forward is not required, R_R is connected to V_{ref}, and in Equation (8.10) $V_{refmin} = 2.35$ V is inserted for V_{RRmin}.

The complete pulse diagram is shown in Figure 8.38 and the mode of operation of the feed-forward in Figure 8.39.

Phase comparator/synchronization

Where the circuit is operated without external synchronization (18) must be connected to 7). The VCO then runs at its nominal frequency f_0. Other TDA4700s can be synchronized

Characteristic data $V_s = 11$ to $30\,V$, $\theta_a = -25$ to $85\,°C$		Test conditions	Lower limit B	typ.	Upper limit A	
ON, OFF *undervoltage* K6						
Switching voltage	V		$V_{Ref} - 30\,mV$		$V_{Ref} + 30\,mV$	V
Dynamic current limiter K7						
Common mode range	V_{IC}		0		4	V
Turn-off delay	t			250		ns
Error detection time	t			50		ns
Overvoltage K5						
Switching voltage	V		$V_{Ref} - 30\,mV$		$V_{Ref} + 30\,mV$	V
Supply undervoltage						
Turn-on wave for V_s rising	V_s	$0° < \theta_a < +70\,°C$	8.8		11	V
					10.5	V
Turn-off wave for V_s falling	V_s	$0° < \theta_a < +70\,°C$	8.5		10.5	V
					10	V
Input C_{filt}						
Mains voltage for mains frequency	V_{NCS}			4		V
Frequency approximately proportional to voltage in the range	V_{NCS}		3		5	V
Voltage for open synchronized input	V_{CS}			1.6		V

Fig. 8.34 Characteristic data for the TDA4700(A)IC over the complete range of operating voltage and temperature.

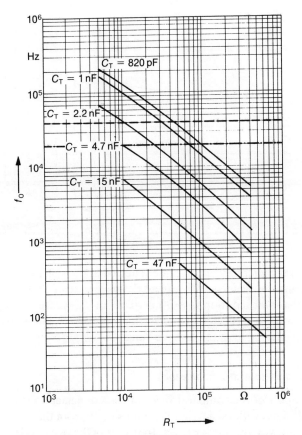

Fig. 8.35 Variation of oscillator frequency f_0 with resistance R_T, with capacitance C_T as parameter (Siemens)

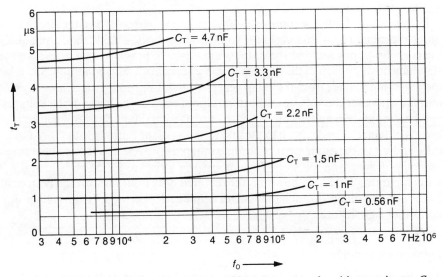

Fig. 8.36 Variation of dead time t_d with oscillator frequency f_0, with capacitance C_T as parameter (Siemens)

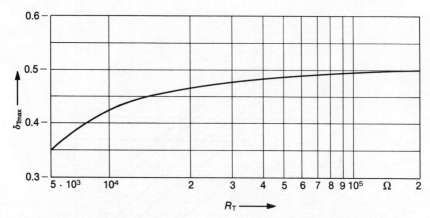

Fig. 8.37 Variation of maximum duty cycle δ_{Tmax} with R_T (Siemens)

via the output (18). Synchronization can be effected at the synchronizing input (18) with a voltage of rectilinear waveform with any mark/space ratio (TTL level). The capture range $\Delta f / f_0$ of the phase-locked loop is ± 30 per cent of f_0, where $\Delta f = f_{synch} - f_0$.

Synchronization is not effected in a phase-locked manner, but with a deviation of up to $\pm 120°$ varying linearly over the capture range.

Comparator K_2 (pulse-width modulator)

The two plus inputs of the comparator K_2 are so connected that the lower plus level is compared at all times with the level at the minus input from the ramp generator. As soon as the leading-edge voltage on C_R exceeds the lower of the two plus levels, both outputs are blocked through the pulse-inhibit flip-flop. These two plus inputs are brought out to input (14), connected to (15), and (8), which is connected to earth through a capacitor C_A (about $1 \mu F$).

C_A is provided for the purpose of a soft start, and is charged, when the circuit is switched on, at a current of $6 \mu A$. So long as the voltage on C_A is below 1.5 V, the output of K_4 is low, the fault flip-flop is set, and the outputs are enabled, provided that a reset signal is not simultaneously applied to the flip-flop in consequence of a fault condition. Since, however, the minimum voltage from the ramp generator is 1.8 V, and the caparator K_2 blocks the outputs whenever the voltage at the minus input exceeds the lower of the voltages on the two plus inputs, the voltage on C_A has to rise to something over 1.8 V before a low duty cycle is obtained. The delay up to the first enabling of the outputs with $C_A = 1 \mu F$ is about 0.3 s, and up to the attainment of the maximum duty cycle about 1 s. These times are in general sufficient for a soft start. The voltage on C_A rises to 5 V and is held at this level by the comparator K_3. Since, however, the required duty cycle has by then been established by the output voltage through the control amplifier K_1 ($V_{14} \leqslant 5$ V), the voltage on C_A is of no further significance.

Pulse-inhibit flip-flop

This flip-flop is set by the synchronizing pulse and reset by a 'low' signal from comparator K_2 (the output assumes low potential when V_{CR} is greater than the lower of the plus

Pulse diagram

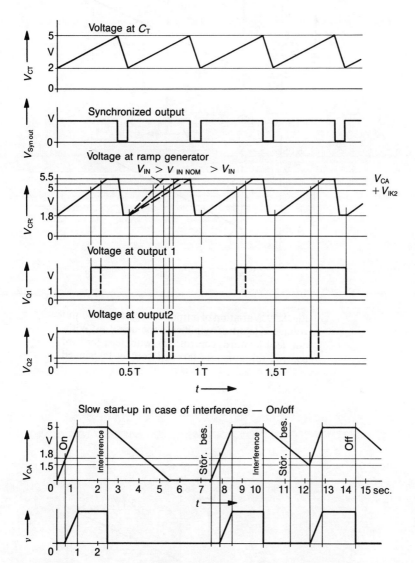

Fig. 8.38 Pulse diagrams for various functional blocks and starting diagram
for the TDA4700(A) IC (Siemens)

potentials) or by comparator K_7 (dynamic current limiting). This causes the output pulse (Q_1 or Q_2: active low) to be interrupted instantaneously. The required duty cycle δ_T is thus determined for each individual output pulse.

Fault flip-flop

If a fault is present, a 'low' signal is applied to the reset input of this flip-flop, and the outputs are instantaneously blocked. Faults may arise through overvoltage (K_5), undervoltage (K_6), IC undervoltage (K_8) or V_{ref} overload.

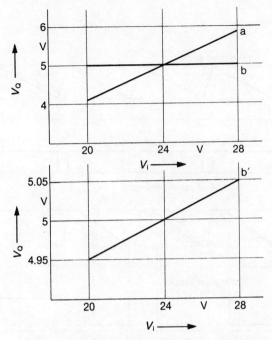

Fig. 8.39 Variation of output voltage V_0 with input voltage V_i (control amplifier inoperative). (a) Without feed-forward control, (b) with feed-forward control, (b) with feed-forward control (expanded scale)

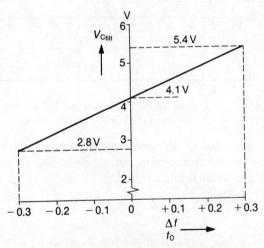

Fig. 8.40 Relationship between relative frequency f/f_0 and voltage V_{Cfil} (Siemens)

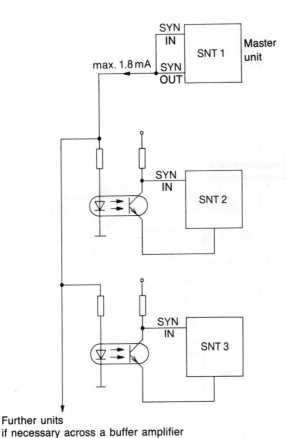

Fig. 8.41 Parallel operation of switched-mode power supplies using a free-running master equipment

Comparator K_5 (overvoltage)

If the voltage on (21) exceeds V_{ref}, the output of K_5 is driven high, which, inverted to a 'low' signal, resets the fault flip-flop instantaneously. At the same time C_A is discharged in about 1 s (with $C_A = 1\,\mu F$). If the overvoltage fault persists, the outputs remain blocked. If the fault is removed, the circuit resumes operation gradually by virtue of the voltage on C_A. Although the overvoltage monitoring can be applied to both the input and the output voltage of the power supply unit, it is useful in most cases to monitor the output voltage. The best method is to take the overvoltage monitoring signal from the voltage divider which is provided for adjustment of the output voltage is such a way that the circuit responds at about 10 per cent overvoltage (divider ratio 1:10—see Figure 8.42).

The output (20) can also be connected to (21), but the voltage to be monitored must then be fed back to (21) through a high resistance (about 4.7 kΩ). A consequence of connecting (20) to (21) is that outputs remain blocked until either the input (21) is connected briefly to earth or the supply voltage is interrupted.

Comparator K_6 (undervoltage)

This comparator responds to undervoltage. Thus if V_{19} falls below V_{ref}, the output of K_6 assumes low potential and blocks the outputs through the fault flip-flop. Operation is

290

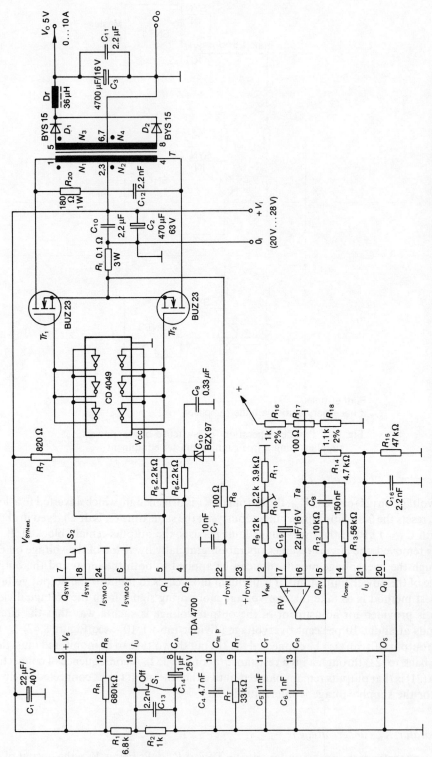

Fig. 8.42 D.C./d.c. converter from 24 V to 5 V/10 A using the TDA4700 IC and the BUZ23 SIPMOS FET (Siemens)

restored with a soft start. If δ_{Tmax} is prescribed at V_{imin} (e.g. as in Figure 8.37), the switch-off threshold must be just below V^*_{imin}. In the case of an inductor-coupled converter, if δ_{Tmax} is well within the available range, and not just on the ill-defined value of 0.5 with a push–pull control circuit, the margin between V^*_{imin} and V_{imin} can be made somewhat greater. In matching the operating voltage V^*_{imin} to V_{ref}, it is convenient to fix one resistance in the voltage divider and calculate the other (designations as in Figure 8.42):

$$R_1 = R_2 \frac{V^*_{imin} - V_{ref}}{V_{ref}} \quad (\Omega) \tag{8.11}$$

where V^*_{imin} is the operating voltage of the undervoltage comparator (V) and V_{ref} is the reference voltage = 2.5 V (allow for tolerances).

Comparator K_7 (dynamic current limiting)

Since both inputs are brought out, the circuit can be chosen without restriction. If $V_{22} > V_{23}$, the output of K_7 assumes low potential and the pulse-inhibit flip-flop switches off the operative output in less than 250 ns. Thus each pulse from the switching transistor is checked for excess and, if necessary, limited. For direct primary-side current limiting the voltage drop across the current-limiting resistor is set at 0.1 to 0.2 V; V_{23} must then be biased to this value from V_{ref}. With primary current limiting through a current transformer (23) can in some cases be connected directly to V_{ref}. In all cases, a filter with a time constant of about 1 μs should be provided at (22) so that the current-limiting circuit does not respond to spikes. The primary current limiting can be supplemented by secondary current limiting as shown in Figure 6.22. The voltage divider consisting of R_{11}, R_{12} and R_{13} (in Figure 6.22) is so adjusted that at the plus input of the additional operational amplifier (OP) the low negative voltage (about 0.1 V) derived from V_{RS} (the voltage across the current-sensing resistor in Figure 6.5) is just balanced. If V_{RS} then increases due to a secondary overcurrent, the output of OP assumes a low potential and δ_T is reduced. The operation of this system is illustrated in Figure 6.13.

Start–stop circuit

The switching-off of a power-supply unit incorporating the TDA4700 can be effected by switching off the power supply to the IC, through the operation of K_6 [connection of (19) to (1)] or through the short-circuiting of C_A [connection of (8) to (1)]. In the latter case a current-limiting resistor of 100 Ω in the turn-off circuit is recommended. When this operation is discontinued, restarting is effected in all cases with a soft start.

Output stages

The two outputs Q_1 (5) and Q_2 (4) are active-low, i.e. they switch the appropriate collector towards earth (1). Each output can be loaded to a maximum of 20 mA. The residual voltage with a transistor turned on is in the region of 0.7 to 1 V.

Balancing

The TDA4700 affords the possibility, by means of an additional balancing circuit, of controlling the two outputs in such a way that unbalanced currents in the push–pull

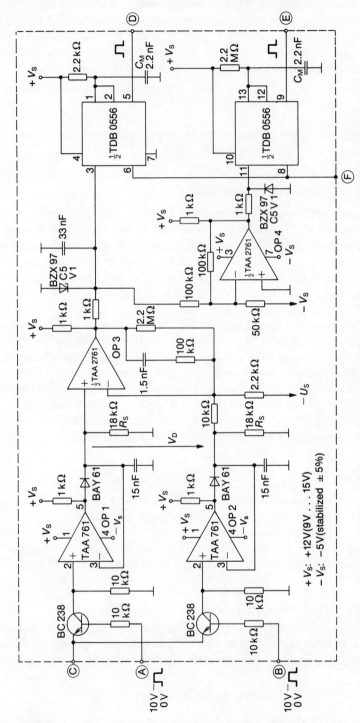

Fig. 8.43 Balancing circuit for push–pull switched-mode power supplies using the TDA4700 IC (Siemens)

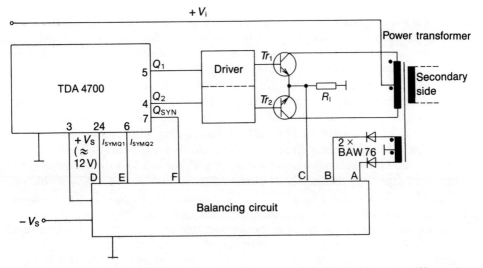

Fig. 8.44 Connection of the balancing circuit into a push–pull switching stage (Siemens)

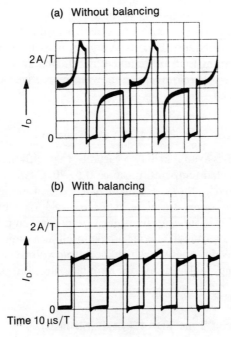

Fig. 8.45 Primary current in the power
transformer of a switched-mode power
supply (a) without balancing circuit and
(b) with balancing circuit (Siemens)

stages are avoided. Such imbalance can arise from partial saturation of the transformer at full load, caused by unequal storage times and voltage drops in the switching transistors. With bipolar transistors this can lead to destruction (through second breakdown). MOSFET transistors are less sensitive in this respect, since they do not exhibit second breakdown. In small power-supply units with MOSFET power transistors, all precautions against transformer saturation can be dispensed with (see Figure 8.42). In medium-sized power supplies saturation can be prevented by a capacitor C_{ba1} in the transformer lead (see Figure 6.4 or 6.5). The cost of the additional balancing circuit shown in Figure 8.43 is to be recommended only in very highly rated power supplies. The incorporation of the balancing circuit into the basic circuit is illustrated in Figure 8.44; Figure 8.45 shows the primary current waveform without and with the balancing circuit.

Since the outputs Q_1 (5) and Q_2 (4) are switched off instantaneously by a voltage in excess of 2 V on the balancing inputs (24) or (6), these must be connected to earth when the balancing circuit is not used.

It should be noted in the circuit of Figure 8.43 that in addition to a positive supply voltage of about 12 V a (negative) auxiliary supply of -5 V is required.

Resistor R_s (18 kΩ) is calculated for an oscillator frequency of 40 kHz. For other frequencies, the value is given by

$$R_s \approx \frac{11}{f_0 \times 15 \times 10^{-9}\,\text{F}} \quad (\Omega) \tag{8.12}$$

where f_0 is the oscillator frequency (Hz).

The value of C_M must be determined experimentally, such that with $V_D = 0$ the pulse width at (D) and (E) is approximately half the correction range provided. Since in many applications all the functions of the TDA4700 are not required, it would be uneconomic to use this large IC in all cases. For this reason there is available a range of 'slimmed-down' versions, which may be mentioned briefly. A common feature of all types in the TDA47xx series is that the functions that are retained are all similar, and similar to those in the TDA4700. It is necessary therefore only to show the respective block diagrams and to refer briefly to the differences.

Integrated control circuit Type TDA4718 (A) (Siemens)

This type (Figure 8.46) is available in two variants, that with the appended code 'A' being specified for the restricted temperature range 0 to 70 °C, while the variant without the 'A'—the letter 'B' is used for the types with further-reduced specifications—is rated for the extended temperature range from -25 °C to $+85$ °C. Compared with the basic TDA4700, the balancing inputs (two terminals), the control amplifier (three terminals) and the output of K_5 (one terminal), six terminals in all, are omitted. The internal control amplifier is not required in circuits in which the output voltage deviation is transmitted by an optocoupler, being replaced by an amplifier in front of the optocoupler (see Figure 5.67).

Integrated control circuit Type TDA4714A/B (Siemens)

Additional omissions in this IC are the entire synchronizing facilities (three terminals) and the undervoltage comparator, a total of four terminals. This is not a very serious lack,

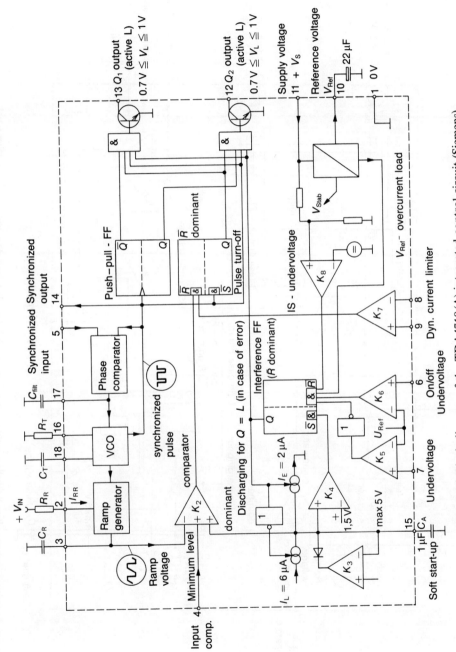

Fig. 8.46 Block diagram of the TDA4718(A) integrated control circuit (Siemens)

296

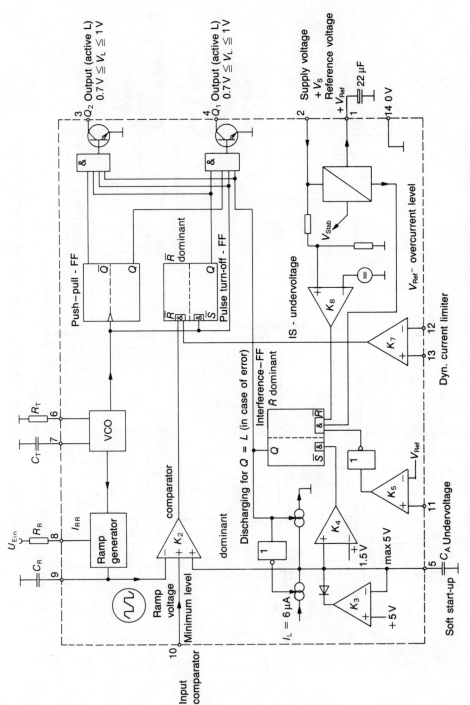

Fig. 8.47 Block diagram of the TDA4714A/B integrated control circuit (Siemens)

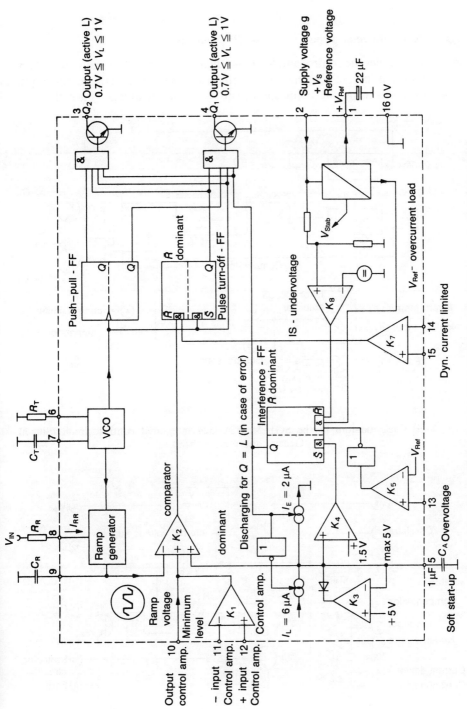

Fig. 8.48 Block diagram of the TDA4716A/B integrated control circuit (Siemens)

since it is rarely required to synchronize a number of ICs. The block diagram is shown in Figure 8.47.

Integrated control circuit Type TDA4716A/B (Siemens)

This variant includes the control amplifier and therefore has two terminals more than the 4714. The output of the amplifier is internally connected directly to the input of comparator K_2, saving one connection. This is no disadvantage, since the control

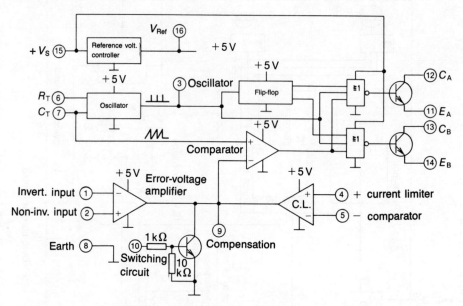

Fig. 8.49 Block diagram of the SG1524/2524/3524 integrated control circuit (Silicon General and many others)

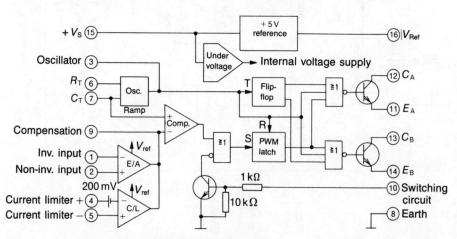

Fig. 8.50 Block diagram of the SG1524A/2524A/3524A integrated control circuit (various manufacturers)

	Conditions	SG1524 SG2524			SG3524			Units
		Min	Typ	Max	Min	Typ	Max	
Reference voltage Oscillator		4.8	5.0	5.2	4.6	5.0	5.4	V
Max. frequency	$C_T = 0.001\ \mu F$, $R_T = 2\ k\Omega$		300			300		kHz
Output voltage Oscillator	(3) $\theta_a = 25\,^\circ C$		3.5			3.5		V
Error-voltage gain								
No load gain	$\theta_a = 25\,^\circ C$	72	80		60	80		dB
Common mode range	$\theta_a = 25\,^\circ C$	1.8		3.4	1.8		3.4	V
Limiting frequency	$A_V = 0dB$, $\theta_a = 25\,^\circ C$		3			3		MHz
Output voltage	$\theta_U = 25\,^\circ C$	0.5		3.8	0.5		3.8	V
Comparator								
Switching ratio r_T		0		45	0		45	%
Threshold voltage	$r_T = 0$		1			1		V
Threshold voltage	r_{Tmax}		3.5			3.5		V

	Conditions	SG1524 SG2524			SG3524			Units
		Min	Typ	Max	Min	Typ	Max	
Current limiting comparator								
Threshold voltage		190	200	210	180	200	220	mV
Common mode range		−1		+1	−1		+1	V
Output stage (in each case)								
V_{CE}		40			40			V
V_{CEsat}	$I_C = 50\,mA$		1	2		1	2	V
V_E	$V_S = 20\,V$	17	18		17	18		V
Quiescent current	$V_S = 8\ldots40\,V$		5	10		5	10	mA
Switching voltage	$\theta_j = 25\,°C, R_c = 2\,k\Omega$	0.5	0.7	1.0	0.5	0.7	1.0	V

Fig. 8.51 Principal characteristic values for the SG1524/2524/3524 (Silicon-General and many others).

amplifier output is practically always connected to the comparator input. The block diagram is shown in Figure 8.48.

An illustration of the terminal designations such as that of Figure 8.32 has not been provided, since the numbering system begins in all cases at the notch in the housing. The connection numbers are indicated in the block diagrams.

The following descriptions cover briefly a further integrated push–pull control circuit—the SG1524...3524, which is made by a large number of manufacturers under licence, so that several suppliers can be nominated, as is frequently required in industrial applications. Since this circuit has been referred to in previous chapters, its characteristic features will be considered.

Integrated control circuit Type SG1524/2524/3524 (Valvo/Signetics, Texas Instruments, Silicon General and others)

This very widely available control circuit is also supplied by other manufacturers under similar designations, e.g. LT1524/2524/3524 from Linear Technology (Metronik-Vertrieb), CA1524/2524/3524 from RCA, XR1524/2524/3524 from Exar or UC1524/2524/3524 from Unitrode-Metronik. In the following discussion these differently designated but electrically identical circuits will be referred to simply as 'SG1524...3524'. In addition, some manufacturers supply the circuit with the code 'A' appended, to denote a more accurate reference voltage, a higher maximum frequency, a higher collector voltage for the output transistors and a higher maximum collector current. Although these are of somewhat more complex internal construction, all types are interchangeable. The block

Absolute limiting values	
Supply voltage V_S	40 V
Collector current	100 mA
Reference current	50 mA
Current in C_T	−5 mA
Power loss, $\theta_a = +25\,°C$	1000 mW
Thermal resistance R_{thjs}	100 k*W
Power loss $\theta_c = +25\,°C$	2000 mW
Thermal resistance R_{thjc}	60 K/W
Ambient temperature	−55 °C...+150 °C

Recommended operating values	
Supply voltage V_S	8 V...40 V
Current from V_{ref}	0...20 mA
Current in C_T	−0.03 mA...−2 mA
Resistance R_T	1.8 kΩ...100 kΩ
Capacitor C_T	0.001 μF...0.1 μF
SG1524	−55 °C...+125 °C
SG2524	−25 °C...+85 °C
SG3524	0 °C...+70 °C

Fig. 8.52 Recommended operating and absolute limit values for the SG1524/2524/3524 integrated control circuit. (Silicon General and many others)

Characteristic values SG1524A…3524A

	Conditions	SG1524A SG2524A			SG3524A			Units
		Min.	Typ.	Max.	Min.	Typ.	Max.	
Reference voltage	$\theta_j = 25\,°C$	4.95	5.00	5.05	4.90	5.00	5.10	V
Oscillator								
Min. frequency	$R_T = 150\,k\Omega$, C_T $= 0.1\,\mu F$			120			120	Hz
Max. frequency	$R_T = 2.0\,k\Omega$, C_T $= 470\,pF$	500			500			kHz
Output (in each case)								
V_{CE}	$I_C = 100\,\mu A$	60	80		50	80		V
I_{CEO}	$V_{CE} = 50\,V$		0.1	20		0.1	20	μA
V_{CEsat}	$I_C = 20\,mA$ $I_C = 200\,mA$		0.2 1	0.4 2.2		0.2 1	0.4 2.2	V V

Fig. 8.53 Characteristic values for the SG1524A/2524A/3524A integrated control circuit (otherwise similar to those listed in Figure 8.51) (various manufacturers)

Absolute limiting values

Supply voltage (V_S)	40 V
Collector voltage (V_C)	
SG1524A, SG2524A	60 V
SG3524A	50 V
Collector current	200 mA
Ambient temperature	$-55\,^{\circ}\mathrm{C}\ldots+125\,^{\circ}\mathrm{C}$

Fig. 8.54 Absolute limiting values for the SG1524A/2524A integrated control circuit (otherwise similar to those listed in Figure 8.52) (various manufacturers)

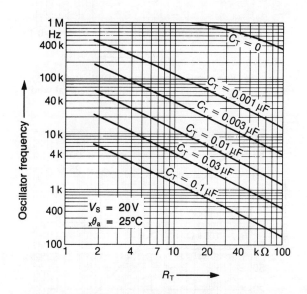

Fig. 8.55 Variation of oscillator frequency (Hz) with resistance R_T; capacitance C as parameter (Silicon General and many others)

diagram of the SG1524/2524/3524 is shown in Figure 8.49, while Figure 8.50 shows the block diagram of the 'A' version for comparison.

Figure 8.51 lists the principal data for the standard SG1524/2524/3524; the recommended operating values and the absolute limits are given in Figure 8.52.

The improved values of the 'A' version, where they differ from those of the normal version, are given in Figures 8.53 and 8.54.

Since this circuit has been used in previous examples (see Figures 2.25, 3.1 and 4.1), it is of interest to illustrate the relationships for the calculation of oscillator frequency and lead time. These are shown in Figures 8.55 and 8.56.

In contrast to the control circuits considered previously, in this case the gain of the error voltage amplifier is adjusted not in the feedback path, but by means of a resistor R_F between (9) and earth (8). In many cases the d.c. gain is set very high and the a.c. gain relatively low, using an RC combination (see Figure 8.57).

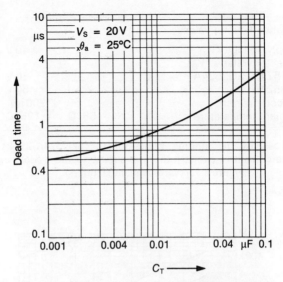

Fig. 8.56 Variation of dead time (μs) with frequency-determining capacitance C_T (Silicon General and many others)

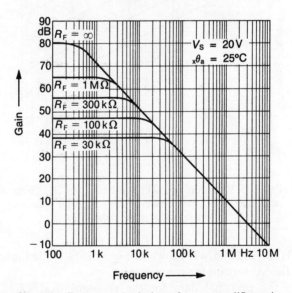

Fig. 8.57 Frequency variation of error-amplifier gain (dB) with resistance R_F as parameter (Silicon General and many others)

8.3 SINGLE-ENDED CONTROLLERS WITH INTEGRATED DRIVER STAGES

Integrated control circuit Type TDA4600-2 (-2D) (Siemens) [23]

This control circuit differs from all the circuits discussed previously in two important features: it includes an integrated driver stage and there is no fixed-frequency operation.

The duty cycle varies as in other circuits with varying operating voltage; i.e. the maximum duty cycle δ_{Tmax} is associated with the minimum input voltage V_{imin}, and similarly δ_{Tmin} with V_{imax}. Whereas, however, in fixed-frequency operation the duty cycle has to alter with changes in load (flyback converter operating with triangular current waveform—see Equation (7.7) and Figures 7.43 and 7.44), in this case it remains constant; instead, the frequency increases as the load decreases. Thus if the product of secondary load and frequency is approximately constant, from Equation (7.7) the duty cycle must remain constant. Figure 8.58 illustrates the situation with different combinations of input voltage and secondary load.

With low mains voltage and nominal load, Section 4 applies, with $\delta_T = 0.42$ and $f = 18\,\text{kHz}$. Similar values were adopted as a starting point for the circuit of Figure 7.32 ($\delta_{Tmax} = 0.5$ and $f = 20\,\text{kHz}$). At nominal and high mains voltage, both with full load (Sections 2 and 1), the frequency changes only by a small amount, while the duty cycle decreases to 0.33 and 0.2 respectively. If, however, at nominal mains voltage, the load is considerably reduced (Section 3), the frequency rises at a constant duty cycle (0.33) to about double. The internal block diagram is shown in Figure 8.59.

Apart from the base current amplifier, all the functional blocks are supplied at the internal reference voltage $V_{ref} = 4\,\text{V}$. The high no-load output voltage enables MOSFET power transistors to be driven as well as bipolar transistors. The following considerations

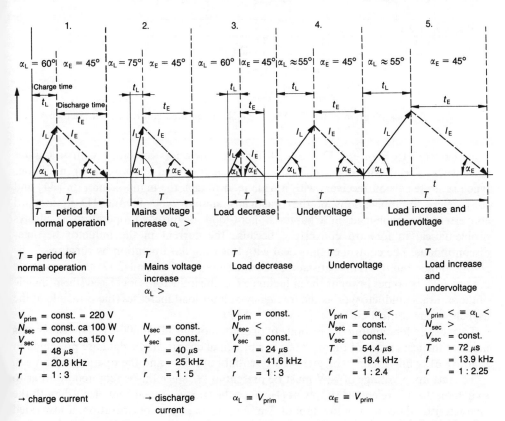

Fig. 8.58 Switching characteristics of the TDA4600-2 (− 2D) integrated control circuit showing the frequency and duty cycle with varying mains frequency and load (Siemens)

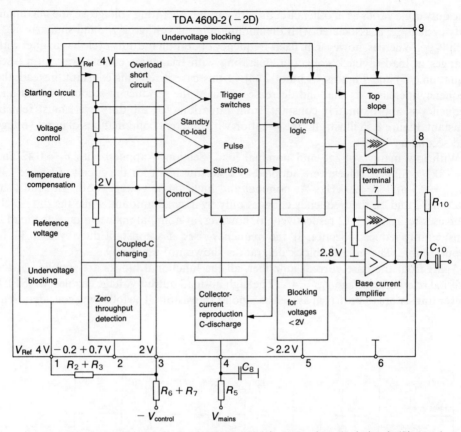

Fig. 8.59 Block diagram of the TDA4600-2 (-2D) integrated control circuit (Siemens)

will be related to Figures 7.32 and 7.34 (flyback converters operating with triangular current waveform). Although the TDA4600-2 is mainly used in converters of the latter type (e.g. in television receivers with a wide load range), the manufacturer, in [34] and [35], refers to applications of the practically similar type TDA4601D in forward converters. As explained in Chapter 2, light-load or no-load operation is always problematical in forward converters, because the current in the inductor becomes discontinuous. The control circuits deal with this irregular operation by turning off the control pulses, but this is not satisfactory. From Equations (2.3) and (2.12), the permissible current I_{0min} becomes lower as the inductance L of the inductor is increased (the principle of the saturable inductor) or as the frequency at light load increases (the principle of the asynchronous control circuit, e.g. Type TDA4601).

To assist understanding of the mode of operation of the TDA4600-2 (-2D), reference is made to the test circuit of Figure 8.60. The principal characteristic data are presented in Figure 8.61, while Figure 8.62 gives the essential limit data and the operational ranges.

To start up, a voltage of 12 V must be present at (9), since this is required in order to switch on the 4 V reference supply at (1). Since the starting-up current, according to the characteristic data, is very low (about 3 mA) up to the point of operation, a low-rated auxiliary power supply, through either series capacitors (Figure 7.32) or a high-value resistor (10 kΩ in Figure 7.34), is sufficient. The auxiliary winding for the IC supply is

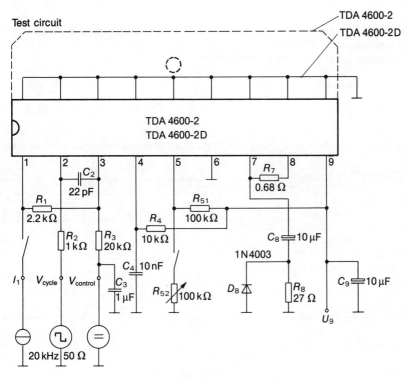

Fig. 8.60 Test circuit for the TDA4600-2 (-2D) integrated control circuit
(Siemens)

connected in the forward-converter mode, so power is supplied to (9) at the first turning-on
of the switching transistor. A voltage of between $-0.2\,$V and $+0.7\,$V must be present at
(2), and this is obtained from an auxiliary winding N_3, as shown in Figure 7.32, producing
about 10 V, and so poled that with the switching transistor conducting the voltage on (2) is
positive. The output (8) for base drive is then enabled. If the switching transistor is turned
off by the control amplifier (because V_0 is higher than the desired value), the voltage on (2)
is negative, and output becomes available at (8) only when this voltage has passed through
zero and become positive again. The control amplifier input (3) is at a voltage of
approximately 2 V under operating conditions, obtained from V_{ref}. If, for example, V_0 rises
as a result of an increase in the mains voltage, V_3 must be reduced, in order to maintain the
desired value of V_0. Either this can be effected by means of an inverting-input control
voltage from the auxiliary winding used for detecting zero voltage on (2) (Figure 7.34), or
an optocoupler can be used to increase the current in the transistor, as in Figure 7.34. The
low-pass filter consisting of $100\,\Omega$ and 8.2 nF (time constant about $1\,\mu$s) removes steep-
fronted overvoltage spikes. The leading edge of the base current from (8) is programmed
by means of the RC network connected to (4), which results indirectly in a reproduction,
and hence a limiting, of the collector current. For the dimensioning of the RC network, see
Equation (7.50). Terminal 5 can be used to switch off the IC; the output is blocked with a
voltage in excess of 2.2 V. Terminal 8 delivers an impressed current, through an electrolytic
coupling capacitor, of a maximum of 1.5 A. The magnitude of this current is determined by
the feedback resistance between (7) and (8), which can be from $0.33\,\Omega$ to $2.2\,\Omega$. Typical
current and voltage waveforms are shown in Figure 8.63.

Characteristic data
$\theta_a = 25\,^{\circ}C$, for test circuit

		min	typ	max	
Starting conditions					
Current consumption (V_1 not yet switched on)					
$\quad V_9 = 2\,V$	I_9			0.5	mA
$\quad V_9 = 5\,V$	I_9		1.5	2.0	mA
$\quad V_9 = 10\,V$	I_9		2.4	3.2	mA
Turn-on point for V_1	V_9	11	11.8	12.3	V

Normal conditions
($V_9 = 10\,V$; $V_{Control} = -10\,V$; $V_{Cycle} = \pm0.5\,V$; $f = 20\,kHz$; Switching ratio 1:2) as follows
Turn-on

		min	typ	max	
Current consumption $V_{Control} = -10\,V$	I_9	110	135	160	mA
$\qquad R_{Control} = 0\,V$	I_9	55	85	110	mA
Reference voltage $I_1 < 0.1\,mA$	V_1	4.0	4.2	4.5	V
$\qquad I_1 = 5\,mA$	V_1	4.0	4.2	4.4	V
Reference voltage temperature coefficient	TC_1		10^{-3}		1/K
Reverse coupled voltage	$V_2{}^*$		0.2		V
Control voltage $V_{Control} = 0\,V$	V_3	2.3	2.6	2.9	V
Collector current reproduction					
Voltage $V_{Control} = 0\,V$	$V_4{}^*$	1.8	2.2	2.5	V
$\qquad V_{Control} = 0\,V/-10\,V$	$\Delta V_4{}^*$	0.3	0.4	0.5	V
Blocking input voltage	V_5	5.5	6.3	7.0	V
Output voltages $V_{Control} = 0\,V$	$V_{97}{}^*$	2.7	3.3	4.0	V
$\qquad V_{Control} = 0\,V$	$V_{98}{}^*$	2.7	3.4	4.0	V
$\qquad V_{Control} = 0\,V/-10\,V$	$\Delta V_{98}{}^*$	1.4	1.8	2.2	V

Fault conditions
($V_9 = 10\,V$; $V_{Control} = -10\,V$; $V_{Cycle} = \pm0.5\,V$; $f = 20\,kHz$; switching ratio 1:2)

		min	typ	max	
Current absorption ($V_5 < 1.8\,V$)	I_9	14	22	28	mA
Turn-off voltage ($V_5 < 1.8\,V$)	$V_{\times 7}$	1.3	1.5	1.8	V
	V_4	1.8	2.1	2.5	V
External blocking input					
$\quad$ Release voltage $V_{Control} = 0\,V$	V_5		2.4	2.7	V
$\quad$ Reverse voltage $V_{Control} = 0\,V$	V_5	1.8	2.2		V
Supply voltage for V_8					
reversed $V_{Control} = 0\,V$	V_9	6.7	7.4	7.8	V
Supply voltage for V_1 onwards					
(for further stages from V_9)	ΔV_9	0.3	0.6	1.0	V

*Equal partions only.

Fig. 8.61 Characteristic data for the TDA4600-2(-2D) integrated control circuit under starting, normal operating and fault conditions (Siemens)

To obtain a satisfactory turn-off characteristic in the switching transistor and low turn-off losses, an inductance must be inserted into the base lead (see Figure 7.32).

Under overload conditions, up to a short circuit, the logic produces a re-entrant characteristic (see Figure 7.13), so that the power developed is quite small. On no load, the 22 pF capacitor between (3) and (2) enables a more regular free-running operation to take place at $f_{max}/2$ and a very low duty cycle.

Integrated control circuit Type TEA1001SP (Thomson CSF) [36]

This control circuit similarly contains a driver stage with a maximum available base current for the switching transistor of 3 A, but is designed for quasi-saturated operation of

Limiting data

Supply voltage	V_9	20	V
(Base current turn-off)	V_7	V_9	V
(Base current amplifier output)	V_8	V_9	V
(Base current turn-off)	I_{97}	1.5	A
(Base current amplifier output)	I_{98}	-1.5	A
Thermal resistance:			
Junction–Surroundings TDA4600-2	R_{thJU}	70	K/W
Junction–Encapsulation TDA4600-2	R_{thEN}	15	K/W
Junction–Surroundings TDA4600-2D	R_{thJU}	60	K/W
Junction temperature	θ_j	125	°C

Operating range

Supply voltage	V_9	7.8 to 18	V
Encapsulation temperature TDA4600-2	θ_{EN}	0 to 85	°C
Ambient temperature	θ_a	0 to 70	°C

Fig. 8.62 Limiting data and operating range for the TDA4600-2 (-2D) integrated control circuit. (Siemens).

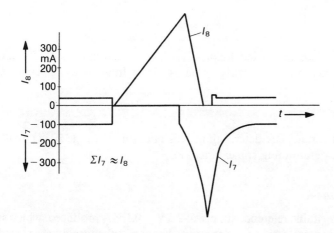

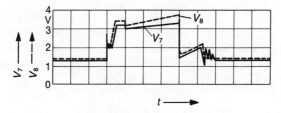

Fig. 8.63 Typical current and voltage waveforms at the output (7) and (8) of the TDA4600-2 (-2D) integrated control circuit (Siemens)

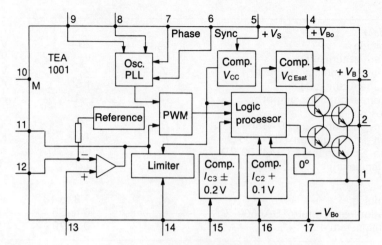

Fig. 8.64 Block diagram of the TEA1001SP integrated control circuit
(Thomson CSF)

the switching transistor. The block diagram is shown in Figure 8.64. Characteristic values are given in Figure 8.65 and the limiting values in Figure 8.66. The special housing is not like the normal DIL package, and the connection designations are shown in Figure 8.67. The circuit will be discussed with reference to the application shown in Figure 7.31.

Oscillator

The oscillator runs at a fixed frequency which is determined by R_T and C_T, up to a maximum quoted as $f_{max} = 50$ kHz. The oscillator frequency is calculated from

$$f_0 = \frac{1.85}{R_T C_T} \quad \text{(Hz)} \tag{8.13}$$

where R_T is the frequency-determining resistance (in the range 10 kΩ to 40 kΩ) and C_T the frequency-determining capacitance (F).

Reference voltage

A temperature-stable reference supply of 2.5 V ± 0.1 V is produced, which supplies all the internal comparators. V_{ref} is connected through an internal resistor of 1 kΩ to the inverting input (12) of the error voltage amplifier.

Synchronization

If it is required to synchronize the oscillator to an external frequency source, this can be done with an input to (6) at a level of between 1.5 V and V_S and a pulse duration of about $5 \mu s$. The resistance R in Figure 8.68 is calculated from

$$12 \text{ k}\Omega < R \frac{T}{T_s} \times 2.3 \text{ k}\Omega \quad (\Omega) \tag{8.14}$$

where T is the period of the set frequency and T_s is the period of the synchronising frequency.

Electrical characteristic values for $\theta_a = 25\,°C$

	Test conditions	Terminal	min.	typ	max.	
Supply voltage		V_{5-10}	6.5	10	14	V
Positive auxiliary voltage		V_{3-10}	4			V
Negative auxiliary voltage		V_{1-10}	-5			V
Supply current in operation	$V_S = 10\,V$	I_5		18	24	mA
V_S threshold voltage for V_s rising		V_{5-10}		6.5	V	
V_S threshold voltage for V_s falling		V_{5-10}		6		V
Primary current limiting threshold		V_{15-10}	±0.18	±0.2	±0.22	V
Secondary current monitoring threshold		V_{16-10}	0.075	0.1	0.125	V
Secondary current monitoring input current	$V_1 = 0\,V$	I_{16}		1	10	μA
OP amplifier gain			100000			
Input impedance			1 kΩ to V_{ref}			
Internal reference voltage		V_{12-10}	2.4	2.5	2.6	V
Oscillation frequency					50	kHz
Optimum current in R_T		I_9		0.5		mA
Synchronizing pulse voltage		V_{6-0}	1.5		V_s	V
Sychronizing input impedance		Z_6		1000		Ω
V_{Tmax} (internal)	$V_S < 6.5\,V$			0		%
	$V_S = 6.5...14\,V$		40		50	
	$V_S > 14\,V$			0		
V_{Tmax} (external)	$V_{14\ 10} = (0.375...0.625)V_S$			90		%
	$V_{14\ 10} = 0.625 V_S$			from $V_{14-10} =$ to 90		
	$V_{14-10} > 0.625 V_S$			0		

Fig. 8.65 Principal electrical characteristic values for the TEA1001SP integrated control circuit (Thomson CSF)

Thermal characteristic values

Thermal resistance Junction/Surroundings	$R_{th(jsr)}$	35	k/W
Thermal resistance Junction/encapsulation	$R_{th(jen)}$	3	k/W

Absolute limiting values

	Terminal		
Operating voltage	V_{5-10}	15	V
Positive auxiliary voltage	V_{3-10}	15	V
Negative auxiliary voltage	V_{1-10}	− 7	V
Voltage between terminal 3 and substrate	V_{3-1}	20	V
Output current	I_2	± 3	A
Current at terminal 4	I_4	10	mA
Primary current limiter input current	I_{15}	± 5	mA
Secondary current monitoring input current	I_{16}	± 5	mA
Junction temperature	θ_j	− 40...+ 150	°C

Fig. 8.66 Thermal characteristic values and absolute limiting values for the TEA1001SP integrated control circuit (Thomson CSF)

If a number of TEA1001SP circuits are to be operated at the same frequency, one only is provided with R_T and C_T; all the others have their Terminals 8 connected together and their Terminals 9 connected to V_S.

Comparators

Four comparators are provided, with the following functions.

V_s comparator (5)

This comparator enables the positive output stage so long as the operating voltage V_S is within the range 6.5 V to 14 V. The lower limit includes a hysteresis of 0.5 V, so that the circuit continues to work until the voltage falls to 6 V.

V_{CEsat} comparator (4)

Although the power transistor is not driven into 'hard' saturation, by reason of the anti-saturation circuit, the collector-emitter voltage is relatively low (about 1.8 to 2 V). In

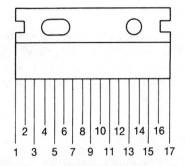

2	4	6	8	10	12	14	16
1	3	5	7	9	11	13	15	17

1,17	Substrate (negative voltage supply for the output stage)
2	Output
3	Positive voltage supply for the output stage
4	V_{CEsat}-Recording
5	V_{CC}
6	Synchronization
7	Phase comparator
8	C_T
9	R_T
10	Earth
11	OP AMP output
12	OP AMP inverted input
13	OP AMP non-inverted input
14	Maximum modulation
15	Primary current limiter
16	Secondary current monitoring

Fig. 8.67 Plan view of the housing and pin allocations of the TEA1001SP integrated control circuit (Thomson CSF)

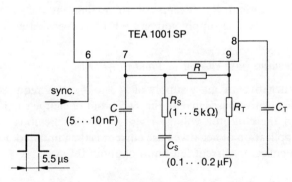

Fig. 8.68 Synchronizing facility (part circuit) for the TEA1001SP integrated control circuit (Thomson CSF)

normal operation, when the positive base current is switched off by the control amplifier, the collector voltage rises. When a threshold of 5 V is reached, the negative base current is switched on and the switching transistor is turned off. In the event of an inadvertent desaturation under fault conditions ($V_{CE} \geqslant 5$ V), the positive base current is switched off instantaneously and the negative base voltage subsequently applied.

Primary current comparator (15)

This comparator, with an operating threshold of ± 0.2 V, is normally made to respond to the voltage drop produced across a resistor by the primary current, and switches off the base current instantaneously through the logic processor. It should not be forgotten that an RC filter with a time constant of a few microseconds is necessary in the lead to (15) (see Figure 7.31).

Secondary current-monitoring comparator (16)

The secondary current-monitoring comparator, with a threshold voltage of 0.1 V, detects a voltage from an auxiliary winding, which is positive during the non-conducting phase. This voltage is proportional to the change in magnetic flux, and approaches zero when the flux in the transformer core has decayed. So long as the voltage exceeds 0.1 V, the turning-on of the switching transistor is inhibited. This monitoring circuit is intended to prevent saturation of the transformer under overload conditions. If the overload current is excessive, i.e. the load resistance is too low, the magnetic energy in the transformer core cannot be completely discharged (the time constant L_2/R_L is too long), and the core may be driven into saturation in a few cycles. In these circumstances the turning-on is inhibited until the magnetic energy is completely dissipated.

Output stage for positive base current (2) + (3)

The logic processor turns on the output stage when there is no fault indication and the enable signal is supplied by the pulse-width modulator (PWM). A short-duration base-current peak (duration about 1 μs) is delivered initially to ensure a rapid and low-loss turn-on. A base current of such magnitude is then supplied, according to the collector current, that the switching transistor operates in the quasi-saturated mode (self-regulating drive circuit). Control of the base current is effected through diode D_7 and resistor R_8 (Figure 7.31). The base resistor R_6 prevents control oscillations; the voltage drop across it should be less than 1.5 V. The supply voltage $+ V_B(3)$ is specified as 4 to 6 V.

Output stage for negative base current (2) and (1)/(17)

For this a further negative auxiliary supply of -3 to -5 V is required. Following the switching-off of the positive base current, the sweep-out of charge carriers from the base region begins, and the collector voltage rises. As in the previous case of the V_{CEsat} comparator, the application of negative base current is then initiated. By this means it is possible to obtain collector current fall times of about 0.1 μs.

Limitation of maximum duty cycle (14)

If (14) is not connected, the maximum duty cycle is between 0.4 and 0.5. Externally-connected components enable $\delta^*_{Tmax} > \delta_{Tmax}$ to be adjusted to suit the requirements of the application. The higher the voltage V_{14}, the lower is δ^*_{Tmax}; with $V_{14} = V_S/2$, $\delta^*_{Tmax} = 0.5$. The same tendency applies to the output of the control amplifier, and hence to its input, since the control quantity is applied to the non-inverting ($+$) input. The control loop must thus be so configured that an increased output voltage V_0 results in an increased current in the LED of the optocoupler and hence an increase in its collector current. This means, in Figure 7.31, an increased voltage on R_{11} and therefore a reduced duty cycle. The control loop thus functions correctly. The same applies to the circuit of Figure 7.35, with either secondary-side or primary-side control.

Light-load operation

In the flyback converter the duty cycle is reduced with decreasing load, and with it the conduction period. To ensure the discharge of the RCD protective network, the minimum conduction period is limited to $2\,\mu s$. If the loading conditions demand a still lower conduction period, the frequency of the drive pulses (synchronous with the oscillator) is reduced.

9 Interference suppression and protection circuits

Although the necessity for interference suppression has not been mentioned in connection with all the examples discussed, it is in fact practically always required. Whether or not protective circuits are also necessary depends upon the specified requirements. The scope of this book permits only a brief discussion.

9.1 MEASURES AGAINST INTERFERENCE

Since the operation of all switched-mode power supplies in the frequency range of about 20 to 100 kHz involves current pulses with the steepest possible wavefronts, the presence of very small, but never wholly avoidable, stray inductances leads to the generation of high induced voltages. These so-called 'spikes' must be removed from both the output and the input (mains) as effectively as possible. In addition, the circuit itself radiates high frequencies, for a current pulse with a fall time of, say, $0.1\,\mu s$ contains high-frequency components up to 10 MHz and higher. While the following comments relate to the so-called primary-switched converters (forward and flyback converters with transformers), they apply equally to inductor-coupled converters.

Satisfactory circuit layout

This entails making the shortest possible connections from the positive side of the mains electrolytic capacitor through the converter, the switching transistor and earth (the negative side of the capacitor). If this is not possible, a low-inductance capacitor of a few microfarads should be connected directly from the transformer to earth, i.e. in parallel with the electrolytic capacitor. Also the loop traversed by the circuit current should be kept as small as possible.

A further proven measure is the incorporation of screening foils in the converter transformer. Where possible, two insulated foils should be applied, one connected to the input and the other to the output (see also Figures 5.18, 5.64 and 7.10).

To reduce the radiation of interference from the often quite voluminous heatsink provided for the switching transistor and the rectifier diodes, isolation of the components in the manner shown in Figure 9.3 is recommended.

The RCD protective networks for the switching transistor and the RC networks for the output rectifier diodes should be soldered directly with the shortest possible connections to the components to be protected. In working out the circuit layout, close attention must

316

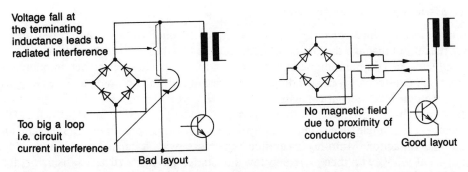

Voltage fall at
the terminating
inductance leads to
radiated interference

Too big a loop
i.e. circuit
current interference

Bad layout

No magnetic field
due to proximity of
conductors

Good layout

Fig. 9.1 A guide to the most satisfactory arrangement of filter capacitor, transformer and switching transistor

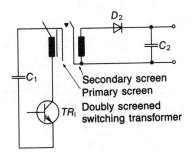

Secondary screen
Primary screen
Doubly screened
switching transformer

Fig. 9.2 Recommended screening in the main transformer

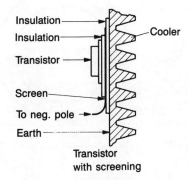

Insulation
Insulation
Transistor
Screen
To neg. pole
Earth

Cooler

Transistor
with screening

Fig. 9.3 A suggested arrangement for insulating the cooler to reduce radiated interference

be paid to short connections, large-area earth conductors, etc. (see also Figure 2.26). In many instances even the most satisfactory arrangement of the circuit based on high-frequency techniques is not sufficient, and the whole circuit has to be fitted into a screening enclosure. The better the planning of the layout, however, the less stringent are the screening requirements.

318

Interference suppression at the output

Although in a forward converter, in theory, no interference voltages should be able to reach the output through the low-pass filter comprised by the relatively high inductance of the inductor in combination with the large output capacitance, in fact interference currents are conducted by the winding capacitance of the inductor and produce voltages across the non-ideal output capacitor (ESR and L_C) (see Figure 7.15). This is still more critical in flyback converters. A remedy is provided by an 'after-filter' with a small inductor—e.g. a single turn with a suitably dimensioned ring core (Figure 7.32)—and an additional capacitor. Multi-layer ceramic capacitors (with a capacitance range up to about 4.7 μF/50 V), with their extremely low inductance, have proved very satisfactory for this purpose. Figure 7.15 shows the output interference voltage without and with an after-filter. The ferrite beads which can simply be slipped over interference-sensitive leads are also very effective.

Interference suppression at the input

In this case it is necessary to incorporate an interference filter directly in the mains lead. The choice and dimensioning of such a filter may be considered here in somewhat more detail. Filters of every kind are available up to the highest suppression grades from industrial firms, and the purchase of a suitable filter constitutes the simplest solution to the problem. If this is not appropriate, or it is desired to manufacture the filter in-house, the

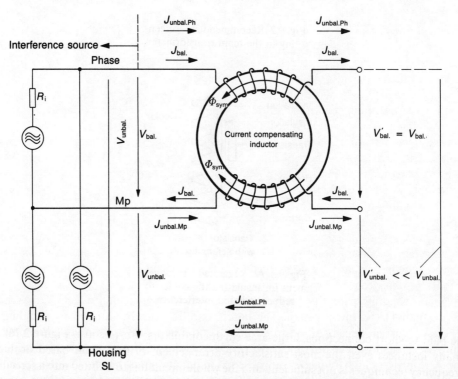

Fig. 9.4 Arrangement of a current-compensated inductor, showing interference currents and voltages (Siemens)

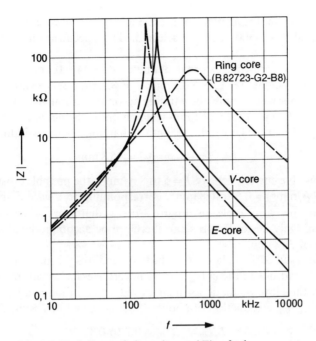

Fig. 9.5 Variation of impedance $|Z|$ of the current-compensated inductor with frequency f with various core constructions (Siemens)

best solution is to use a so-called 'current-compensated' inductor, in which a ferrite core carries two *identical* windings in which the input current (d.c. in inductor-coupled converters or 50 Hz mains current in primary-switched converters) flows in such a way that the magnetic field produced is exactly neutralized [37] [38].

The magnetic flux generated by the input current thus flows only to a very small extent in the core, and mainly in the air as stray flux. It is thus possible to use a core without an air gap and maintain a high inductance from the point of view of interference currents without any appreciable saturation effects due to the input current. Since the input filter is required to be effective over the wide frequency range from about 20 kHz to 30 MHz, a core shape with wide-band impedance characteristics is preferred; from Figure 9.5 this means using a ring core, which is the practice in most commercial filters.

The winding of the ring core is subject to the following restrictions. Since the ferrite material is electrically conducting, the winding cannot be applied directly to the core. Either a ready-insulated core must be used or insulation must be applied (e.g. powder-coated plastic, shrunk-on film or a winding of film strip). Such insulation, however, reduces the important internal diameter of the ring. In addition, according to VDE, the separation between the two windings must be at least 3 mm, which further restricts the available winding space. Furthermore the losses should not result in a temperature rise of more than 115 °C in operation. The latter condition can be satisfied by avoiding an excessive current density ($J = 5$ to 8 A/mm^2) and if necessary resin-casting the inductor into the screening enclosure which is in any case required. Since the inductor is required to operate in a very high frequency range, its self-inductance must be very low. This points to a *single-layer* winding. To illustrate the effects of these restrictions and conditions, a design for an inductor of this type will be presented. It is useful to choose a core material with a high permeability. From the available types the R34/12.5 ring core of material N30, with $A_L = 5000\,\text{nH}$ and $\mu_i = 4300$ may be selected. The specification of the inductor is $L = 2.7\,\text{mH}$ at $I_a = 4\,\text{A}$.

The data for this core are: external diameter $d_e = 34 \pm 0.7\,\text{mm}$; internal diameter $d_i = 20.5 \pm 0.5\,\text{mm}$; height $h = 12.5 \pm 0.3\,\text{mm}$; $l_e = 82\,\text{mm}$; $A_e = 83\,\text{mm}^2$; weight $= 33\,\text{g}$. The internal diameter which limits the winding accommodation amounts, with the lower tolerance, to only 20 mm. A total of 1 mm is taken up by the necessary insulation. There thus remains $d_i' = 19\,\text{mm}$. The winding length remaining for one winding is then

$$l = (d_i'\pi - 6\,\text{mm})/2 = 26.8\,\text{mm}$$

Given the permissible current density $J = 5$ to $8\,\text{A}/\text{mm}^2$, the possible cross-sectional area $A = 4\,\text{A}/(5\text{ to }8\,\text{A}/\text{mm}^2) = 0.5$ to $0.8\,\text{mm}^2$, corresponding to a wire diameter $d = 0.8$ to $1.01\,\text{mm}$. From the wire table of Figure 2.10, a diameter of 0.9 mm is selected; $d_{\text{max}} = 0.965\,\text{mm}$. Only 26 to 27 turns can therefore be accommodated in the available winding length of 26.8 mm.

Although as a result of the balancing of the two windings the magnetic lines of force lie mainly in the air and not in the core, there is nevertheless a certain magnetic bias, and allowance must therefore be made for a small reduction in inductance owing to I_a. In accordance with [38], a reduction of about 30 per cent will be assumed. Thus

$$L/L_0 = \mu_\Delta/\mu_i = 0.7 \text{ to } 0.8 \tag{9.1}$$

where L_0 is the inductance without magnetic bias, L is the inductance with magnetic bias, μ_Δ is the incremental permeability from Figure 9.6 and μ_i is the initial permeability (from the data sheet for the material used).

From Figure 9.6, with $\mu_\Delta/\mu_i = 0.7$, or $\mu_\Delta = 0.7 \times 4300 = 3010$, a field strength of approximately 25 A/m can be read off (taking the mean between the two curves, since the

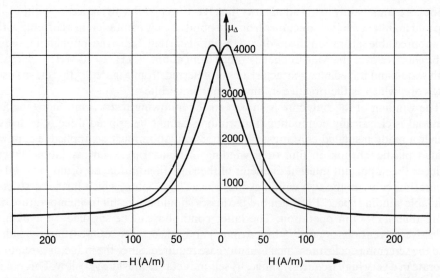

Fig. 9.6 Incremental permeability μ_Δ of Ferroxcube 3C11 (N30) as a function of d.c. field strength (measured on ring cores at $\Delta B \leqslant 0.2\,\text{mT}$) (Valvo)

current is alternating). The maximum number of turns is then given by

$$N \leqslant \frac{H l_e}{k_\sigma I_a} \tag{9.2}$$

where H is the field strength from Figure 9.6 (or dashed boundary curve in Figure 2.9) (A/m), l_e is the effective magnetic path length (m) and k_σ is the leakage coefficient, from [38]: approximately 0.7 to 0.8 per cent

$$N \leqslant \frac{25 \, \text{A/m} \times 82 \times 10^{-3} \, \text{m}}{8 \times 10^{-3} \times 4 \, \text{A}} \leqslant 64$$

These 64 turns are permissible, but cannot be accommodated. If the practicable number is put into the rearranged Equation (9.2), H is found to be

$$H = \frac{26 \times 8 \times 10^{-3} \times 4 \, \text{A}}{82 \times 10^{-3} \, \text{m}} = 10.15 \, \text{A/m}$$

From Figure 9.6 this produces hardly any reduction in μ_A, which can be estimated as approximately 4100. The reduction in permeability is thus only about 5 per cent. Since the inductance factor A_L is proportional to the permeability $\mu_A(\mu_e)$, from Equation (2.16), A_L is also reduced by only 5 per cent.

$$A_L = 0.95 \times 5000 \, \text{nH} = 4767 \, \text{nH}$$

The inductance, from Equation (2.17), is then

$$L = N^2 A_L = 26^2 \times 4767 \times 10^{-9} \, \text{H} = 3.22 \, \text{mH}$$

The required inductance is thus easily obtained with 26 turns; 24 turns would in fact suffice.

As is apparent from Figure 9.4, the balanced-current inductor reduces the asymmetrical, but not the symmetrical interference voltages. To satisfy the suppression requirements, the inductor must be used in conjunction with capacitors; the capacitors between the main conductors and the protective conductor or the housing are referred to as 'Y' capacitors, those between the main conductors as 'X' capacitors. To avoid personal danger in the event of a possible break in the protective conductor, the possible leakage

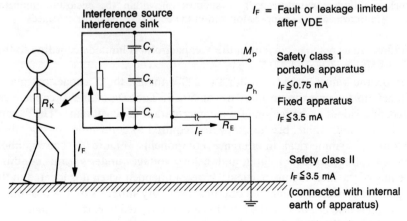

Fig. 9.7 X and Y capacitors and leakage current (Siemens)

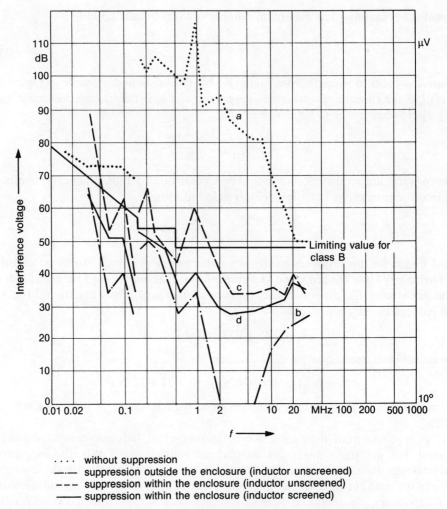

.... without suppression
—·— suppression outside the enclosure (inductor unscreened)
- - - suppression within the enclosure (inductor unscreened)
——— suppression within the enclosure (inductor screened)

Fig. 9.8 Interference levels in a switched-mode power supply with various suppression measures (Siemens).... without suppression, —·—·— suppression outside the enclosure (inductor unscreened), ----suppression within the enclosure (inductor unscreened), ———suppression within the enclosure (inductor screened)

current (flowing as reactive current in the Y capacitors) is limited according to the safety class (test voltage at least 1500 V a.c.).

Based on the values stated here for the 220 V mains, the Y capacitors may have a capacitance up to a maximum of 10 nF or 50 nF. The size of the X capacitors is not restricted, and values in the region of some microfarads are customary. The X capacitors need to be of a high value, because they are required to contribute considerably to the suppression of symmetrical interference components. Figure 9.8 shows, above, the interference voltage without a filter, and, below, voltages under various conditions.

Screening of the filter is thus particularly recommended when it is located in the same housing as the converter circuit. To obtain satisfactory suppression the mounting of the screening can is also important; the filter must be connected to earth through a large-area conductor to be effective [39].

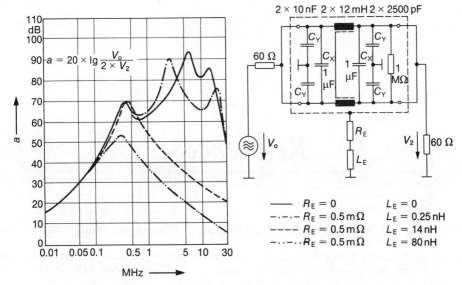

Fig. 9.9 π filter and earth connection ($60\,\Omega$ insertion resistance) (Siemens)

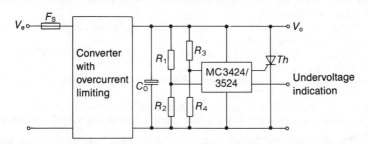

Fig. 9.10 Overvoltage protection (crowbar) for the output of a power-supply unit with undervoltage indication, using the MC3424/MC3524 IC (Motorola)

9.2 PROTECTIVE CIRCUITS

In the illustrations of the various control circuits, overvoltage and undervoltage indication circuits have also been shown, integrated into the control IC. However good an indication circuit may be in detecting impermissible deviations in, say, the output voltage, it is of no benefit if an excessive voltage appears at the output owing to a defect in the circuit, which cannot be altogether ruled out. Some control circuits include control facilities for the avoidance of excessive output voltage, but if the whole IC fails these circuits also are of no avail. The destruction of voltage-sensitive components (e.g. TTL circuits, CMOS circuits, etc.) can be very expensive. If, however, at a prescribed voltage (e.g. 6 V in a circuit with a 5 V output) a thyristor in parallel with the output is triggered, the output voltage is limited to the forward voltage drop of the thyristor (1.5 to 2 V) in about 1 μs, and the electronic components supplied by the power-supply unit can suffer no damage. The whole system admittedly ceases to function, but the same applies in the event of a mains failure. Such a protection and indication circuit, with its associated thyristor, is shown in Figure 9.10. The power-supply unit is of course protected by current limiting and suitable fuses.

References

[1] Siemens: Schaltnetzteile 4 'Bemessungsgrunglagen'. Siemens AG, Bereich Bauelemente, Balanstr. 73, 8000 München 80.

[2] Vacuumschmelze: Firmenblatt Z013, Weichmagnetische Bauelemente 'Speicherdrosseln für geschaltete Stromversorgungsgeräte'. Vacuumschmelze Werk Hanau, 645 Hanau, Grüner Weg 37.

[3] Valvo: Technische Informationen für die Industrie Nr. 770801 'Transformatoren, Speicherdrosseln sowie Ein-und Ausgangskondensatoren in Schaltnetzteilen', 1977. Valvo, Burchardstr. 19, 2000 Hamburg 1.

[4] J. Wüstehube, u.a.: *Schaltnetzteile-Grundlagen, Entwurf, Schaltungsbeispiele*', 2. Auflage. 7031 Grafenau 1, Expert-Verlag, 1982.

[5] Silicon General: 'Regulating pulse with modulator SG1524/SG2524/SG3524'. Neumüller, Eschenstr. 2, 8021 Taufkirchen/München.

[6] O. Macek, *'Schaltnetzteile, Motorsteuerungen und ihre speziellen Bauteile'*. 1. Auflage, Heidelberg: Dr. A. Hüthig-Verlag, 1982.

[8] O. Kilgenstein, *'Passive Bauelemente'*, 1. Auflage. 1977. Stuttgart: Frech-Verlag.

[9] Unitrode: 'Applikationsschriften für Schaltnetzteile'. Metronik GmbH, Vogelsgarten 1, 85 Nürnberg.

[10] Thomson-CSF: Datenbuch Dioden, Z-Dioden, Gleichrichter, Thyristoren 1975/76. Thomson-CSF, Fallstr. 42, 8000 München 70.

[11] O. Kilgenstein, *'Grundlagen aktiver Bauelemente'*, 1. Auflage. Stuttgart: Frech-Verlag, 1979.

[12] Siemens: Datenbuch Ferrite 1982/83. Siemens AG., Bereich Bauelemente, Balanstr. 73, 8000 München 80.

[13] G. Roespel and M. Zenger, 'Ferritkernformen für die Leistungselektronik'. *Bauteile-Report* Siemens **17** (1979). Issue 5, S. 209–214.

[14] Valvo: Technische Informationen Nr. 820402 'Elkos für industrielle Anwendungen'. Zweigbüro Nürnberg 10, Bessemerstr. 14.

[15] J. M. Peter, J. C. Baudier, J. Redoutey, B. Maurice and K. Rischmüller, 'Leistungstransistoren im Schaltbetrieb'. Thomson-CSF, Fallstr. 42, 8000 München 70.

[17] F. Böhner, 'Schaltnetzteil (Durchflußwandler) 5 V/60 A'. Diplomarbeit SS 1983, Fachhochschule Nürnberg.

[16] K. Rischmüller and D. Henke, *'Schaltnetzteil-Grundlagen'* Nr. 38/79. Thomson-CSF, Fallstr. 42, 8000 München 70.

[18] Valvo: Technische Informationen für die Industrie Nr. 780131 'Epitaxial-Gleichrichterdioden für Schaltnetzteile'. Valvo, Burchardstr. 19, 2000 Hamburg 1.

[19] Thomson: 'Handbuch Schalttransistoren', 1. Auflage 1979. Thomson-CSF, Fallstr. 42, 8000 München 70.

[20] M. Herfurth, 'Ansteuerschaltungen für SIPMOS-Transistoren im Schaltbetrieb'. *Siemens Components* **18** (1980), Issue 5, 218–224.

[21] Applikationsbericht über ein Sperrwandler-Schaltnetzteil Valvo-Applikationslaboratorium Nr. 403/70-1.

[22] O. Kilgenstein, *'Stabilisierte und geregelte Spannungsquellen'*, 1. Auflage. Stuttgart: Frech-Verlag, 1978.

[23] Valvo: Technische Information Nr. 770415 'Steuer- und Regelschaltung TDA 1060 für Schaltnetzteile'.

[24] Siemens: 'Integrierte Schaltnetzteil-Steuerschaltungen TDA 4700/TDA 4718; Funktion und Anwendung'. Technische Mitteilung aus dem Bereich Bauelemente, Best. Nr. B2332.

[25] D. Glang, 'Schaltnetzteile: Inversund Aufwärtswandler'. Diplomarbeit SS 84, Fachhochschule Nürnberg.

[26] W. Schaub, 'Schaltnetzteil unter Verwendung von Leistungs-MOS-FET's und hoher Taktfrequenz für 5 V/15 A'. Diplomarbeit WS 1980/81, Fachhochschule Nürnberg.

[27] A. Fuchs, 'Gegentakt-Durchflußwander-Schaltnetzteil in Halbbrückenschaltung mit bipolarem und MOS-FET-Leistungstransistor bei $U_0 = 12\,\text{V} \pm 10\%$ einstellbar und $I_0 = 1,5\ldots15\,\text{A}$. Diplomarbeit SS 1984. Fachhochschule Nürnberg.

[28] Valvo: Technische Informationen für die Industrie 'Schalt-Netzteile mit Transistoren der Reihe BUX 80'; Best. Nr. 761027.

[29] Siemens: Schaltbeispiel 117 V_{eff}/220 V_{eff}, 5 V/20 A; Schaltnetzteil nach dem Eintakt-flußwandlerprinzip mit TDA 4718 und SIPMOS-FET. Siemens-AG, Bereich Bauelemente, Best. Nr. B/3031.

[30] Siemens: Schaltbeispiel Durchflußwandler-SN mit meheren Ausgangsspannungen (5 V/10 A, $\pm$ 12 V/2 A) mit TDA 4718 und SIPMOS-FET. Siemens-AG, Bereich Bauelemente, Best. Nr. B1-B3030.

[31] R. Grünzinger, 'Sperrwandler im Dreiecksbetrieb mit bipolarem und MOSFET-Leistungstransistor und verschiedenen Steuerschaltungen bei $U_0 = 24\,\text{V}$ und $I_0 = 0 \cdots 2\,\text{A}$'. Diplomarbeit WS 1984/85, Fachhochschule Nürnberg.

[32] L. Metze, 'Sperrwandler im Dreiecksbetrieb mit bipolarem und MOS-FET-Leistungstransistor und verschiedener Art der Rückführung bei $U_{01} = +15\,\text{V}$; $I_{01} = 0\ldots3\,\text{A}$ und $U_{02} = -15\,\text{V}$; $I_{02} = 0\ldots-3\,\text{A}$'. Diplomarbeit WS 1984/85, Fachhochschule Nürnberg.

[33] E. Paulik and G. Peruth, 'Schaltnetzteile mit der IS TDA 4600'. Technische Mitteilung aus dem Bereich Bauelemente. Siemens-AG, Bereich Bauelemente, Best. Nr. B/2439.

[34] W. Rössler, 'Flußwandler-Schaltnetzteil mit Minimalaufwand für 5 V/10 A'. *Siemens Components* **22** (1984), Issue 6, 259–264.

[35] W. Rössler, 'Flußwandler-Schaltnetzteile mit Minimalaufwand'. *Siemens Components* **23** (1985), Issue 1, 11–13.

[36] J. Rates, 'Schaltnetzteil mit TEA 1001 SP'. Technische Information Nr. 44, Thomson-CSF-Bauelemente, Fallstr. 42, 8000 München 70.

[37] E. Ristig, 'Neue Bauformen stromkompensierter Drosseln'. *Siemens Components* **18** (1980), Issue 6, 294–297.

[38] Vqlvo: Technische Informationen für die Industrie Nr. 790910 'Stromkompensierte Drosseln mit Ferroxcube-Kernen'. Valvo UB Bauelemente, Burchardstr. 19, 2000 Hamburg 1.

[39] R. Schaller, 'Elektromagnetische Verträglichkeit durch den Einsatz von Entstörfiltern'. Siemens Technische Mitteilung aus dem Bereich Bauelemente, Best. Nr. B/2418. Siemens AG, Bereich Bauelemente, Balanstr. 73, 8000 München 80.

Index